GLOBAL ECOLOGY
in
Human Perspective

Poverty and prosperity in Bangkok. Two decades of rapid economic growth have turned Thailand's capital into a construction boomtown dominated by scores of new skyscrapers, but many newcomers to the city are forced to live in squalid slums along the fabled Chao Phraya River. Air pollution induced by traffic congestion and water pollution beneath their dwellings are part of their daily environment. *(Photo by Steve Raymer, © National Geographic Society, 1993, by permission)*

GLOBAL ECOLOGY
in
Human Perspective

Charles H. Southwick
University of Colorado, Boulder

New York Oxford
OXFORD UNIVERSITY PRESS
1996

OXFORD UNIVERSITY PRESS

Oxford   New York
Athens   Auckland   Bangkok   Bombay
Calcutta   Cape Town   Dar es Salaam   Delhi
Florence   Hong Kong   Istanbul   Karachi
Kuala Lumpur   Madras   Madrid   Melbourne
Mexico City   Nairobi   Paris   Singapore
Taipei   Tokyo   Toronto

and associated companies in

Berlin   Ibadan

Published by Oxford University Press, Inc.,
198 Madison Avenue, New York, New York 10016

Library of Congress Cataloging-in-Publication Data
Southwick, Charles H.
Global ecology in human perspective / Charles H. Southwick.
p.   cm.
Includes bibliographical references and index.
ISBN 0–19–509867–6
1. Environmental sciences. 2. Man—Influence on nature.
3. Pollution.   I. Title.
GE105.S68   1996
304.2—dc20   95-30198

3 5 7 9 8 6 4 2

Printed in the United States of America
on acid-free paper

PREFACE

Are we facing a worldwide boom in prosperity and expansion or an ecological collapse of the global environment? This book is an introduction to some of the complex dilemmas represented by broad questions of this type.

The book is not solely another volume on the environment. Although our environment is a central concern, the book deals with broader issues. It did not originate from environmental activism. Rather, its roots are in the natural and social sciences, primarily ecology, the behavioral sciences, and geography. This orientation is shaped by 45 years of personal scientific work in both industrialized and developing nations around the world.

The following chapters cover introductory material on the scope and meaning of global ecology, a brief review of ecological principles relevant to global concerns, the meaning of global change, human impacts on earth, population growth and regulation, the interactions of economics and ecology, world health, and prospects of human futures. The central theme deals with the ways in which we, as the human species, are altering planet earth, and how, in turn, it is affecting us.

The book is intended for undergraduate students in all fields of the arts and sciences, for adult education courses, and for professional students in many areas including business and financial services, public policy, land management, agriculture, law, medicine, public health, and international affairs. No formal prerequisites are assumed except a reasonable background in the physical and biological sciences, represented by good high school and/or introductory college courses.

Boulder, Colorado   C. H. S.
May 1995

ACKNOWLEDGMENTS

Any book that represents the thoughts and experiences of a lifetime owes a debt of gratitude to many individuals and organizations. For my professional education in biology and ecology, I am indebted to teachers and colleagues at the College of Wooster in Ohio, Hamilton College in New York, and the University of Wisconsin, Ohio University, Johns Hopkins University, Oxford University, Aligarh Muslim University, the University of Calcutta, Stanford University, and the University of Colorado. I was especially privileged to receive much of my early training in ecology from John Emlen, Joe Hickey, and Robert McCabe at Wisconsin; and Charles Elton, Mick Southern, Dennis Chitty, and Peter Crowcroft at Oxford. This was later enhanced by collaboration with Fred Bang, Dwain Parrack, George Schaller, Bill Sladen, Ed Gould, Carleton Ray, and John Oppenheimer at Johns Hopkins; M. Rafiq Siddiqi, M. Farooq Siddiqi, M. B. Mirza, J. L. Bhaduri, R. K. Lahiri, K. K. Tiwari, R. P. Mukherjee, Don Lindburg, Iqbal Malik, and S. M. Mohnot in India; D. D. Bhatt, M.K. Chalise, and Daniel Taylor-Ide in Nepal; Frank Cadigan and Illar Muul in Malaysia; Zhang Yongzu, Jiang Haisheng, and Liu Zhenhe in China; Jatna Supriatna, Jito Sugardjito, and Joe Erwin in Indonesia; Nick Collias, C. R. Carpenter, James Zetek, and Bill Mason in Panama; Paul Heltne and George Whitesides in Bolivia; Matt Kessler, Rich Rawlins, and Janis González-Martinez in Puerto Rico; Jean Baulu in Barbados, Vernon Reynolds, Irv DeVore, and Sherwood Washburn in Stanford; Dave Armstrong, Al Bartlett, Marc Bekoff, Wilson Crumpacker, Carl and Jane Bock, Michael Breed, Mike Grant, Jim Halfpenny, Ruth and Steve Bernstein, Bill Lewis, and many others in Colorado. I am also indebted to many students over the years, undergraduate and graduate, who have provided friendship, excitement, and new ideas.

I am further grateful for close friends in professional societies, too many to list individually, in scientific organizations including the American Institute for Biological Sciences, American Association for the Advancement of Science, Animal Behavior Society, Ecological Society of America, American Society of Primatology, International Primatological Society, American Society of Mammalogy, International Society for the Study of Aggression, and the American Society of Zoology. These organizations, along with the National Geographic Society, especially the NGS Committee for Research and Exploration, the National Science Foundation, National Institutes of Health, the National Academy of Sciences, and the National Research Council count in their memberships many individuals dedicated to science and human affairs. The scientific community is enriched by their loyalty to scientific and humanitarian ideals. Neither my work, nor that of science in general could function well without their hard work, expertise, and support.

I would like to express my special thanks to several colleagues who have read early drafts of this book and commented constructively on the entire manuscript—especially Peter Raven, Syd Radinovsky, Charles Stine, and two anonymous reviewers; and David Pimentel, Richard Williams, Harvey Nichols, Carol Wessman, Donella Meadows, and Henry Wright, who read selected chapters. They have done much to improve the book, but they are, of course, in no way responsible for conclusions expressed in the book or for any errors which may remain.

For practical assistance with word processing, I am grateful to Phyllis O'Connell, Linda Bowden, and Marissa Wilson, and for assistance with graphics, Diane Lorenz. Pat Gasbarro of Oxford University Press has handled the arduous task of permissions. Linda Carlson has assisted with proofreading. Outstanding editorial guidance has been generously provided by Kirk Jensen, Bob Rogers, and Colby Stong of Oxford University Press.

Finally, I owe the greatest debt of gratitude to my wife, Heather, who has been my most steadfast companion for nearly a half-century, and who has worked and traveled with me in more than 60 countries around the world. Our children, Steven and Karen, now professionals in engineering and medicine, respectively, and their families, have also shared with us the joys and concerns of an international outlook on the world of the twentieth and twenty-first centuries.

CONTENTS

PART II.
BASIC ECOSYSTEM ECOLOGY   *31*
Chapter 4
Ecological Principles: Ecosystems   *33*

Some Basic Principles of Ecology
Ecosystem Structure
Abiotic Substances
Producer Organisms
Consumer Organisms
Decomposer Organisms and Detritivores
Incomplete Ecosystems
Cybernetic Control in Ecosystems
Systems Analysis and Modeling in Ecology
Summary and References

Chapter 5
Ecosystem Organization and Function   *48*

The Hydrologic Cycle
The Carbon Cycle
The Nitrogen Cycle
The Phosphorus Cycle
The Sulfur Cycle
Summary and References

Chapter 6
Energy Flow and Trophic Structure   *57*

The Source of Energy
Food Chains and Trophic Structure
Ecologic Pyramids
Ecologic Efficiencies
Energy and Human Activities
Summary and References

Chapter 7
Ecosystem Homeostasis, Succession and Stability   *68*

Homeostasis
Homeostasis in Ecosystems
Global Implications of Ecosystem Homeostasis
Trophic Structure and Ecosystem Stability
Ecosystem Development and Succession
The Conservation of Diversity
The Controversy over Old-Growth Forests
Summary and References

PART III.
HUMAN IMPACTS ON PLANET EARTH   *81*
Chapter 8
Our Global Condition: A Clash of Concepts   *83*

Global 2000
Opponents of Global 2000

INTRODUCTION

This book is about 6 billion people and a single planet.[1] It's about coexistence and quality—common words destined to become the critical issues of the twenty-first century.

In many ways, these issues are critical already, though we tend to ignore the bigger picture in our daily preoccupations with more immediate problems. The big question is, Can we live together without committing violence to ourselves and our planet?

As we try to answer this question, the arithmetic rolls on. Despite the encouraging decline in the birth rates of many nations, the world has a net increase of over 85 million people every year—20 times the population of Los Angeles—enough new people to add a city the size of Denver, Colorado, every two days. These 85 million more people every year are competing for jobs, housing, and a reasonable standard of living. Every year, 70,000 square kilometers of farmland (approximately 24,000 square miles) are abandoned because the soil is exhausted (Hammond, 1994). Every year more than 30 local wars cause untold misery because people cannot coexist peacefully with their neighbors. Every year an estimated 4,000 to 40,000 species of plants and animals become extinct because we have not yet learned to live with nature.

Daily our television screens show us scenes of crime and deprivation in the midst of affluence; over 5,000 violent crimes are committed daily in the United States alone (*Uniform Crime Report,* 1994). Newspapers, radio, and television remind us of more circumstances in the world where the quality of life has eroded to a struggle for security and survival.

Throughout the 1970s and 1980s, we defined coexistence, security, and survival largely in terms of the Cold War. If we could just break the back of Soviet communism, we could launch a "new world order." We defined progress with economic statistics and corporate profits.

Now we know the world is not quite so simple, not so reassuring. Defeating

1. The world's population is estimated at 5.8 billion in mid-1996, and is projected to surpass 6 billion within three years (Population Reference Bureau, Washington, DC).

Soviet communism and spreading capitalistic enthusiasm around the world has not necessarily ensured our well-being. Perhaps in our political fervor we overlooked ecological realism and adequate knowledge of our life support systems—realism and knowledge that considers the interrelationships of ourselves, our environments, and the limits of planet earth. These are the subjects of global ecology.

Global ecology, concerned with ecological principles on a worldwide scale, emerged in the 1970s and hopefully may come of age in the 1990s. At least the field has achieved academic and political recognition, while still searching for its own scientific identity. That identity is hard to define. Realistically, the world is so large, so varied, and so complex, that it seems naive to embrace all of its features in one scientific discipline called "global ecology." On the other hand, so many ecological forces are now impacting the entire planet, we must think in terms of one world, at least scientifically. This book is an effort in that direction.

Wendell Wilkie wrote a book about global issues entitled *One World* in 1943 but the term "global ecology" was not commonly used for many years. In 1971 John Holdren and Paul Ehrlich edited a volume with the title *Global Ecology,* and in the 1980s and early 1990s, several other books appeared with this title (Corson, 1990; Rambler, Margulis, and Fester, 1989; Southwick, 1985). Use of the term grew rapidly, and by 1993 the University of Colorado library listed 70 volumes under the subject heading "global ecology." The subject received more widespread public attention in the 1990s with prominent international meetings.

In June of 1992, the United Nations "Earth Summit" met in Rio de Janeiro, 20 years after the world's first environmental conference in Stockholm in 1972. In Rio, the world's environmental, scientific, and to a certain extent political attention, focused on one of the largest international meetings ever convened. More than 20,000 delegates, 100,000 visitors, 118 heads of state, and representatives of 180 countries met to consider problems of the global environment. Media attention was extensive, perhaps not as much as for the Superbowl or the Olympics, but still remarkable for an international scientific meeting.

The agenda of the Earth Summit, also known as UNCED (the United Nations Conference on Environment and Development), included a wide range of global issues: air and water pollution, climate change and global warming, agriculture and soil erosion, deforestation and desertification, biodiversity and conservation, poverty and health, economics and development, education and minority relations. But one of the overriding issues remained in the background—the issue of population (Sitarz, 1993).

Population emerged front and center in the UN conference in Cairo in September of 1994. Here delegates from 180 nations met to consider the ramifications of global population growth. For the first time, the United States was aligned with the United Nations in recognizing the dimensions of global population growth—growth that has taken world populations from less than 2 billion people to nearly 6 billion in the lifetime of many of the Cairo delegates. Still, there were cries of backlash against the "myths of the population crisis." Early in the proceedings, rifts developed between the majority of nations, who advocated family planning and birth control, and the Vatican hierarchy and Islamic fundamentalists, who rejected any efforts at population planning as unacceptable threats to religious beliefs and moral values. The conference was almost derailed on the issue of abortion, but with concerted effort and compromise it returned to the main topic of family planning through improved education, better health care, support of women's rights, and higher living standards. The great advance of the 1994 Cairo conference was its focus on population as a global issue. Two years previously in Rio de Janeiro, this topic could scarcely be discussed. The most notable achievement of the Cairo meeting was a resolution signed by delegates of nearly 180 nations to support family planning and to recognize the vital link between population policies and development. This is a remarkable step forward in global ecology.

Other conferences, books, articles, films, and scientific achievements, especially the exploration of space, have fostered the growth of global ecology. The concepts of "only one earth," of the earth as "an oasis in space or a fragile lifeboat in a hostile universe" have caught the collective imagination of humanity. Global ecology has blossomed as a concept, even if its scientific boundaries remain vague.

Although the field of global ecology has grown, it is not without skeptics and dissenters. The science of global ecology has often been criticized with some justification, and even ridiculed. Some proponents of global ecology in the early 1990s warned of runaway global warming, but the winter of 1993–94 brought wave after wave of blizzards, ice storms, heavy snowfalls, and record low temperatures in North America. Global ecologists have warned of acid rain and dying forests, but many forests are thriving. They warned of agricultural Armageddon and worldwide famine, while global food supplies continued to increase for most nations. Economic collapse was predicted for many countries, yet financial booms and unprecedented bull markets swept through countries and regions considered ecological basket cases not long ago, including China, eastern and southeast Asia, and several countries of Latin America. Never in human history have so many people in the world enjoyed adequate food, housing, clothing, and the basic necessities of life. Most of us in western nations and increasing numbers of people throughout the entire world live with more comfort and better health than the kings and queens of medieval Europe. Who, then, can we believe—the doomsayers of ecology or the prophets of worldwide prosperity?

The Sierra Club and the Sahara Club are similar in the sound of their names, but they are worlds apart in their attitudes and philosophies. The Sierra Club, founded by John Muir in the nineteenth century, believes in environmental preservation and conservation; we are part of nature and must live within ecological limits. The Sahara Club, founded in the late twentieth century by U.S. citizens fed up with pious environmentalists, insist that the ecological movement is out to take away individual freedoms, eliminate our jobs, and destroy our industrial strength. We are masters of the land, and all its resources are to be exploited for human use. The backlash against environmentalism in times of economic stress and job shortages has been sufficiently severe that both membership in and contributions to environmental groups such as the Sierra Club, Greenpeace, and the Wilderness Society declined sharply in 1994 (Clifford, 1994).

These controversies are old ones, of course, as old as the laws of Abraham and the writings of the Old Testament. The Judeo-Christian concept of land ownership has been blamed for many environmental atrocities. The Biblical injunction to "Be fruitful and multiply, and fill the earth and subdue; and have dominion over the fish of the sea and over the birds of the air and over every living thing that moves upon the earth" (Genesis 1:28) was God's directive to go forth and "have dominion over everything upon the earth." According to this concept, whatever humans consume, God will replenish. It is interesting to note, however that the non-Judaic, non-Christian world has ravaged the earth with equal ferocity, a point to be explored later in this book.

Many of these dilemmas and controversies emerge again and again in both scientific and popular media, and their long history does little to diminish the furor of debate. The broad issues of ecology always have been in dispute scientifically and philosophically, and they will continue to be controversial. The recent prominence of global ecology, along with increasing competition for the earth's resources, highlights these disparities and leads to some skepticism about the entire field of ecology.

Among the disputes at the 1992 Earth Summit and the 1994 population conference there were fundamental disagreements about the value and outcome of the conferences. Some declared UNCED a failure because it did not come to grips with the tough problems and did not impose standards of international regulation necessary to reverse global deterioration. Others claimed UNCED a great success in that so many nations and so many different factions were discussing environmental issues. Apart from the politics and media hype of the conference, many basic scientific issues, such as the reality, timing, extent, and direction of climate change, remained unresolved. Skeptics pointed out that only 30 years ago scientists insisted we were headed for a new Ice Age. Then came the 1980s with the hottest temperatures of the twentieth century, and fears of runaway global warming were raised. In the early 1990s the spector of a new Ice Age emerged again. Can ecological science even come close to dealing with such complex, variable, and large-scale problems? Are the ideas of global ecology a help or a hindrance to human betterment?

The Cairo population conference also produced its share of skeptics. Whereas the majority of delegates felt that the momentum of population growth will cause unprecedented social and economic problems, a vocal minority insisted there is no such thing as a "population crisis." Vernon Jordon, former U.S. Ambassador to the United Nations under President Reagan, expressed the minority view by stating there is no real population problem. Proponents of this view note that some of the world's most densely populated nations, such as Denmark, do not have the social and environmental problems of less crowded countries. For example, Denmark, with four times the population density of the United States, has a violent crime rate less than 1 percent of ours. They also note that vast areas of the world are unpopulated and undeveloped. The problem, Mr. Jordon stated, is one of development and education, not overpopulation.

An issue of *U.S. News and World Report* (September 12, 1994) expressed the typical view of technological optimists that there is no reason why the world cannot support a population of 10 billion people. In contrast, neo-Malthusians predict that such growth will lead to ecological disaster and social chaos.

The controversies about population and environment are both ideological and scientific. Take, for example, the question of preservation, utilization, or exploitation. Do we use the earth's resources to the fullest and harvest the last old-growth forests because they are there for our use, or do we save them because they are not ours to destroy? Do we convert virgin forests into managed timber plantations because rapid-growth forests may absorb more carbon dioxide, or do we preserve old-growth forests as repositories of biodiversity?

Questions such as these obviously have economic consequences and lead to social divisions: jobs vs. endangered species, corporations vs. environmentalists, rich vs. poor, north vs. south, the political right vs. the left, and so on. Who is responsible for environmental degradation: the developing nations hacking their way through tropical forests to survive or the industrial nations using the earth's resources to promote an extravagant lifestyle?

Virtually everyone agrees on the need for development, or at least human betterment, but how and at what cost? The buzzword of the 1990s is "sustainable development," or "sustainable growth," but is this possible? Perhaps sustainable growth is an oxymoron, a contradiction of terms (Bartlett, 1994).

Amid these questions and controversies the need is more apparent than ever for a return to facts and principles. These principles may be either scientific or ethical. In this work, we will deal first with scientific principles, but moral judgments are also involved in all human endeavors. Science, the environment, and human affairs in all respects must consider ethical values. Ecology is no exception. In exploring global ecology, we will focus on these questions: (1) What principles are we dealing with? (2) What facts do we have relating to these principles? (3) What conclusions can we reach? (4) What options do we have for thought and action?

These are challenging questions we can ask in all areas of human activity, including our social and environmental relationships. Hopefully, this book will help the reader decide not so much what to think but how to think: how to relate facts, processes, and beliefs; how to assess outcomes and draw conclusions; and finally, how we might view and treat the one world in which we live.

REFERENCES

Bartlett, A. 1994. Reflections on sustainability, population growth, and the environment. *P(p ilation and Environment: A Journal of Interdisciplinary Studies* 16: 5–35.

Clifford, F. 1994. Hard times for environmentalism. *Boulder Camera,* Sept. 25, 1994, p. 1F.

Corson, W. H. (ed.). 1990. *The Global Ecology Handbook.* Boston: Beacon Press.

Hammond, A. L. (ed.). 1994. *World Resources, 1994–95, a Guide to the Global Environment: A Report by the World Resources Institute, the United Nations Environmental Programme, and the United Nations Development Programme.* New York: Oxford University Press.

Holdren, J., and P. R. Ehrlich (eds.). 1971. *Global Ecology.* New York: Harcourt Brace Jovanovich.

Rambler, M. B., L. Margulis, and Rene Fester (eds.). 1989. *Global Ecology: Towards a Science of the Biosphere.* Boston: Academic Press.

Sitarz, D. (ed.). 1993. *Agenda 21: The Earth Summit Strategy to Save Our Planet.* Boulder, CO: Earth Press.

Southwick, C. H. (ed.). 1985. *Global Ecology.* Sunderland, MA: Sinauer Associates.

Uniform Crime Report. 1994. FBI: Washington, DC.

Wilkie, W. 1943. *One World.* New York: Simon and Schuster.

Introduction to Global Ecology

Chapter 1

THE MEANING AND
SCOPE OF
GLOBAL ECOLOGY

Global ecology is the study of ecological principles and problems on a worldwide basis. It consists of the environmental, biological, and behavioral sciences relating to planet earth. It involves structure, process, and change.

Global change has many components. Some changes are local and regional phenomena that accumulate and eventually assume global importance, such as deforestation, overgrazing, soil erosion, and coastal pollution. Other events in global ecology may not be seen locally, but they affect planetary ecology on a broad scale, such as depletion of the ozone layer in the stratosphere or accumulation of carbon dioxide in the atmosphere. These changes, too, may be the accumulation of small events, but their total significance can be far greater than we might suspect from merely adding up their separate impacts.

Global ecology is eminently multidisciplinary. It involves many areas of the physical, biological, and social sciences. Conventionally, global ecology focuses on the earth sciences: atmospheric, geological, geographic, and oceanographic. Much of the conventional study centers on climatology, meteorology, landforms, and oceanic processes. The physical sciences, especially chemistry and physics, are essential in these areas.

Now there is increasing recognition of the role of biological communities in atmospheric and oceanic processes, and appreciation of the interplay between plants, animals, and microorganisms and their physical environments.

Global ecology also has a behavioral and social dimension; that is, it involves human activities. For some time human populations have been a physical and biological force on earth (Thomas, 1955–56; Goudie, 1994; Turner, 1990). Now it is obvious that we are a dominant factor on earth, shaping the landscape, altering atmospheric and oceanic conditions, and severely modifying biotic communities throughout the world. This does not mean we are divorced from nature's

3

forces. Quite the contrary, it is more apparent than ever that we are not exempt from natural laws. As the German poet Goethe once said, "God may forgive, and Man may forgive, but Nature never forgets." Not only are we still very much at the mercy of physical actions such as volcanoes and earthquakes, hurricanes and tornadoes, droughts and floods, but in many of these, except the first two, we now have an influencing role. The possibility exists, for example, that human activities such as air pollution are altering the frequency or severity of violent weather. Certainly patterns of human land use can exacerbate floods, droughts, and the effects of storms. We even increase the economic impact of earthquakes by concentrating some of our populations is seismically active areas.

These possibilities lead to necessary considerations of the numbers, distribution, and actions of human populations on the earth. Such evaluations bring the social sciences to the study of global ecology; the fields of economics, political science, and sociology. By such means, poverty, war, and world health all become subtopics of global ecology.

Thus the incredible span of global ecology. To make sense of this multidisciplinary field we need to define *ecology* and *environment,* and then proceed from these definitions to some basic considerations of the position, qualities, and components of planet earth.

A FEW BASIC DEFINITIONS

Ecology is usually defined as the study of the interrelationships of living organisms and their environment. In short, ecology is the study of biological interactions. This involves the full range of living organisms: bacteria, protozoa, algae, fungi, plants, animals, and human beings. Viruses and subviral particles are also ecological factors, although they do not have the complete characteristics of living organisms (Raven and Johnson, 1995).

Ecology comes from two Greek words, *oikos,* meaning "house" or "home," and *-logy,* meaning "discourse" or "science." Hence, *ecology* literally means "the science of our house or home." The term *economics* has the same stem, *oikos,* and the suffix *-nomics,* meaning "law" or "management." Hence, *economics* literally means "laws of the home" or "management of the home." The common origin of these two terms suggests that, although the fields of ecology and economics are often at odds in our modern world, they must finally come together again in our consideration of world futures. Ultimately, good economics will have to involve sound ecology.

THE ENVIRONMENT

The word *environment* comes from the French verb *environner,* which means "to surround, encompass or encircle." Thus the environment refers to the surroundings of ourselves or any organism. These surroundings can be considered in three categories: physical, biological, and social.

Figure 1.1 Global ecology spans the physical, biological, and social sciences. In the savannah grasslands of Ethiopia, the Oromo people herd cattle, and seek greater autonomy over their land. Their dependence upon water, soil, and biota is clearly evident. City-dwellers have the same dependence involving more circuitous routes. Both pastoral herders and urban inhabitants experience direct linkages between ecology, economics, and politics. (Photo by Robert Caputo © copyright National Geographic Society)

The physical environment of an organism includes the air, water, inorganic chemicals, and physical structures around that organism. This includes terra firma, buildings, automobiles, and fabricated goods.

The biological environment includes all living organisms of other species— microorganisms, plants, animals, and the entire living community of other species that exist around a given individual.

The social environment includes all individuals of the same species in the surroundings of the individual under consideration.

Thus, if we consider the environment of a child in the inner city of New York, her or his physical environment consists of urban air and water, the houses, schools, factories, stores, streets, playgrounds, cars, trucks, bikes, and so on in the neighborhood, as well as every manufactured product around that individual.

The biological environment includes sparrows, pigeons, starlings, rats, dogs, cats, cockroaches, flies, mosquitos, trees, shrubs, weeds, disease organisms, soil organisms (if soil is present), and any other living organisms in the area.

And finally, the social environment consists of the child's family, schoolmates, friends, teachers, police officers, fire fighters, and other human beings with whom she or he might come in contact. Media such as television, movies, radio, audio- and videocassettes, newspapers, and magazines could be considered part of the social or the physical environment. Although these are physical objects, a case could be made that they are really part of the social environment in the sense that they involve communication. All of these components of the environment are

Figure 1.2 A model of economic growth and prosperity, Hong Kong overlooks scenic Victoria harbor. Beneath the water's surface lie vast quantities of sewage, industrial waste, and agricultural chemicals. Recognizing potential threats to human health, Hong Kong has launched a $1.55 billion cleanup. (Photo by Steve Raymer © copyright National Geographic Society)

part of an individual's surroundings and are actual or potential influences on that individual. Certainly each individual interacts with all three aspects of the environment.

This definition of the environment is broader than that used in most ecology books, but this broad approach is essential to understanding the true scope of modern ecology. It would be fruitless to try to interpret what is happening to the earth without considering human factors.

Since our purpose is to look at the earth as an ecological system, it is helpful to consider planet earth in terms of its position in the universe. In chapter 2, we will consider some properties of planet earth and then the four major components of our planet: the lithosphere, hydrosphere, atmosphere, and biosphere. Our goal is to have some understanding of the physical and biological properties of the earth.

PLANET EARTH IN THE UNIVERSE

In the dimensions of the universe, planet earth seems relatively insignificant. Physically, the earth is only the fifth largest planet in the solar system of an average star. Our sun, classified as a yellow dwarf star, is just one of 100 billion stars in the Milky Way galaxy. We don't know how many galaxies exist in the universe, but there are estimated to be 100 billion (Sagan, 1985). Most galaxies are estimated to be 50,000 to 150,000 light-years in diameter, they typically have 100 billion stars in each galaxy, and galaxies average 1 to 2 million light-years apart. Thus, the number of stars in the cosmos is unknown and incomprehensible.

Forty years ago Professor Harlow Shapley of Harvard University estimated that at least 10^{20} stars (the exponent indicates the number of zeros after 1) were visible to us with modern telescopes, and that vast numbers existed beyond the reach of our most powerful observatories. The Hubble space telescope will tell us more about the dimensions and structure of the universe.

The number of stars with solar systems is completely unknown, but also incredible in magnitude. If even a small percentage of stars have solar systems, the total number of such systems would be enormous.

These statistics emphasize the minute size of our solar system in relation to the universe, but other figures also convey the impressive distances and gigantic nature of our solar system. The sun is 864,950 miles in diameter and approximately

---

BOX 1.1
Coming of Age

How old is the science of global environmental change? I would date it from 1864, when George Perkins Marsh published his *Man and Nature: or, Physical Geography as Modified by Human Action*. For those who focus on climate change, however, the relevant date might be 1896, when Svante August Arrhenius identified the burning of fossil fuels as a factor that could alter the radiative balance of the Earth. And for those who would start with the formal organization of the science, it might be 1957 (when the International Geophysical Year began) or 11 years ago in Ottawa (when the International Council of Scientific Unions created the International Geosphere-Biosphere Program to study human-induced global environmental change in depth.)

While the grand design that emerged at Ottawa was good, it was disappointing in one crucial respect: It failed to promote the collaboration between natural scientists and social and behavioral scientists necessary to understand the interactions of nature, society, and technology—to address the complex questions of human actions and sustainability. Such collaboration, of course, would have to confront both attitudinal and structural barriers. Attitudinally, natural scientists tend to see social science as "soft," while social and behavioral scientists denigrate the "naive determinism" of natural science. Structur-

ally, disciplinary identification and the reward system of the social and behavioral sciences discourage research on nature and society together. Thus, there is a real shortage of scholars willing to invest the necessary time and energy on interdisciplinary studies.

Even so, there has been substantial progress since Ottawa. At the first open meeting of the Human Dimensions of Global Change Community (organized by the Social Science Research Council's Committee for Research on Global Environmental Change and cosponsors and held at Duke University in early June), more than 120 presentations focused on the human factor in global environmental change—the sources of that change, its impacts on people and society, and society's responses. These presentations recognized that environmental change is more than climate change and that global change is more than simply environmental change. Included were analyses of the great demographic, economic, social, cultural, institutional, and technological changes under way in the world and the resulting changes to land, water, and biota as well as climate. For all those present, there was a sense that a truly interdisciplinary global environmental science is coming of age.

—Robert W. Kates

From *Environment* 37(6): Inside cover, July–Aug. 1995.

330,000 times the mass of the earth. One million earths could fit inside a hollow sphere the size of the sun. The sun contains 99 percent of the mass of our solar system. Every second, the sun converts 4 million tons of matter into energy through the process of nuclear fusion, yet this is a negligible portion of its total mass. The sun has an estimated life of 4 to 10 billion years remaining to provide solar energy for the earth's ecosystems.

The amount of solar energy striking the earth is small in relation to the sun's total output, but unbelievably large in relation to our own energy needs. Solar energy striking the earth's surface averages 9 million calories per square meter per day, equal to 36 billion calories per acre per day, assuming 10 hours of sunshine. The total amount of solar energy striking the United States in 20 minutes on an average summer day is sufficient to meet our country's entire energy needs for one year, providing it could be harnessed. Solar energy powers virtually all global ecosystems except those surrounding deep-sea thermal vents. Stored solar energy in the form of wood, coal, and petroleum powers most human domestic, industrial, and transportation activities. Chapter 6 will briefly review energy flow in ecological systems.

To further illustrate the position of earth in our solar system, the earth is the third planet in distance from the sun, approximately 149,600,000 kilometers (92,960,000 miles) from the sun. This distance is only 8.3 light-minutes, but it would require a spaceship traveling 25,000 miles per hour (the speed required to escape earth's gravitational force) over five months to travel from the earth to the sun. The outermost planet in our solar system, Pluto, at the most distant point on its orbit, is approximately 4.5 billion miles from earth, a distance that would take 13 years, 10 months and two weeks in a space craft traveling 37,000 miles per hour for a one-way trip (37,000 mph would be the minimum speed necessary to reach Pluto's orbit from the earth).

These facts of astronomy are humbling to say the least, and may give the impression that planet earth does not matter in the cosmic scale of the universe. It is always helpful in science to gain perspective, however, and to look at topics from as many points of view as possible. When our focus is on the earth itself, considering its own dimensions and qualities, quite a different impression is gained. The astronauts of the world who have had the opportunity of seeing planet earth from space have seen this with remarkable clarity. They have viewed earth's magnificent blues and greens and browns surrounded by a swirling atmosphere of white clouds—a striking contrast to the pale and lifeless moon on which the Apollo crews landed and the vast distances of space that we have barely started to explore. For all astronauts space flight has been an awesome experience, and for many a deeply spititual one as well. Chapter 2 takes a closer look at this planet that is our home.

SUMMARY

Global ecology is the study of planet earth in physical, biological, and sociobehavioral terms. This definition is broader than conventional definitions, which usually focus on physical and biological factors. It is based on the reality, however, that behavioral and social interactions of human populations are major, often dominant, factors in global relationships.

REFERENCES

Goudie, A. 1994. *The Human Impact on the Natural Environment.* 4th ed. Cambridge, MA: MIT Press.

Raven, P. H., and G. B. Johnson. 1995. *Biology.* 4th ed. St. Louis, MO: Mosby–Year Book.

Sagan, C. 1985. *Cosmos.* New York: Ballantine.

Thomas, W. L. (ed.). 1955–56. *Man's Role in Changing the Face of the Earth.* Chicago: University of Chicago Press.

Turner, B. L. (ed.). 1990. *Earth as Transformed by Human Action.* New York: Cambridge University Press.

Chapter 2

PROPERTIES AND
COMPONENTS OF
PLANET EARTH

On the cosmic scale, planet earth is infinitely small; on the human scale it is finitely large. We are fond of saying, "What a small world this is!", a statement both true and false. Earth is small in the sense of communications, crowding, jet transportation, and finite physical dimensions. We can phone or fax London, New Delhi, or Hong Kong in a matter of minutes. Earth is vast in terms of complexity, diversity, and functional capabilities. If we tried to canoe up the Amazon River, walk across the Sahara, or sail a small boat across the Pacific, we would find out how huge the earth is. In the next few chapters, we should keep these opposing impressions in mind as we look at some of its physical and functional properties.

The earth has a circumference of 40,075 kilometers (24,902 miles), and an average radius of 6,373 km (3,960 miles). Its surface extent is 510,230,000 km^2 (197,000,000 square miles), of which only 29 percent is land. Of the land surface, only 11 percent is arable under our present-day systems of agriculture (Miller, 1992). The rest is too mountainous, too cold, too dry, too devoid of soil, or otherwise unsuitable for growing crops. Hence, we depend upon about 3 percent of the earth's surface for terrestrial food production. Earth is the only planet known to have liquid water, which covers approximately 70 percent of its surface.

Although the earth may seem insignificant in astronomical terms relative to the entire universe, many of its properties are uniquely favorable for life. The presence of liquid water, for example, is the basis for all life on earth, and both atmospheric concentrations and temperature conditions are especially favorable for living organisms as we know them. On Mars, water occurs only as ice and water vapor, and on Venus, only as vapor. Mars is too cold for life as we know it, and Venus is too hot.

The earth is remarkable in the moderation of its temperature regimes. Although temperatures vary greatly on the surface of the earth, from a recorded high temperature of 58°C (136°F) to a recorded low of -105°C (-127°F), the average surface air temperature of the earth is 15°C (59°F) which is right in the favorable range for most life forms existing on earth. One can say, of course, that this is solely because life has evolved in adaptation to these temperatures, but the fact remains that most chemical and physical processes that are the basis of biological functions perform best at the temperature conditions found on earth.

In contrast, the average surface temperatures on Mars are well below freezing (-23°C or -9.4°F), whereas those for Venus are excruciatingly hot (460° to 480°C), well above the boiling point of water. Furthermore, on Venus the average surface atmospheric pressure is 94 times that of earth. These extreme conditions on Mars and Venus do not totally eliminate the possibility of some form of life on these planets, for we have recently learned through research on oceanic thermal vents that organisms can live and function in seemingly inhospitable environments. But these physical facts do indicate that life on other planets in our solar system, if it exists at all, would have to be very different from that on earth. No evidence of such life has yet been found. The numerous *Mariner* spacecraft expeditions past Mars did not detect any life forms, nor were any found by the *Viking Lander* which reached the surface of Mars and conducted specific experiments designed to detect the presence of life forms.

Atmospheric conditions represent another vitally important set of environmental conditions on earth favorable for living organisms. These will be discussed in a subsequent section, but here we can note that the atmospheres of Mars and Venus are predominantly carbon dioxide with only trace amounts of oxygen. It is clear, therefore, that planet earth, a modest planet in astronomical terms in a modest solar system, is a statistical miracle according to our present knowledge.

These unique features of our planet bring us to consider its four major components: the lithosphere, hydrosphere, atmosphere, and biosphere.

THE LITHOSPHERE OR GEOSPHERE

The lithosphere refers to the solid and molten portions of the earth—the land, including that under the oceans, and those portions of the earth consisting of molten rock. The lithosphere can be divided into the crust, mantle, and core. The crust, the portion most familiar to us, consists of rocks, alluvium, and a thin veneer of soil in some regions. It is actually a set of crustal plates that fit together like a jigsaw puzzle; the visible portions of these plates we recognize as continents. Seven large continental plates (North American, South American, Antarctic, Eurasian, Indo-Australian, and Pacific) average about 40 km in thickness (24.8 mi.), but they vary in thickness from 0 at the site of an active lava flow up to 100 km. The earth's crusts under the ocean are usually thinner, averaging only 8 km (de Blij and Muller, 1993).

The large continental plates, plus a number of small ones, such as the Arabian and Philippine, provide the basic form of the visible lithosphere (Fig. 2.1).

The crustal plates, or tectonic plates, rest upon the mantle, thought to be molten rock, which extends to depths of nearly 3,000 km. According to the theory of

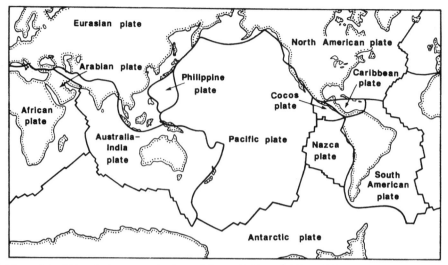

Figure 2.1   Lithospheric or tectonic plates of the world (from Gubbins, 1990).

plate tectonics, which is now well established, lithospheric plates formed a single large land mass, known as Pangaea, 200 to 250 million years ago. By approximately 135 million years ago, this large mass had broken into several smaller masses, dominated by a large area in the northern hemisphere known as Laurasia and large areas in the southern hemisphere known as Gondwana. Segments of these continued to drift apart in the process known as continental drift, in which the tectonic plates may be thought of as floating or riding on the mantle of molten rock. By 65 million years ago, the present continents were taking shape, though North America was still closely attached to Eurasia, and India was adrift in the Indian ocean, not yet tied to the rest of Asia. The modern configuration of continents occurred within the past 65 million years (Fig. 2.2), and India has become a part of mainland Asia only in the last 10 million years.

Our present landforms of mountains and plains, valleys and canyons are the combined result of crustal movements, or continental drift, and surface actions such as erosion. The Himalayan mountains, for example, formed from uplift as the Indian plate moved northward and collided with the Eurasian plate. This movement is still occurring, virtually imperceptibly, but to a measurable extent, and some mountains of the Himalayas increase in height a few inches per decade. Continental drift has had a major influence on the distribution and evolution of plants and animals throughout the world. Crustal movements are evident today in fault lines, earthquakes, volcanic activity, and thermal vents (Fig. 2.3).

The inner core of the earth is thought to consist of an outer layer of molten iron and an inner solid portion. This structure is speculative, of course, as is the detailed structure of the mantle, since we have never directly sampled either one except as the mantle is sampled by volcanic activity.

Several scientific projects have been launched to drill all the way through the crust to obtain a chronological history of the crust and also to tell us more about

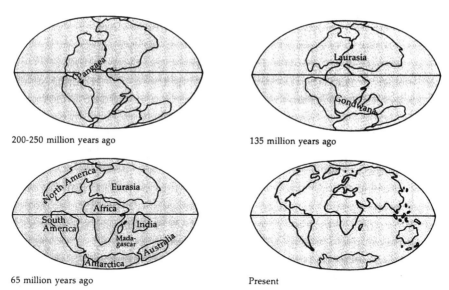

200-250 million years ago

135 million years ago

65 million years ago

Present

Figure 2.2 Continental drift. Two hundred to 250 million years ago, the earth's land surfaces were locked in a supercontinent named Pangaea. About 180 mya, Pangaea began to split into northern and southern land masses, Laurasia and Gondwana, which later separated into modern continents (from Campbell, 1987).

the true structure of the mantle. One such project in the United States, known as Project Mohole, was started in 1961 by the National Science Foundation. It was begun in oceanic crust off the California coast, then shifted in 1964 to the Puerto Rican coast in the Caribbean, and in 1965 to an oceanic area 100 miles north of Hawaii where the crust beneath the ocean is only 5 km in thickness. The purpose was to reach the earth's mantle and learn more about the composition of both crust and mantle. Unfortunately, the project was abandoned in 1966 because of skyrocketing costs and a lack of funds. Our present direct knowledge of the crust goes to a depth of about 3 km, and our knowledge of the mantle comes only from volcanic emissions.

Figure 2.3 Cross-sectional diagram of oceanic and continental plates, showing sea-floor spreading and seismically active zones of earthquakes and volcanoes (from Sullivan, 1991).

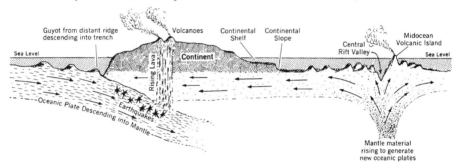

THE HYDROSPHERE

The hydrosphere consists of the water on planet earth: oceans, estuaries, lakes, ponds, rivers, streams, springs, ground water, snow, ice, glaciers, clouds, and atmospheric water vapor, although the last two forms are also a part of the atmosphere. The earth is appropriately called the "water planet" because 71 percent of the earth's surface is covered by water and ice. Oceans constitute 97 percent of the hydrosphere, polar ice caps and glaciers form approximately 2 percent, and lakes, rivers, streams, springs, ground water, and water residing in vegetation account for about 1 percent of the total hydrosphere. Technically, water in vegetation is a part of the biosphere.

A significant aspect of the hydrosphere is its dynamic motion and change. With few exceptions, the water on earth is constantly flowing, mixing, evaporating, and seasonally melting, rising, and falling. This is obvious in streams and rivers and at the edges of lakes and oceans, where one can see waves and tides, but it is also true of open oceans where there are great currents such as the Gulf Stream and Humboldt Current. In the northern hemisphere, major oceanic currents, such as the Gulf Stream in the Atlantic and the Japanese or Kuroshio Current in the Pacific, move in a clockwise fashion when viewed from above, whereas in the southern atmosphere, the Humboldt and South Atlantic Currents move in a counterclockwise direction (Fig. 2.4). These patterns result from the earth's rotational forces, and they have major influences on regional climates. For example, the mild climate of Great Britain, at the same latitude as northern Newfoundland

Figure 2.4   Major ocean currents that influence climates of adjacent land areas (from Kemp, 1994).

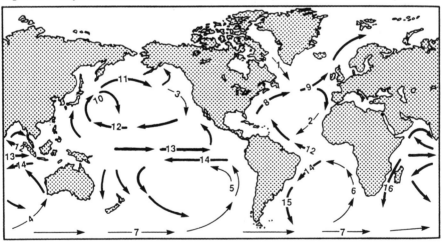

COLD CURRENTS		WARM CURRENTS	
1 Labrador	5 Humboldt	8 Gulf Stream	13 Equatorial Counter
2 Canary	6 Benguela	9 N. Atlantic Drift	14 S. Equatorial
3 California	7 West Wind Drift	10 Kuroshio	15 Brazil
4 West Australia		11 N. Pacific Drift	16 Agulhas
		12 N. Equatorial	

and Labrador, results from the Gulf Stream bringing warm tropical waters from the Caribbean.

We also know from recent satellite photography that the oceans of the world have numerous swirls and eddies, often of gigantic size, encompassing hundreds of miles. There are also strong vertical movements, as for example at the Antarctic convergence, where pronounced updrafts bring nutrient-rich water to the surface at the edge of the Antarctic seas.

There are also vast pools of relatively warm water that appear periodically in the mid-Pacific, the so-called El Niño southern oscillations (ENSO), which influence global weather patterns, especially by producing droughts in the western Pacific and excessive rains in the eastern Pacific (Gulnaraghi and Kaul, 1995). These will be discussed in Chapter 18.

We cannot see the hydrospheric processes of evaporation and transpiration, but we are well aware of their subsequent stages—cloud formation, fog, and precipitation.

The major exceptions to the statement that the hydrosphere is in constant motion are glaciers and ground water, especially deep aquifers, where some motion occurs but on a much slower timetable of years, decades, and centuries rather than minutes, hours, and days.

A remarkable fact stemming from recent research on the hydrosphere is that the total amount of water on earth has probably remained constant for 1 billion years or more. There have been shifts between various components of the hydrosphere, especially between ice and liquid water. During Pleistocene ice ages, a larger percentage of the hydrosphere was frozen into continental ice sheets, producing cold climates and lower ocean levels. We do not fully understand the geological and astronomical forces producing these changes, whether they were entirely due to changes in earth's rotational and orbital patterns or whether atmospheric forces played some role, but they obviously occurred without human influence. Now, of course, there is increasing evidence that human factors are changing atmospheric and hydrospheric conditions and that many fundamental processes, including climate, will change as a result. We will discuss these in Chapters 17 and 19 on air and water pollution.

THE ATMOSPHERE

The atmosphere is a multilayered envelope of gases around the earth that not only provides many of the basic components necessary for life (e.g., oxygen, nitrogen, carbon dioxide, water vapor) but also moderates temperatures and serves as a protective shield from lethal dosages of excessive ultraviolet radiation and cosmic rays from outer space. In structure, the atmosphere consists of several layers with a total thickness, or height above the surface of the earth, of about 80 kilometers, approximately 50 miles (Fig. 2.5). These layers are not distinct and sharply defined; rather they represent a gradual thinning of atmospheric gases and gradual changes in average temperatures. The portion of the atmosphere closest to the earth, the troposphere, is about 16 km, or 10 miles, in thickness and contains over

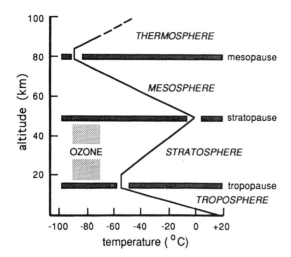

Figure 2.5 Vertical structure of the atmosphere and its associated temperature profile (from Kemp, 1994).

75 percent of all gaseous molecules in the atmosphere. It covers the highest mountain peaks, and at its outer boundary temperatures are below −50°C. Oxygen concentrations at this boundary are too low for normal human respiration. Most earthly weather occurs within the troposphere, and some long-distance commercial airline flights approach its outer boundary of approximately 50,000 feet.

The gaseous concentration of the atmosphere near the surface of the earth is approximately 78 percent nitrogen, 21 percent oxygen, and 1 percent miscellaneous gases, such as argon (0.93 percent), carbon dioxide (0.03 percent), and a wide variety of other gases including oxides of sulfur and nitrogen, ozone, chlorides, and inert gases such as neon, helium, and krypton. Both atmospheric pressures and oxygen concentrations decline markedly with altitude. At 30,000 feet oxygen constitutes only about 12 percent of the atmosphere.

The stratosphere lies beyond the troposphere, from 16 to 50 km in altitude (10 to 30 miles). The stratosphere is bounded by the ozone layer, formed when molecules of oxygen (O_2) are dissociated by solar radiation and then recombine into O_3. On earth's surface, ozone is a powerful oxidant and air pollutant, but at the outer edge of the stratosphere it protectively absorbs ultraviolet radiation. There is concern that some air pollutants of human origin that migrate to the upper stratosphere, such as chlorofluorocarbons used in air conditioners and refrigerators, are reducing the stratospheric ozone layer with potentially serious consequences. Strangely, temperatures at the outer layer of the stratosphere are higher than at the lower levels, probably due to the absorption of solar energy by the ozone layer. Diminution of the ozone layer will increase both UV radiation and temperature at the earth's surface (see Chapter 17). Approximately 99 percent of the atmospheric gases lie below the outer boundary of the stratosphere.

The layer beyond the stratosphere, the mesosphere, extends from approximately 50 km to 80 km (30 to 50 miles). It has gradually decreasing temperatures and ozone concentrations. At the outer limits of the mesosphere, temperatures reach −90°C.

Beyond the mesosphere, the outermost zone of the atmosphere, known as the thermosphere, again shows slightly rising temperatures, to about −40°C as the atmosphere merges gradually with the hydrogen and helium atoms in outer space.

Many of our satellites and spacecraft orbit the earth beyond the atmosphere at altitudes ranging from approximately 200 km (125 miles) for the early manned spacecraft to 900 km (560 miles) for Landsat and earth resource satellites to 35,800 km (22,300 miles) for many existing communication satellites. At this highest altitude, satellites have the same orbital speed as the earth's rotational speed, so they remain in a stationary position in regard to any location on earth's surface.

ATMOSPHERIC WEATHER

Just as the oceans are in constant motion, so also is the atmosphere, with complex interactions of high- and low-pressure areas, prevailing winds, local storms, and high-altitude jet streams. Weather is basically driven by the differential heating effects of solar energy. In equatorial regions, air warmed by tropical sunshine rises and starts to flow toward the poles. As it rises and cools, precipitation forms, so that many tropical regions also have high rainfall (providing they have a vegetative cover, as we will see later). As upper-level air flows toward the poles, it cools, sinks, and becomes reheated, producing complicated patterns of pressure, precipitation, and airflow. These air movements are known as Hadley cells. Desert regions accumulate high-pressure areas with an outflow of dry air; their relative lack of vegetation tends to ensure continued aridity. We will see later that vegetative cover over the land surface, especially forests, creates humidity and attracts clouds and precipitation. Likewise, lack of vegetation creates arid conditions, a lack of precipitation. Major river, lake, and forested basins generate atmospheric moisture and attract low-pressure areas with an inward flow of moist air. Boundaries, or fronts, between high- and low-pressure areas and between warm and cold air masses tend to generate storms. The greater the pressure and temperature differentials, the more violent the storms tend to be. Hurricanes, typhoons, severe thunderstorms, and tornadoes are all products of atmospheric energy and pressure differentials.

CLIMATE

Climates and seasons reflect long-term weather patterns. These patterns result from solar radiation, day length, temperature regimes, precipitation, cloud cover, and wind—factors that contribute to tropical, temperate, or cold climatic regimes. Seasons are predominantly a function of the axis tilt of the earth, the tilt determining the direction and intensity of sunlight on the northern and southern hemispheres. Sunlight shining vertically on a portion of the earth's surface delivers much more energy than sunlight shining obliquely, both because of the area receiving the sunlight (much larger in the case of oblique rays), and the filtering effect of the atmosphere (also much greater for oblique rays). Thus, as the northern and southern hemispheres present different angles to the sun at different times of the year, the seasons change. The most oblique sunlight occurs in winter, the most direct in summer.

Climate is a function primarily of latitude, altitude, and topography. Altitude and latitude are broadly related to the extent that 100 meters of elevation is approximately equivalent to 1 degree of latitude, or approximately 110 kilometers. In other words, ascending a mountain in the western United States from 1,800 meters (5,200 feet, the elevation of Denver) to 4,500 meters (14,000 feet, the approximate elevation of Mount Evans, 50 km west of Denver) is the climatic equivalent of going 27° north, that is, from 40° north latitude to 67° north latitude. This is comparable to traveling from Denver, Colorado, to Fairbanks, Alaska.

In both cases, one ascends from a temperate climate with a grassland and temperate forest habitat to an arctic or alpine climate with a tundra habitat. Topography also influences climate by altering wind patterns, cloud formation, and producing rain shadows on leeward sides of mountains. The climates of continental interiors are often very different from those of coastal areas—a collective influence of topography, the hydrosphere, and the atmosphere.

THE BIOSPHERE

The biosphere is that portion of planet earth where living organisms exist. It is both an abstract concept and a striking reality: abstract because it combines and includes many parts of the atmosphere, hydrosphere, and lithosphere and is not separate from them; real because we are always aware of the presence or absence of living organisms on earth. Furthermore, the earth is the only planet in the universe that we know for certain contains living organisms. Statistical theory and modern astronomy indicate that it is extremely likely that other planets in other solar systems have living organisms, but we cannot prove this at the present time. Nor can we prove that any other planet in our own solar system is endowed with life.

The biosphere is such a remarkable and overwhelming aspect of planet earth, and such a central feature to global ecology, that the next chapter is devoted to some of its characteristics.

COMPONENT INTERACTIONS

Although we have been discussing the biosphere, atmosphere, hydrosphere, and lithosphere separately, in fact they all interact and have innumerable mutual influences. Many of these interactions are obvious, but some are more subtle, and we are just beginning to appreciate the magnitude of certain interactions.

Several of the more obvious interactions have already been mentioned. The atmosphere shields the earth from excessive ultraviolet radiation and moderates temperatures on the earth's surface. Without the protective shield of ozone, and our atmosphere of nitrogen, oxygen, and small amounts of carbon dioxide and other gases, planet earth would be either a raging inferno like the planet Venus or a frozen wasteland like Mars. In either case, it would be unsuitable for life as we know it.

The hydrosphere is also an environmental moderator, contributing moisture to the atmosphere, forming clouds, and creating precipitation which enables plant and animal life to flourish. The hydrosphere both absorbs and releases heat to moderate climate and generate weather over the earth's surface.

The atmosphere, hydrosphere, and lithosphere all contribute essential elements and compounds for the existence of life. Some obvious examples are oxygen from the atmosphere for respiration of plants and animals, carbon dioxide from the atmosphere for photosynthesis in plants, and oxygen from photosysnthesis back into the atmosphere for respiration in all aerobic organisms including ourselves. The bumper sticker that asks, "Have you thanked a green plant today?" alludes to the important role of plants in biochemical interactions, as well as a source of food.

Other biochemical interactions between the earth's components are numerous, continual, and complex. The atmosphere supplies the nitrogen for soil and water that becomes the basis for protein and nucleic acid synthesis in living organisms. Phosphorus from the lithosphere ultimately meets the needs of ATP synthesis within the biosphere. The list could go on and on, and other examples of essential component interactions will be given throughout this book.

We do not fully understand the overall importance of many of these interactions. How significant is the uptake of carbon dioxide and the formation of oxygen in photosynthesis in maintaining the global balance of the atmosphere? Only in the last 25 years have we learned that forests and grasslands help to attract and maintain atmospheric moisture and precipitation necessary for their own survival—a case of mutual interdependence between a physical process and a biological one which we will discuss more fully under the hydrologic cycle in Chapter 5. We are also just beginning to learn how recycling processes in natural ecosystems stabilize and detoxify pollutants. We are beginning to appreciate some of the mechanisms of ecosystem homeostasis that maintain, renew, and reproduce favorable environments on earth.

The topic of interactions is really the heart of ecology. The following chapters will provide many examples of interactions between the biosphere, atmosphere, hydrosphere, and lithosphere which will give us a better understanding of the interlocking nature of these global components. Our search for understanding will begin with the biosphere.

SUMMARY

In conventional scientific terms, planet earth may be studied in four components: lithosphere, hydrosphere, atmosphere, and biosphere. The composition of these components and the presence of the biosphere make the earth unique among the planets about which we have direct knowledge. These components interact in many ways, and understanding the nature of these interactions is the heart of global ecology.

REFERENCES

Campbell, N. 1987. *Biology.* 1st ed., Menlo Park, Ca: Benjamin Cummings.

deBlij, H., and P. O. Muller. 1993. *Physical Geography of the Global Environment.* New York: John Wiley and Sons.

Gubbins, D. 1990. *Seismology and Plate Tectonics.* New York: Cambridge University Press.

Gulnaraghi, M., and R. Kaul. 1995. Responding to ENSO. *Environment* 37: 16–20, 38–44.

Hopson, J. L., and N. K. Wessels. 1990. *Essentials of Biology.* New York: McGraw-Hill.

Kemp, D. D. 1994. *Global Environmental Issues: A Climatological Approach.* 2nd ed. London and New York: Routledge.

Miller, W. T. 1992. *Living in the Environment.* 7th ed. Belmont, CA: Wadsworth.

Odum, E. P. 1993. *Ecology and Our Endangered Life Support Systems.* 2nd ed. Sunderland, MA: Sinauer Associates.

ReVelle, P., and C. ReVelle. 1992. *The Environment: Issues and Choices for Society.* 3rd ed. Boston: Jones and Bartlett.

Sullivan, W. 1991. *Continents in Motion: The New Earth Debate.* New York: American Institute of Physics.

Chapter 3

THE BIOSPHERE

From the rocky coasts of the high Arctic to the cold Antarctic shores, planet earth possesses an incredible profusion and diversity of life. No one knows how many species of plants, animals, and microorganisms exist on earth; the number is generally thought to be between 5 million and 30 million, most of which have yet to be named or studied (Raven and Johnson, 1995). Since the time of Linnaeus in the eighteenth century, less than 2 million species have been described sufficiently to be given scientific names. The abundance, variety, and assemblages of all living organisms on earth constitutes the biosphere.

The earth is the only planet we know that is endowed with life. The statistical probability is that life exists on other planets, perhaps even in our own solar system, but if not here, most probably on unseen planets in other solar systems. We can only speculate about these possibilities, however, for we have yet to see other solar systems, let alone know anything about them. The eminent astronomer Harlow Shapley was one of the first modern astronomers to write convincingly about the unbelievable numbers of stars in the universe. Shapley estimated that 100 pentillion (10^{20}) stars are visible to present-day telescopes. Furthermore, he suggested the likelihood that many of these have planetary systems with physical conditions favorable for life on at least some of their members. Our notions of the universe continue to expand into concepts of living systems beyond planet earth, but even today we lack evidence of their existence, and we remain entirely in the realm of mathematical probability concerning extraterrestrial life. Hence, earth provides the only planetary ecosystem with living organisms that we know about with certainty. Although it may be timeworn to say so, planet earth is our only home, our total life-support system, and our "grand oasis in space," as phrased by Christensen (1984).

But how much do we really know about the earth? We have a vast storehouse of knowledge about its physical characteristics and its biological inhabitants, but only recently with the advent of space science have we seen the earth in its entirety. We are just beginning to appreciate some of its complexities, and we aren't at all sure how most of these complexities function. We don't know, for example, how earth's oxygen and carbon dioxide balances are maintained. We can't fully understand or predict many of the earth's most basic processes—climatic changes, ocean levels, volcanic activities, and crustal movements. We are just beginning to decipher some key processes of physical and biological change on the earth, but we are a long way from understanding or predicting these changes. Global science and global ecology as scientific disciplines are still in their infancy.

EXTENT OF THE BIOSPHERE

The biosphere can be thought of as a thin and discontinuous film over the surface the earth, including the hydrosphere. It varies considerably in thickness and is incomplete in surface coverage. It extends above the earth to elevations that may reach as high as 10,000 meters when insects or microorganisms are carried aloft by wind currents and updrafts. On land, the biosphere extends below ground to the deepest roots of plants and to the chambers of subterranean caverns. Recent studies have shown that the biosphere extends much deeper than previously known, with the discovery of bacteria in crystalline rock aquifers over 1000 m beneath the Land Surface in the Columbia River basin of Washington (Stevens and McKinley, 1995). In oceans, the biosphere extends to the deep-sea regions wherever life may be found, now known to include deep-sea thermal vents at depths of 3,000 meters, and possibly undiscovered vents considerably deeper. Even deep-ocean trenches, such as the Mariana Trench, provide habitat for hetero-trophic organisms over 10,000 meters beneath the ocean's surface. The ocean depths and the high elevations of 10,000 meters (10 km) represent the extreme limits of the biosphere on earth, and they emphasize the need to distinguish be-tween different zones within the biosphere.

Stretching from +10,000 meters to −10,000 meters from the surface of the earth, the biosphere may seem like a broad band, but it is only 12 miles (20 km) thick, whereas the planet itself is nearly 8,000 miles (12,800 km) in diameter. Hence, the biosphere at its maximum thickness represents less than two-tenths of 1 percent of the diameter of the earth.

This broad definition of the biosphere includes any place on earth or in the atmosphere where life may be found. If we consider only those places where living organisms actively reproduce, the extent of the biosphere is more limited. With few exceptions, the zone of biological production is much narrower. Its greatest extent on land is in a tall forest, such as a redwood grove or primary tropical forest, where the zone of biospheric primary production, or BPP,[1] may be 100 meters thick, reaching from the canopy of the forest to the depth of the roots. At the other extreme on land, the zone of biospheric primary production in a rice field or potato field is only 1 or 2 meters thick; in a mown lawn, only a few centimeters thick; and in the case of lichens covering a rock, only a few millimeters thick.

In aquatic environments, the zone of BPP may be over 100 hundred meters

---

1. BPP (biospheric primary production) is used here in a comparable sense to GPP (gross primary production) known in traditional ecology; that is, it refers to total organic synthesis in plants by photosynthesis and in microorganisms by chemosynthesis.

thick in a very clear ocean or lake where sufficient light can penetrate to support photosynthesis well below the surface. This would represent unusually clear water, occurring in some marine and freshwater environments. Conversely, a typical lake or ocean coastal zone would have photosynthesis extending only a few meters below the surface, and in turbid waters, only a few centimeters.

The major exceptions to the above statements are deep-sea thermal vents, discovered in 1977 at depths up to 3,000 meters. These vents are cracks in the earth's crust on the bottom of the sea where hot gases and heated water emerge, often several hundred degrees centigrade. Such vents have an incredible assemblage of marine life—uniquely adapted clams, marine worms, and crustacea—capable of living in complete darkness, under very high pressures, and at extreme temperatures. In such habitats, photosynthesis does not occur; rather, primary production is accomplished by bacteria capable of synthesizing organic compounds from hydrogen sulfide (Childress, 1995). This is a totally different world from the biosphere as we know it on the surface of the earth. It operates on different energy sources and different metabolic pathways than life on the surface, and its existence radically alters our concept of where and how living organisms can exist. The organisms in sulfidic environments have developed special physiological mechanisms to deal with highly toxic, high temperature, high pressure, but energy-rich habitats (Scott and Fisher, 1995). Figure 3.1 portrays in a diagrammatic fashion the vertical dimension of the biosphere. It suggests the great variation of habitats from high to low in the biosphere and in its lateral extent as well.

Despite the extreme environments in the deep-sea thermal vents where living organisms occur, there are many places on earth where the biosphere does not exist, or if living organisms are present, they are so transient or so sparsely distributed that they do not constitute a permanent biotic community. For practical purposes, the biosphere does not exist in the extreme parts of the polar regions, over vast areas of the driest deserts, on most of the highest mountain peaks having an environment of permanent ice and snow, in some land and water areas highly polluted with toxic wastes, and throughout some of the deepest oceanic spaces other than thermal vents and upwellings. Such areas may have transient life forms, but they do not contribute significantly to the total picture of biospheric production.

The relative thickness of the biosphere can be visualized by an analogy with more familiar structures. If the diameter of the earth is represented by the height of an eight-story building, approximately 100 feet tall, the total thickness of the biosphere would be represented by the thickness of a two-by-four (about 4 cm) on top of the building. On the same scale, the zone of active biological production, excluding deep-sea vents, would be represented by the thickness of a piece of paper (approximately 0.3 mm), and even this thickness would represent the most favorable habitats, such as a clear coral sea or a tropical rain forest.

The point of this descriptive exercise is to emphasize that the biosphere is surprisingly thin. Meaningful terrestrial biosphere occupies less than one-quarter of the earth's surface, and it is continually subject to alteration and insult at the hands of human populations. It is like a delicate veil covering part of the earth's surface. Yet this biosphere is our total life-support system. It generates our oxygen, produces our food, reprocesses our wastes, and makes life possible.

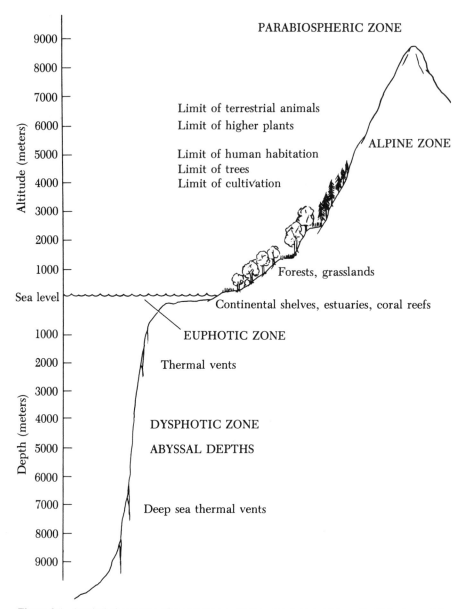

Figure 3.1 Vertical dimensions of the biosphere. Living organisms exist from the highest mountain peaks to ocean depths. Life at the extremes (parabiospheric zone) is rare or transient. The zone of significant biological activity is relatively thin on a global scale (from Southwick, 1976).

FUNCTIONAL PROPERTIES OF THE BIOSPHERE

Although it is an elementary exercise in biology to enumerate what a natural ecosystem does, it is worth listing some of these properties in relation to a nonliving system. What, for example, can a natural grassland accomplish that AstroTurf or a parking lot cannot? Some of these accomplishments are listed in Table 3.1.

Figure 3.2 The biosphere extends vertically from environments of permanent ice and snow in high mountains to ocean depths, but much of the earth's surface is too cold, too dry, or too rocky to support permanent biological communities. Only a relatively thin and sparse veil of productive environments provide the life-support systems for human and animal populations. Helicopter skiing in the Canadian Rockies of British Columbia affords exciting recreation for a fortunate few. (Photo by Dick Durrance II © copyright National Geographic Society)

TABLE 3.1
Simplified comparison of some system properties between a natural ecosystem and a man-made structure (Rodale, 1972).

Natural Ecosystem: Pond, Marsh, Grassland, Forest, etc.	Manufactured System: House (conventional), Factory, Parking Lot, AstroTurf, etc.
1. Captures, converts, and stores energy from the sun	1. Consumes energy from fossil or nuclear fuels
2. Produces oxygen and recycles carbon dioxide	2. Consumes oxygen and produces carbon dioxide
3. Produces carbohydrates and proteins; accomplishes organic synthesis	3. Cannot accomplish organic synthesis; produces only chemical degradation
4. Filters and detoxifies pollutants and waste products	4. Produces waste materials that must be treated elsewhere
5. Is capable of self-maintenance and renewal	5. Is not capable of self-maintenance and renewal
6. Maintains beauty if not excessively disturbed	6. Usually causes unsightly deterioration if not properly engineered and maintained
7. Creates rich soil	7. Destroys soil
8. Stores and purifies water	8. Often contributes to water pollution and loss
9. Provides wildlife habitat	9. Destroys wildlife habitat
10. Supports plant life; grows food	10. Does not support plant life or grow food

These facts are familiar to most readers, but nonetheless they are often forgotten as we forge ahead paving the biosphere with concrete and asphalt, chopping down its trees, washing away its soil, and polluting its waters.

We can also think of the biosphere as a mosaic of biochemical processes forming an infinitely complex biochemical system. This system captures, converts, processes, and stores solar energy through an incredible diversity of organisms. Despite the diversity of hundreds of thousands of species of green plants (at least 250,000), about 30,000 species of algae, unknown numbers of other microorganisms, and a great diversity of animal species, perhaps 5 to 10 million, the fundamental structures of all living organisms are remarkable similar in their basic organization. The patterned structures of DNA, RNA, proteins, lipids, and carbohydrates form a blueprint for all life. Modern biology is stretching our understanding of this living world on both the molecular and the global scale.

As this remarkable biosphere of which we are a part recycles biogeochemical products through itself, the atmosphere, the hydrosphere, and the lithosphere, it counters the physical process of entropy (the increase of disorder and disorganization) by constantly reorganizing, restructuring, and rebuilding the biochemical basis of living organisms. Life itself can be thought of as anti-entropic in its phases of normal growth and reproduction, whereas it is entropic in its processes of catabolism and decomposition. Without life the world would indeed proceed to disorder; with life there is productive reorganization.

The biosphere restructures its components not only in a physical and biological sense; it does so in a behavioral and social sense as well.

Groups of ants, bees, fish, mice, deer, monkeys, and people all tend to organize themselves in behavioral and social systems. Social breakdowns occur, but from each process of social disintegration or entropy, reorganization begins. In this elemental sense, a corporation after Chapter 11 bankruptcy or a nation after war follows the same basic process as a herd of deer or a covey of quail after the hunting season; a corporation or a society must reorganize, achieve a new structure, find new leaders, develop new routes of communication, establish new systems of political process. The order in the living world can be disassembled or shattered, but that order can also be reassembled provided we do not entirely destroy an ecosystem's life-giving ability to do so.

ENVIRONMENTAL BUFFERING AND HOMEOSTASIS

The biosphere can also be thought of as a great moderator or buffer of environmental conditions on earth. One need only compare the summer ground temperatures on a bare earthen field or an asphalt parking lot at mid-day to those in the shade of deep forest to realize how much a plant community moderates temperature. When the temperature is in the high 90s on a hot and sunny July day, how much nicer it is to park the car in the shade of a large tree than on a concrete parking lot. In a similar way, the biosphere moderates humidity, wind, precipitation, oxygen and carbon dioxide balance, and many aspects of atmospheric chemistry. Forests provide moisture to the air through transpiration, helping to maintain the rainfall necessary for their own survival. Aspen forests in Colorado have a significant buffering effect on acid precipitation, and most biotic communities have various capacities to detoxify pollutants.

All of these qualities relate to ecosystem homeostasis, the ability of biotic communities to maintain environmental conditions favorable for the perpetuation of life. When the biosphere is destroyed, physical conditions are more likely to swing to extremes; reasonable balances can no longer be maintained. These principles have many ramifications, often of direct importance to human survival. We know that natural drought cycles can be tragically exacerbated when the vegetative cover is destroyed by overgrazing or excessive land misuse. Temperature differentials become much greater when forests are destroyed. Flood conditions become more dangerous when watersheds are denuded. Storms become more violent. The maintenance of equitable climates on earth is intimately associated with intact ecological systems (Schneider and Londer, 1984).

Traditional courses in biology are organized around certain aspects of the biosphere: its complexity, its taxonomy, and its central functions. Some textbooks of biology are even titled *Biosphere* (Wallace, King, and Sanders 1984). Within the study of the biosphere, the focus may be the cell, the organism, the population, or the ecosystem. At each of these levels, the interplay of diversity and unity are impressive themes. At the most comprehensive biological level of all, the biosphere itself, these themes are also logical ones to emphasize, but it is perhaps even more important for us to realize what we are doing to the biosphere. How are we affecting the earth's ability to function as a life-support system? The question is a critical one not only for ourselves but for millions of other organisms. It is easy to see local effects at many points, but how important are these in a cumulative sense? Are numerous arenas of local pollution and countless scars of erosion, deforestation, and desertification altering the function of the global system? Or are they negligible? Data on global increases in CO_2, accumulation of airborne lead in the Greenland icecap, DDT in the penguins of Antarctica, and expansion of the deserts of Africa provide evidence of global influences, but how important are they?

Although the earth is a vast assemblage of infinitely complex and varied environments, we must also think of its total health. Global concepts are more essential than ever before. They can provide guidelines for local action and clues to what we might expect from local or regional developments when seen in broader perspective.

THE GAIA HYPOTHESIS

An ecological idea that has gained momentum in recent years is the concept of the earth as a living organism. The Gaia hypothesis states that the earth has evolved as a living functional unit, not merely as an assemblage of components that seem to work together. The concept was developed by James E. Lovelock, a British astrophysicist, who studied the Martian atmosphere and became increasingly impressed by the unique properties of earth. The term stems from the name of the Greek goddess Gaia, the goddess of earth.

The concept has similarities with certain ancient ideas about the earth. The

Greeks had a goddess to represent the earth, and the Chinese saw the earth as a complex entity with contrasting forces (yin and yang) that could operate in balance or out of balance. If yin and yang become imbalanced in an individual or in the earth, illness develops.

Lovelock defined Gaia as a "single complex entity involving the earth's biosphere, atmosphere, oceans, and soil, the totality constituting a 'feedback' or 'cybernetic' system which seeks an optimal physical and chemical environment for life on this planet" (Lovelock, 1987). Gaia remains a hypothesis, not an established fact, but considerable scientific evidence is accumulating that is compatible with the Gaia concept.

The Gaia hypothesis, which strangely enough was a spinoff of the space program, has many ramifications. If true, it would mean that the biosphere may actually control the atmosphere within limits. Thus anthropogenic influences on the biosphere will ultimately affect the atmospheric balances of carbon dioxide and oxygen. This idea has been refuted by some, but more recent changes in CO_2 levels give greater credence to the concept. In fact, it has been estimated that the earth without living organisms would have an atmosphere of 98 percent carbon dioxide and an average surface temperature of 290°C ± 50°C (Odum, 1993).

The Gaia hypothesis also asserts that any significant changes in the biosphere will change the heat balance of the earth, and there are definite scientific concerns that this may occur within the next 20 to 50 years.

If the earth is indeed a single organismic entity, it becomes valid to think of the earth as having respiratory and metabolic cycles and an overall state of health. This concept invites analogies with what we know about the metabolism and health of individual organisms. An individual organism breathes, ingests, utilizes energy, excretes, ages, and is subject to injury and illness. Most individual organisms also reproduce, and at this point our analogy breaks down, but up to this point a number of interesting comparisons are possible. The earth obviously respires, utilizes energy, produces waste products, ages, on a much broader time scale, and is certainly subject to injury and illness, at least local injury and illness.

Just as an individual may experience repeated or even multiple injuries and illnesses and then recover, so also may the earth. But a point can be reached where recovery is no longer complete or even possible. An individual may survive burns that cover 50 percent of the body, but he or she may be scarred for life; if the burns cover 80 percent of the body, recovery may not occur. If an arm is broken, recovery is usually complete; if the spinal cord is broken, the individual is never again the same. Recovery from blood poisoning may be complete if the liver and kidneys are fully functional, but if not, recovery may be impossible.

One can think of analogies applicable to the earth, if the Gaia Hypothesis is correct. The earth may be able to withstand so much tropical deforestation or desertification, but a point may be reached where recovery will not be possible. Similarly, the earth can withstand a certain amount of air pollution, and some polluted rivers, but if air pollution became too pervasive, for example, if acid rain were to develop on a worldwide basis, or if large amounts of river and coastal pollution seriously affected the health of the oceans, global health would never again be the same, and recovery might not even be possible. The trouble is that we have no idea where ecological thresholds may lie, and usually we have difficulty recognizing when thresholds are approached. The Gaia hypothesis alerts us to these concepts, however, and forms a framework around which research questions can be asked.

SUMMARY

In astronomical terms, the earth is a modest planet in the solar system of a second-rate star. Our sun is one of 100 billion stars in our galaxy, which in turn is one of billions of galaxies in the universe. Despite the incomprehensible magnitude of the cosmos, however, earth is the only planet we know of for certain that has a biosphere. Earth is uniquely endowed in terms of temperature, light, atmospheric conditions, and the presence of liquid water to promote rich and diverse communities of living organisms.

The biosphere is that portion of the earth in which living organisms exist. It can be thought of as a relatively thin and discontinuous film over the land and in the seas, extending to a height of no more than 10 km above the surface and to a depth of approximately 10 km beneath the sea. Its thickness comprises less than two-tenths of 1 percent of the thickness of the earth. Its productive zone, where significant biological reproduction occurs, is even much narrower, no more than 100 meters in thickness, and often much less. The biosphere does not exist to any major extent on the highest mountain peaks, in large areas of the polar icecaps, in the driest deserts, in vast areas of ocean depths, on highways, airport runways, or paved parking lots, and in the earth's most polluted regions. Less than 25 percent of the earth's surface contains terrestrial biosphere with significant levels of productivity.

At once resilient and fragile, the biosphere is our total life-support system. The metaphor of a thin and delicate veil of life is entirely appropriate. At the same time, some forms of living communities are tough and resilient. It is now essential for us to understand the nature and functioning of the biosphere. It is of equal importance for us to realize what we are doing to the biosphere in both positive and negative ways to enhance or damage its ability to survive and function.

The Gaia Hypothesis states that the earth is a highly integrated living entity with internal feedback mechanisms that regulate temperature, moisture, and atmospheric conditions. The hypothesis has many implications for global ecology. While still hypothetical and not a set of proven facts, many aspects of the Gaia Hypothesis are supported by an increasing number of scientific studies.

REFERENCES

Childress, J. J. 1995. Life in sulfidic environments: Historical perspective and current research trends. *Amer. Zoologist: A Journal of Integrative and Comp. Zoology*, 35(2): 83–90.

Christensen, J. W. 1984. *Global Science: Energy, Resources, Environment*. 2nd ed. Dubuque, IA: Kendall-Hunt.

Lovelock, J. E. 1987. *Gaia: A New Look at Life on Earth*. Oxford: Oxford University Press.

Odum, E. P. 1993. *Ecology and Our Endangered Life Support System.* 2nd ed. Sunderland, MA: Sinauer Associates.

Raven, P. H., and G. B. Johnson. 1995. *Biology.* 4th ed. St. Louis, MO: Mosby–Year Book.

Rodale, R. 1972. *Ecology and Luxury Living May Not Mix.* Emmaus, PA: Rodale Press.

Schneider, S., and R. Londer. 1984. *Coevolution of Climate and Life.* San Francisco: Sierra Club.

Scott, K. M., and C. R. Fisher. 1995. Physiological ecology of sulfide metabolism in hydrothermal vent and cold seep vesicomyid clams and vestimentiferan tube worms. *Amer. Zoologist: A Journal of Integrative and Comparative Zoology,* 35(2): 102–111.

Shapley, Harlow. 1964. *Of Stars and Men: The Human Response to an Expanding Universe.* Boston: Beacon Press.

Stevens, T. O., and J. P. McKinley. 1995. Lithoautotrophic microbial ecosystems in deep basalt aquifers. *Science,* 270: 450–454.

Wallace, R. A., J. L. King, and G. P. Sanders. 1984. *Biosphere: The Realm of Life.* Glenview, IL: Scott, Foresman.

Basic Ecosystem Ecology

Chapter 4

ECOLOGICAL
PRINCIPLES:
ECOSYSTEMS

The principles of ecology are relatively easy to state, more difficult to understand, and virtually impossible at this point to appreciate in their full ramifications. At the risk of oversimplification, it is helpful to enumerate some of the most widely recognized ecological principles that have broad applications in global ecology. Subsequent chapters discuss these principles more fully.

First of all, ecological systems can be studied at different levels of organization: individuals, populations, communities, ecosystems, biomes, the biosphere, and the ecosphere. These constitute a hierarchy of increasing complexity.

Ecology can begin at the level of the individual *organism;* the object of study may be an individual plant, animal, or microorganism and its environmental adaptations. These adaptations include the ways an individual organism responds to physical factors such as temperature, light, and moisture, as well as to biotic factors in the environment, that is, all other organisms.

Populations are groups of individuals, members of the same species in a definable area. Populations obviously have properties not possessed by individuals, such as spatial distribution, age and sex ratios, birth rates, and death rates. Sometimes the definition of a population states that individuals should be potentially interbreeding, to distinguish isolated populations of the same species. In common usage, the term "population" may have a biological meaning, as in the population of trout in a lake, or it may have an economic, geographic, or political meaning, as in the population of mule deer in Colorado, or the population of people in Denver.

A biotic *community* consists of all the different populations of living organisms in a definable area, such as the community of different species in a lake or a forest. Communities involve relationships between species, and species may be classified according to their roles in the community. The role of a species in a community is known as a *niche*.

Ecosystems are defined as biotic communities interacting with their physical environment. This interaction involves cyclic exchanges of elements and compounds, such as oxygen, nitrogen, carbon, water, and nutrients. These exchanges are known as biogeochemical cycles, and they involve a flow of energy.

Biomes are large regional ecosystems with characteristic plant and animal associations, usually covering a large geographic area. Some examples are the grassland or prairie biome, the coniferous forest biome, the desert, and the tropical rain forest.

Collectively, all the communities and biomes of the world constitute the *biosphere*. As discussed in Chapter 3, the biosphere is the global assemblage of all living organisms.

Finally, the *ecosphere* consists of the biosphere interacting with all other components of planet earth. In brief, the ecosphere is the global ecosystem.

The principles of ecology have been publicized and reiterated many times, but it is still helpful to review some of them to serve as a guideline for subsequent discussions. All of these concepts are developed more fully in excellent texts by Colinvaux (1993), Odum (1993), Ricklefs (1990), and Smith (1992).

SOME BASIC PRINCIPLES OF ECOLOGY

Ecology is a vast and complicated subject, the broadest and most inclusive of the natural sciences. Many of its principles, however, relate to familiar aspects of general biology. The following list is not final or complete but highlights some of the most basic principles that have relevance to global problems. Some of these principles may seem vague and abstract, others may appear to be the painful elaboration of the obvious, but they have innumerable applications to our world and are often forgotten in human affairs.

1. While it is convenient to think of ecology in terms of levels of organization, all levels overlap and interact; that is, individuals in populations, communities, and ecosystems all have mutual influences.

2. Within ecological systems, virtually everything is related to everything else. Ultimately, nothing stands alone. Every component may potentially influence every other component.

3. The abundance and distribution of living organisms on earth, including ourselves, are affected by many aspects of the physical, biotic, and social environment. Not all aspects of the environment are equal in their influences. Limiting factors play dominant roles; these are factors that limit or regulate the abundance and distribution of organisms. The identification of limiting factors is an important part of ecological research.

4. Biotic communities differ in the numbers and kinds of species, the diversity of species, and in the roles or ecological niches of species in the community. Some communities, typically those in moist tropical environ-

ments, have great numbers of species, whereas other communities, often those in harsh environments with extreme conditions of temperature or dryness, have relatively few species.

5. The species in biotic communities form networks of relationships, or symbioses, which may be either favorable or unfavorable for each species. Examples of such interspecific relationships include cooperation, commensalism, mutualism, competition, predation, and parasitism. These relationships will be discussed in later sections.

6. Different ecological communities have analogous components and may be similar in basic organization even though their constituent species are very different. If the ecosystems are complete, they have identifiable producers, consumers, and decomposers, or recyclers. Microscopic green algae in lakes and giant redwood trees on land have analogous roles as producers; that is, both produce organic compounds through photosynthesis.

7. Biological interactions are usually multiple and cumulative. Rarely, if ever, do interactions occur singly or in isolation. One interaction often triggers another. For example, a decline in bird populations due to pollution or habitat loss may lead to an outbreak of insects which will affect different species of plants and thereby influence many other aspects of the biotic community.

8. Interactive synergisms are frequent. Several factors working in combination often have a result that is more than the sum of their individual effects. An example would be the effects of increased temperature and increased nutrients (such as nitrates and phosphates) on the growth of green algae in a lake or river. Either one of these factors alone would very likely increase algal growth, but together they could cause an explosive growth of undesirable algal blooms.

9. Ecological reactions or interactions are often delayed. The initial result of any one action may be inapparent or undetectable, but a response may occur at a later time, sometimes much later. In the early uses of DDT in the 1940s and 1950s, pest insects were controlled without any undesirable side effects. Many years later, several species of insectivorous and predatory birds died because toxic doses were accumulated through food chains. The true effects of certain ecological changes may not be evident for decades or even centuries. Even though atmospheric carbon dioxide has been increasing since the beginning of the industrial revolution in the nineteenth century, its full impact on global ecology will not be evident until the twenty-first century.

10. Threshold effects are common in ecological systems. Any given stimulus may be inadequate to produce a reaction, but at some point a slight increment beyond a threshold level may produce a pronounced effect. Stimulus-response curves in ecosystems, as in individuals, may be linear, but are often exponential, which means that ecological effects accelerate rapidly after a period of no noticeable response.

11. Human populations are subject to ecological principles, although we artificially dissociate ourselves from living systems. We buy food from grocery stores and supermarkets, but it originates in living ecosystems. All

of our activities and even our advanced technologies are ultimately dependent on nature.

12. Planet earth is a finite and closed ecological system that depends on solar energy. Within a closed ecological system, all life-support materials must eventually be recycled, and no component of the system can expand indefinitely.

These principles certainly do not cover all aspects of ecology, but collectively they represent a short course in ecology. Their importance to global ecology will become more evident as we review the concept of ecosystems.

ECOSYSTEM STRUCTURE

A system is defined in Webster as "an aggregation or assemblage of objects united by some form of interaction or interdependence" or "a group of diverse units so combined by nature as to form an integrated whole." Such definitions can encompass anything from a television set or automobile to a pond or forest. A system with biological components is an ecological system or ecosystem.

More specifically, an ecosystem is a definable biotic community and its nonliving environment functioning as an interacting system. Ecosystem interactions involve the flow of energy through the system, and the cycling of materials between the living and nonliving components. Both will be considered later.

We have already mentioned the fallacy of separating components of interacting systems, but it is the nature of science to take systems apart, classify the parts, and then try to put them back together again. The goal is to understand how they really work, a goal more easily achieved in mechanical systems than in biological ones. Nonetheless, the procedure has certain logical merits in sciences ranging from biochemistry and physiology to ecology.

In the initial analysis of ecosystems it is convenient to consider four major components: abiotic components, producers, consumers, and decomposers, or detritivores.

ABIOTIC SUBSTANCES

The term *abiotic* means "without life" or "nonliving." Inorganic substances such as water, oxygen, sodium chloride, nitrogen, and carbon dioxide are abiotic when they are physically outside living organisms, but once within living organisms they become part of the biotic world. Many elements may be tightly bound in inorganic compounds, such as silicon in sandstone or aluminum in feldspar, and are essentially unavailable to living organisms. Elements that are normally very active in biological processes, such as oxygen, may exist in an abiotic form readily available to living organisms, such as O_2 or CO_2 in air, or they may exist in an inaccessible form such as silicon dioxide (SiO_2) in quartz, a major component of granite. As another example, potassium may be available to plants in the form of potassium chloride (KCl) in soil but relatively unavailable in the form of monoclinic feldspar or orthoclase ($KALSi_3O_8$), one of the commonest of minerals.

An important property of an ecosystem that determines its productivity is the form and composition in which bioactive elements and compounds occur. For example, an ecosystem may have an abundance of vital nutrients, such as nitrates and phosphates, but if they are present in relatively insoluble particulate form, as they would be if linked to ferric ions, they would not be so readily available to plants, as they would be in the form of potassium or calcium nitrate and phosphate. One of the regulatory aspects of ecosystem function is the rate of nutrient release from solids, for this can determine the metabolism of the entire system.

PRODUCER ORGANISMS

Producers are green plants and bacteria that synthesize organic compounds. They are said to be *autotrophic,* or self-productive, because they take inorganic compounds and manufacture organic materials and living protoplasm from them. All green plants, algae, and cyanobacteria (blue-green algae) are producer organisms since they exhibit photosynthesis. Some bacteria are producers if they exhibit chemosynthesis. For example, sulfur bacteria are chemoautotrophic in that they oxidize elemental sulfur (S) to produce sulfate ions (SO_4^-), a basic step in organic synthesis. Some nitrogen-fixing bacteria, such as *Nitrosomas,* oxidize ammonia (NH_3) to nitrites (NO_2^-), and other nitrogen bacteria continue the process to nitrate form (NO_3^-). Deep-sea thermal vents represent a type of ecosystem totally dependent upon chemoautotrohic bacteria, in which bacteria oxidize hydrogen sulfide (H_2S) issuing from hot-water vents in the ocean bottom.

In quantitative terms, the most important producers are green plants and algae which take carbon dioxide and water, and with the energy of sunlight and enzyme systems in chlorophyll, produce carbohydrates. The generalized equation for this reaction is

$$CO_2 + 2H_2O + light\ energy \rightarrow (CH_2O) + H_2 + H_2O + O_2$$

This process, also known as carbon fixation, involves many stages and a variety of intermediate pathways. Some kinds of plants produce intermediate compounds with three carbon atoms (known as C_3 plants), whereas others produce intermediate compounds with four carbon atoms (C_4 plants). Both pathways involve the Calvin cycle, named after Melvin Calvin of the University of California who received the Nobel Prize for his work on the biochemistry of photosynthesis. Many C_4 plants, such as corn and Bermuda grass, have evolved in hot, dry climates, and their biochemical mechanisms of photosynthesis are more efficient at utilizing CO_2. The photosynthetic return for C_4 plants is greater per gram of water used than for C_3 plants which live in moister habitats. C_3 plants, on the other hand, are adapted to moderate temperatures and light regimes and can tolerate shade or cloudy weather more readily.

Ultimately, the end products of photosynthesis are carbohydrates such as glucose and fructose, and in highly oversimplified form, the total chemical reaction may be represented by

$$6 \ CO_2 + 6 \ H_2O + light \rightarrow C_6H_{12}O_6 + 6 \ O_2$$

Ecosystems can be compared according to the rates of carbon fixation or carbohydrate production per unit area per unit time. These rates provide a quantitative way of comparing the relative productivity of diverse natural or artificial ecosystems.

CONSUMER ORGANISMS

Consumers in an ecological sense are organisms that utilize the organic materials manufactured by plants and chemosynthetic bacteria. They are unable to produce their own organic compounds for basic energetic and nutritive purposes. Animals can synthesize certain vitamins and enzymes that are essential in nutrition, but they cannot manufacture their own basic food. Hence, animals are said to be *heterotrophic,* meaning they have different or varied nutritional sources.

Primary consumers or herbivores directly consume the organic compounds of plants. Secondary consumers may be omnivores or carnivores which depend partially or entirely on other animals for food. Tertiary and quaternary consumers may be second- or third-stage predators, such as a muskellunge fish feeding on a small bass that had eaten a mosquito that had bitten a rabbit. At the base of a consumer chain is an herbivore that has fed directly on some plant, algal, or bacterial product.

All the consumer food chains in a community are referred to as the *trophic structure* of the ecosystem. Trophic structure is discussed more fully in Chapter 6.

DECOMPOSER ORGANISMS AND DETRITIVORES

Decomposer organisms are bacteria and fungi that break down organic compounds into simpler substances. Their nutrition is termed *saprophytic* because it is associated with decaying material. In a sense, decomposers are the digestive organisms or the recyclers of an ecosystem. They reduce the complex organic molecules of dead plants and animals to simpler compounds that can be absorbed by green plants as vital nutrients to begin the cycle of life once again. Three basic processes may be involved in the breakdown of carbohydrates: aerobic respiration, which is essentially the reverse of photosynthesis in its results, anaerobic respiration, and fermentation. In highly simplified chemical terms, these processes are represented by the following equations:

Aerobic respiration:
$$C_6H_{12}O_6 + 6 \ O_2 \rightarrow 6 \ CO_2 + 6 \ H_2O + energy$$

Anaerobic respiration (glycolysis):
$$C_6H_{12}O_6 + 2\ NAD^+ + 2\ ADP + 2\ P \rightarrow$$
$$2\ C_3H_3O_3^- + 2\ NADH + 2\ H^+ + 2\ ATP\ (pyruvate)$$

Fermentation:
$$C_6H_{12}O_6 + 2\ ADP + 2\ P \rightarrow$$
$$CO_2 + 2\ C_2H_5OH + 2\ ATP + 2\ H_2O$$

In the above equations, NAD and NADH are coenzymes, known as nicotin-amide adenine dinucleotide and nicotinamide adenine dinucleotide phosphate, which catalyze the reaction; ADP and ATP (adenosine di- and triphosphate) play the key role in energy transfer. The above equations are very simplified formulations and each involves many steps. In glycolysis, for example, the citric acid cycle, or Krebs cycle, involves 10 major biochemical steps.

In fermentation, carbohydrates are broken down into carbon dioxide, alcohol (C_2H_5OH), and water. Fermentation processes also break down more complex organic substances through the action of enzymes produced by molds, bacteria, and yeasts. For example, the fermentation of milk by bacteria changes milk sugar (lactose) into lactic acid, which we recognize as the souring of milk. If we consider the breakdown of innumerable organic compounds such as proteins, lipids, and a variety of carbohydrates, it becomes apparent how complex decomposition processes are. If decomposers were not present in ecosystems, organic compounds would be locked into complex insoluble molecules that could not be utilized as nutrients by plants. Hence, decomposers are necessary for the renewal of life.

Detritivores are larger organisms, including scavengers, that utilize dead organic material and extract energy from the process of decomposition. Many soil invertebrates, such as earthworms and millipedes, are detritivores, as are many bottom-dwelling invertebrates in water. Large scavengers such as vultures and other carrion eaters are also detritivores in that they consume primarily dead organic material.

Ecosystems involve a variety of life forms not specifically mentioned in the preceding paragraphs, even though most components of an ecosystem can be classified as producers, consumers, or decomposers. For example, parasites are specialized consumers. Plant parasites feed directly on plants and are thus herbivores; in effect, they differ from grasshoppers or rabbits only in their attachment to a plant for more prolonged feeding periods. Animal parasites derive their nutrition from living animals and are thus carnivores; they differ from predators only in the fact that they normally do not kill the host and they remain attached to the host for a longer time. Both plant and animal parasites may be either internal (e.g., nematodes) or external (e.g., aphids, ticks).

Scavengers, considered as detritivores above, might also be considered carnivores in the case of vultures or hyenas, which differ from predators in that they feed on an animal after it has died from some other cause. Lions and coyotes are both carnivores and scavengers; they sometimes kill their own prey, but they also consume dead animals that they may come upon. The point of this type of classification is not necessarily to fit each species into a single precise or rigid classification, although many do have a single function, but rather to have some understanding of the role of each species in its ecosystem. Figure 4.1 is a general diagram of major ecosystem components, and Figure 4.2 portrays familiar producers, consumers, and decomposers in a terrestrial ecosystem.

INCOMPLETE ECOSYSTEMS

Most ecosystems contain all four basic components discussed above, although incomplete ecosystems exist in which one or more components are lacking. An example of an incomplete system lacking producers is a cave where consumers and decomposers exist, but no green plants or appreciable communities of chemo-autotrophic bacteria are present. Cave communities harbor a variety of organisms, including bats, insects, arachnids, and heterotrophic bacteria. Practically all cave-dwelling animals must depart from the cave for feeding periods, as do bats, or depend on extrinsically produced nutrients that enter the cave by flowing water or seepage.

Many abyssal regions of the oceans away from thermal vents also have communities of heterotrophic organisms but lack producers. The heterotrophs live on fallout detritus from upper layers of the ocean. Predators, such as some of the bizarre deep-sea fish, feed upon the detritivores.

The central cores of most large cities might also be considered incomplete ecosystems without producers, at least from the human standpoint. Some green plants obviously exist, and most cities have inner-city parks and tree-lined avenues, but these do not provide significant food production for humans and most

Figure 4.1 Basic components of an ecosystem. The arrows represent some of the major pathways of organic compounds (from Southwick, 1976).

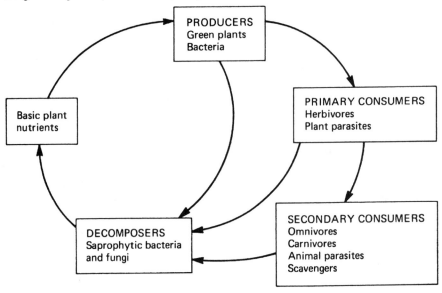

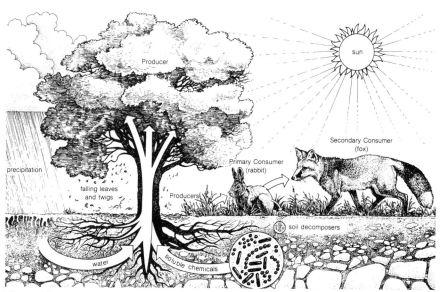

Figure 4.2 Diagrammatic representation of a familiar terrestrial ecosystem showing producers, consumers, and decomposers (from Miller, 1982).

of the vertebrate animals living within the inner city, including dogs, cats, pigeons, starlings, and sparrows. For all of these, the inner city requires extrinsic production and imported food. The only other alternative is for the inhabitants to leave the city and feed in surrounding areas. This occurs with starlings, creating an ecological situation analogous to bats in a cave. The lack of production in an inner city is not due to a lack of light but to a lack of soil and suitable substrate. Inner-city gardening is an exception to this general scenario, but even the most productive gardens provide only a small portion of the food requirements of all urban inhabitants.

In other ways, cities can be considered incomplete ecosystems, dependent upon the surrounding landscape. Not only do they import food, but they must also import fresh air and water. At the same time, cities must export waste products: sewage, solid waste, carbon dioxide, sulfur dioxide, and a host of other pollutants that cannot be decomposed or recycled within the city. Hence, all cities have "sanitary landfills," formerly called dumps, and wastewater treatment plants where the waste products and detritus of the city can be processed. If cities were cut off from their surrounding environments, they would soon perish from thirst, starvation, asphyxiation, or the accumulation of waste products. In exchange for this life support, cities, of course, provide a great many cultural and economic benefits: jobs, housing, transportation, manufacturing, education, entertainment, and so forth. So the relationship between city and landscape is vital in both directions, but it is particularly important to remember, as cities expand, that they cannot sustain themselves. They are not complete ecosystems.

Incomplete ecosystems could also exist in specialized cases where producers and decomposers are present without consumers. A theoretical example is a massive bloom of algae in an aquatic ecosystem where a toxic metabolite of the algae creates unfavorable conditions for zooplankton, fish, and all other consumers. In such a case, excessive production and decomposition go hand in hand. This highly unstable and undesirable situation has been known to occur in limited areas of severe red-tide outbreak, caused by dinoflagellates such as *Gymnodinium brevis,* or in small stagnant ponds with dense populations of the toxin-producing algae *Microcystis.*

Agriculture is basically an attempt to maintain relatively incomplete ecosystems dominated by producers, so that the production and yield can be harvested for human consumption. In efficient agricultural practice, all feasible efforts are made to reduce extraneous consumption and decomposition. Crops are sprayed with insecticides to reduce insect herbivores, seeds are treated with fungicides to reduce mold and decomposition, and vertebrate pest-control programs are employed to reduce rats, mice, and birds that might consume production. Hence, from the human standpoint, a corn field or an orchard is an incomplete ecosystem tightly managed for production.

CYBERNETIC CONTROL IN ECOSYSTEMS

Ideally, in complete ecosystems a balance is maintained between the major components. A system of feedback controls normally regulates component abundance. For example, if an increase in production occurs due to favorable weather, consumers will benefit and will also increase in number, which then has the effect of reducing the number or quality of producers. In this case, the increase in producers is said to create positive feedback on consumers, whereas the increase in consumers creates negative feedback on producers. An excessive increase in either one will ultimately lead to an increase in decomposition. Hence, an ecosystem is normally a network of checks and balances, or interplays of positive and negative feedback relationships. This interplay is the essence of a cybernetic or self-regulating system.

The fact that ecosystems are self-regulating does not mean they are simple or necessarily stable, however. Simplicity rarely prevails because most natural ecosystems have innumerable actors in each major component, and feedback interactions involve many factors, such as numbers, biomass, rates of metabolism, reproduction, and mortality. Stability rarely prevails because ecosystem feedback processes often involve lags, thresholds, cumulative effects, and synergisms, as discussed earlier. They do not work with precision. Furthermore, ecosystems often change slowly over time, through an ecological process known as *succession* in which populations replace each other and gradually change community structure. Thus a field of grasses and forbs may change to shrubs, pine forest, or deciduous forest in the course of a century depending upon local soil and climate. A pond or shallow lake may fill in with aquatic vegetation and eventually become a wet meadow or bog.

Ecosystems may be characterized as young, transitional, or mature, and each stage has certain definable traits. Young or pioneer stages usually have low species diversity, low total biomass, low nutrient storage, and high rates of production in relation to biomass. Mature or climax stages typically have higher species diversity, high biomass, higher nutrient storage, and low rates of production in relation to biomass (Odum, 1993). The course of ecosystem development depends upon many factors such as topography, soil, climate, and frequency and degree of disturbances.

SYSTEMS ANALYSIS AND MODELING IN ECOLOGY

Ecosystems are sufficiently complex and dynamic that they cannot always be studied by traditional techniques of individual research and scholarship. Nor can they be subjected to laboratory analysis and simple experimentation. Broadly based team research, systems analysis, and modeling are often necessary. Although these terms have a certain mystique about them, they are basically nothing more than methods of looking at phenomena and problems in a systematic and quantitative manner. This is, of course, more easily said than done. While it is important to be comprehensive, it is not possible to evaluate every conceivable factor influencing an ecosystem, so it becomes especially important to identify key variables and regulating factors.

Systems analysis is a logical scientific method that approaches complex phenomena in several stages: (1) *systems measurement,* in which the objectives are outlined and data are obtained on important variables relating to these objectives; (2) *data analysis,* in which statistical relationships between variables are examined, and key variables most important in regulating the system are identified; (3) *systems modeling,* in which functional or mathematical models are composed to provide a theoretical basis for interpreting the variables; (4) *systems simulation,* in which variables are manipulated mathematically to test the model and predict the consequences of changes within the system; and (5) *systems optimization,* in which the best strategies for achieving the objectives are selected.

This approach is basically similar to standard scientific procedure which involves description, classification, hypothesis formation, experimentation, and testing. Systems analysis, however, can be applied to very complex systems that require computerized data storage and analysis and do not lend themselves to direct experimentation. One cannot easily or intentionally experiment with a large ecosystem such as a river, a city, an ocean, or a range of mountains. Thus, mathematical modeling and computer simulation is sometimes the only feasible experimental approach in large-scale ecological research. Team research is also a usual component of systems analysis. One individual rarely possesses all the skills necessary for a complete systems approach.

Often we do have the opportunity to observe the effects of inadvertent experiments in large-scale ecosystems, for example, when we clearcut a forested watershed, overgraze a grassland, or pollute a lake; or when a forest fire destroys a stand of timber, or an industrial accident creates a huge oil spill. Such incidents provide unintended ecological experiments, but they are usually uncontrolled experiments and are not specifically designed for scientific study.

The stages of systems analysis can be illustrated with the common problem of evaluating pollution and improving water quality in a river or lake. The first stage, *systems measurement,* involves recording water quality and river characteristics (e.g., volume, depth profiles, flow rates, current patterns, temperatures, turbidity or clarity, dissolved oxygen, nutrient levels, trace-element concentrations, bottom characteristics) and biological characteristics (e.g., bacterial, algal, zooplankton, insect, fish, amphibian and waterfowl populations, use of the river or lake by terrestrial vertebrates). In systems measurement, scientists make a detailed inventory of existing conditions to come up with an accurate description of the river or lake in its present state. Finally, systems measurement identifies the major influences on the stream or lake. These can include the watershed, land-use patterns, geological formations, weather, domestic inputs, industrial operations, and commercial and recreational uses. It may take a team of five or ten scientists several years to obtain baseline data and understand the prevailing situation. It is important as well to gain some idea of seasonal and annual variability in the parameters studied.

Stage two, *data analysis,* involves computer and statistical analysis of relationships between variables. How do land-use patterns, industrial operations, and domestic inputs interact with topography, weather, and stream-flow characteristics to influence the physical and biotic qualities of the river? Some specific questions in this area would be: What are the sources of the nutrients? How do the nutrient levels influence the biotic communities? Are oxygen and temperature levels within satisfactory ranges? If not, why not? Are toxic chemicals present in the steam? If so, what are their origins and their effects, and what happens to them? Because many questions of this type can be asked, multivariate analyses, as well as other statistical procedures, are often utilized.

The third stage, *systems modeling,* is undertaken when the investigators feel they have a lead on the origins and relationships of the important variables within the system. They can then propose a mathematical relationship between various factors influencing the prevailing conditions within the river. Data analyses might suggest, for example, that pesticide runoff from agricultural operations is poisoning the river and is responsible for the absence of certain organisms. The analyses might also suggest that excessive nutrients from domestic sewage are creating high bacterial populations which are reducing dissolved oxygen below satisfactory levels. From these and other relationships implied by data analysis, mathematical models can be constructed to examine the effects of eliminating pesticide runoff from a certain percentage of the watershed or of reducing domestic sewage input from a certain number of towns in the watershed.

The fourth stage, *systems simulation,* consists of computer runs with different variables modified. The model could then be used to project changes in other variables. For example, systems simulation could project changes in the river if agricultural practices were altered, if sewage treatment plants were installed, or if industrial operations were curtailed. Modern computer techniques are essential at several stages in systems analysis, especially in data analysis and systems simulation. Ideally, an integrated computer technology program should be utilized throughout the entire systems research if possible, from data collection, storage, and analysis to systems optimization.

If the ultimate purpose is management to solve a problem, a fifth stage, *systems optimization,* is used to determine the best course of action to achieve the goals. This process relates the biological and physical data with economic feasibility studies. Alternatives and options could then be examined in realistic terms. Scientifically, optimization is not necessarily a part of systems analysis, but in the practical world of management, conservation, and restoration, it is an appropriate goal. Figure 4.3 summarizes the stages of systems analysis that can be used in any ecologic problem.

Systems analysis is also essential in working in the opposite direction from that outlined above, in determining, for example, the effects of a new industrial operation or land-use modification on existing conditions in an ecosystem. What would be the effects, for example, of constructing a series of dams on the Mekong River

Figure 4.3   Basic stages of systems analysis and management (from Southwick, 1976).

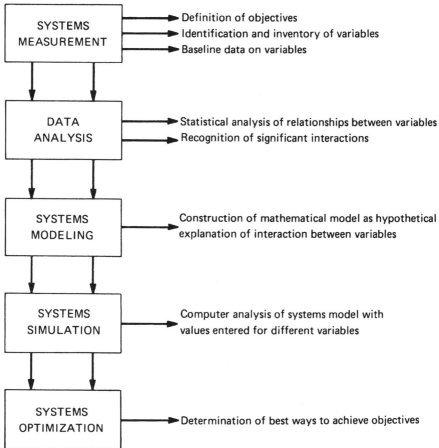

in Southeast Asia or of basing all new electric power plants on coal and petroleum combustion if nuclear power is deemed too risky? Virtually all human activities and construction projects have environmental impacts on air and water quality or agriculture or wildlife populations, and it is usually difficult to predict the true nature of these impacts. Broad-spectrum system analyses are needed for many environmental and natural resource problems throughout the world. Mathematical modeling and systems analysis may not provide the final answer to environmental problems, but in many situations they make it possible to understand the relationships among the complex variables that exist in ecologic problems. They provide the only reasonably objective way of evaluating the cost-benefit relationships of environmental modifications, assuming that accurate and complete data can be obtained. There are many problems, of course, in relating economic and ecologic values; it is not always possible to assign numerical cost or benefit figures to some ecological considerations.

It is also essential to remember that systems analysis initially depends upon accurate descriptive data gathered by careful field study. The most sophisticated computer technology and mathematical models are of little help if the basic ecologic data are incorrect or grossly incomplete. The field biologist still plays an essential role in the most elaborate ecosystem analysis. In some circumstances, just as in weather prediction, a naturalist or field ecologist who has never seen a computer might have knowledge and experience of more value than an incomplete or hastily composed systems analysis program that uses the most advanced computer technology. Aldo Leopold never used computers, yet he provided some of the most accurate insights available on the ecology and management of America's environmental problems. In modern times, however, ecological modeling is an advanced and rapidly moving field of science (Robertson, 1991).

SYSTEMS ANALYSIS IN GLOBAL ECOLOGY

Systems analysis and ecological modeling are absolutely essential in trying to understand environmental problems affecting the entire world. The complex issues of global warming, for example, require that we consider many factors: air pollution and atmospheric chemistry, patterns of solar radiation, the dynamics of weather and climate, cloud formation, ocean circulation, global and regional precipitation, terrestrial vegetation, land use, human population growth, regional and global temperature regimes, and even astronomical factors such as the orbital and axial patterns of planet earth's rotation. With this complexity of variables, it is no wonder that there is uncertainty and controversy about the reality, extent, and timing of global warming. We will look at some relevant data and discuss these issues in Chap. 18.

Many other global problems are similarly complex. Human population growth is affected by a great variety of ecological, economic, sociological, and medical factors. Agricultural production is influenced by weather, climate, soil, energy resources, human activities, pollution, and so on. Biodiversity is affected by population pressures, global warming, deforestation, desertification, and agricultural expansion. All these of topics will be considered in later chapters. At this point, we will simply emphasize the necessity of broad-scale systems analysis and ecological modeling to achieve some scientific understanding of these problems.

SUMMARY

The concept of the ecosystem provides the central theoretical framework for ecology. This concept focuses on the interactions and exchanges between living organisms and their nonliving environment.

Ecosystems can be analyzed in terms of four basic components: abiotic factors, producers, consumers, and decomposers, or detritivores. In complete ecosystems, all components interact to produce an exchange of materials and a flow of energy. Incomplete ecosystems exist which may lack one or more components.

Newer studies are quantifying different components of ecosystem structure and function, so that interdependent relationships are now understood more precisely. It is increasingly important for us to comprehend the dynamics of ecosystems and their patterns of diversity, change, and stability. It is especially important for us to understand the impact of human activities on ecosystem structure and function. We are now aware that human activities influence many global processes including atmospheric conditions. There is no longer any doubt that human populations are a dominant ecological force on planet earth.

Systems analysis and ecological modeling offer the best prospects for evaluating large-scale ecosystem change. Such studies involve large databases and multidisciplinary approaches.

REFERENCES

Colinvaux, P. 1993. *Ecology 2*. New York: John Wiley and Sons.
Miller, G.T. 1982. *Living in the Environment*. 3rd ed. Belmont, CA: Wadsworth.
Odum, E. P. 1993. *Ecology and Our Endangered Life Support Systems*. 2nd ed. Sunderland, MA: Sinauer Associates.
Ricklefs, R. E. 1990. *Ecology*. 3rd ed. New York: W. H. Freeman.
Robertson, D. 1991. *Ecologic-Logic-based Approaches to Ecological Modelling*. Cambridge, MA: MIT Press.
Smith, R. L. 1992. *Elements of Ecology*. 3rd ed. New York: HarperCollins.
Southwick, C. H. 1976. *Ecology and the Quality of Our Environment*. 2nd ed. Boston, MA: Willard Grant Press.

Chapter 5

ECOSYSTEM
ORGANIZATION AND
FUNCTION

Ecosystems can be best understood by reference to the cycling of materials and the flow of energy. Elements and compounds cycle and recycle throughout ecosystems in biogeochemical cycles. The term implies a passage of materials from the living to the nonliving world and back again. In contrast, energy flows through the system without returning. It's passage is unidirectional and not circular. Both biogeochemical processes and energy flows have many applications to global ecology. In this chapter we review some principles of the major biogeochemical cycles, and in the next, energy flow.

Many elements and compounds enter and exit living organisms through the processes of photosynthesis, ingestion, metabolism, growth, and decomposition. These processes can be illustrated with the hydrologic, carbon, nitrogen, phosphorus, and sulfur cycles.

THE HYDROLOGIC CYCLE

In the hydrologic cycle, water passes through all four "spheres" (bio-, atmo-, hydro-, and litho-) in a variety of physical and biological processes (Fig. 5.1). Plants absorb water and utilize it in photosynthesis to produce hydrocarbons. It thus becomes a part of protoplasm and basic plant tissues. Water is also retained in plants without chemical modification. From plants, water may be incorporated into animals eating the plants, or it may pass on to the atmosphere as water vapor by either respiration or direct transpiration. Transpiration is a major route for the return of water to the atmosphere. A temperate forest may transpire an average of 2,500 gallons of water to the atmosphere per day (McCormick, 1959), and this moisture is an important factor in maintaining humidity and atmospheric moisture in cloud formation. Water vapor is also returned to the atmosphere by direct evap-

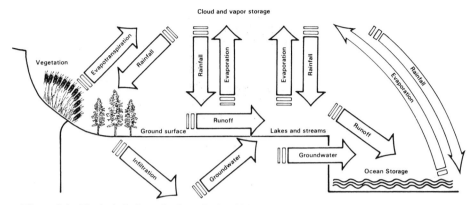

Figure 5.1 The hydrologic cycle (from Cook, 1984).

oration from surface waters, and the two processes together occurring from wet-leaf surfaces are called evapotranspiration.

Water returns to the earth's surface through precipitation, as rain, snow, sleet, hail, and dew, and then re-enters the liquid hydrosphere. The total cycling time may be short, occurring in hours or days, as in July thunderstorms when evaporated water may quickly form clouds and precipitate back to earth, or it may be much longer. Cycling time is probably weeks in cases where water is incorporated into plant tissue. It may be months or years when water enters underground aquifers or benthic depths of lakes and oceans. It may be decades or centuries when it enters glaciers and polar icecaps.

The hydrologic cycle is a topic of great interest in global ecology, in agriculture, in the study of deserts and droughts, water supplies, pollution, and in numerous other areas of ecological interest. There is certainly no shortage of water on earth, but its distribution and quality often present major environmental problems. Most of the earth's water is saline (at least 97 percent), and significant portions of fresh water are frozen or polluted. Only about 1 percent of all water on earth is available for human use (Graves, 1993). Storms, floods, and droughts are common problems as well, and these will be considered in subsequent chapters.

THE CARBON CYCLE

Carbon, a key element in all living organisms, cycles relatively simply between plants, animals, and the inorganic world. Figure 5.2 portrays the general pathways of carbon in an ecosystem. The diagram is greatly simplified. Cycles on land and water are somewhat separate, but they are basically similar and are interrelated at the air-water interface.

Carbon exists in the atmosphere primarily as carbon dioxide, in which form it is incorporated directly into plant protoplasm through photosynthesis. A tropical rain forest can incorporate between one and two kilograms of carbon per square meter per year in organic compounds, equivalent to 30 tons per hectare of phyto-

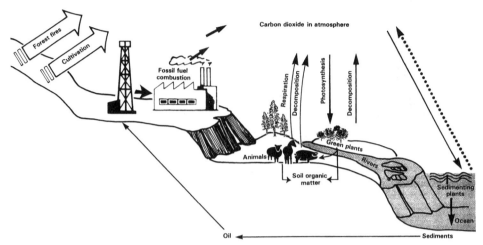

Figure 5.2   The carbon cycle (from Cook, 1984).

mass production, or plant growth (Reichle, 1975). About 70 billion tons of carbon are fixed into organic compounds per year on a global basis, representing approximately 10 percent of the earth's total carbon dioxide (Raven and Johnson, 1992).

From plants, animals, and bacteria, organic carbon can enter animals, where it goes through various stages of digestion and assimilation, and from either plants or animals it can re-enter the atmosphere as CO_2 through respiration. If it remains in plants and animals, it ultimately passes into dead organic material, from which it can return to atmospheric CO_2 by oxidation or decomposition. In some animals, carbon is tied up in hard parts, such as shells, and thus remains in the form of carbonates for a long time. Limestone can result from marine deposits of animal carbonates as well as from inorganic precipitation of carbonates in water. These carbonates in limestone can then return to the living carbon cycle only very slowly through a process of erosion and dissolution. Dissolved carbonates in water may be absorbed by plants. Some aquatic plants, for example, Eurasian watermilfoil *(Myriophyllum spicatum),* can use carbon in dissolved carbonates as a direct carbon source in photosynthesis. Most aquatic plants, or algae, however, are more efficient when using free CO_2 in water as a carbon source.

Carbon can also become locked into organic deposits of coal and petroleum, remaining in this form for millions of years until it is released in combustion. Fossil fuels (coal, petroleum, and natural gas) all come originally from plant and animal remains, which ultimately depended on photosynthesis and solar energy.

The carbon cycle is of special interest in global ecology because of increasing levels of carbon dioxide in our atmosphere. This increase is a result of increasing combustion of coal, petroleum, and wood by human activities, and possibly reduced uptake of CO_2 through photosynthesis due to deforestation. Also, agricultural activities usually produce a net increase of CO_2 because the CO_2 fixed by crops does not compensate for the release of CO_2 during fallow periods, plowing, cultivating, and harvesting (Odum, 1993).

The process of CO_2 increase has long-term consequences for climate, which will be discussed in Chapter 18. At this point, we can note that the carbon cycle demonstrates the importance of basic ecology in global concerns. This point is also emphasized by Clark (1989) in a special issue of *Scientific American* titled "Managing Planet Earth." Clark presented a world map of carbon emissions and discussed the role of the carbon, nitrogen, phosphorus, and sulfur cycles in global environmental change.

THE NITROGEN CYCLE

Nitrogen, another essential element of all protoplasm, has a more complex series of cyclic pathways through the ecosystem, dependent at many stages on bacterial metabolism (Fig. 5.3).

The largest reservoir of nitrogen is gaseous N_2 in the atmosphere, comprising about 78 percent of the air that we inhale. To become available to plants and the

Figure 5.3   A simplified diagram of the nitrogen cycle (from Southwick, 1976).

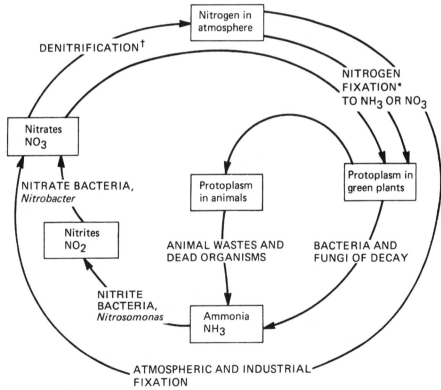

† *Denitrifying bacteria*
   Pseudomonas
   Thiobacillus
   Micrococcus denitrificans

* *Nitrogen fixing bacteria and Algae*
   Free-living: Azotobacter, Clostridium
   Leguminous: Rhizobium
   Blue-green: Nostoc, Anabaena

biosphere, gaseous nitrogen must be "fixed" or incorporated into compounds such as ammonia (NH_3) and nitrates (NO_3). From these nitrogen compounds, plants synthesize proteins, nucleic acids, and other essential nitrogen-containing compounds. Nitrogen fixation is accomplished by both free-living soil bacteria such as *Azotobacter* and *Clostridium* and by symbiotic bacteria such as *Rhizobium* which live in the root nodules of leguminous plants, including clover and alfalfa. If leguminous plants are plowed back into the soil, they can add as much as 300 to 350 kilograms of nitrogen per hectare per year to the soil (about 300 pounds per acre). Several other kinds of plants, other than legumes, also contain nitrogen-fixing bacteria in root modules, including alders, wax myrtle *(Myrica)* and mountain lilacs *(Ceanothus)*.

Other kinds of bacteria, including the cyanobacteria, or blue-green algae, such as *Anabaena* and *Nostoc,* can perform nitrogen fixation. One species of *Anabaena* lives symbiotically with a water fern *(Azolla)* which is purposefully introduced into rice paddies in Asia. It serves the important function of providing nitrogen fixation in the water, thus ensuring soil fertility despite intensive agricultural use. A healthy crop of *Azolla* can add 50 to 450 kilograms of nitrogen per hectare to a rice field (Raven and Johnson, 1992).

Nitrogen fixation also occurs as a physical process in the atmosphere through the ionizing effect of lightning and cosmic radiation and industrially through the Haber and Bosch method. The greatest single source of fixed nitrogen, however, is probably terrestrial bacteria.

After fixed nitrogen is incorporated into protoplasm by amino acid and protein synthesis, organic nitrogen compounds may follow any one of three courses: (1) they may be stored or modified as proteins and nucleic acids within the plant, (2) they may be incorporated into animal protein through consumption and assimilated by animals, (3) they may be decomposed through death and bacterial action. In animals, these compounds can be metabolically decomposed into urea and other excretory products.

In the decomposition cycle through death and decay, ammonia (NH_3) is produced from amino acids by the action of ammonifying bacteria such as *Pseudomonas* or *Proteus*. Under normal conditions, ammonia is quickly converted into nitrite form (NO_2) by nitrite bacteria such as *Nitrosomonas* and into nitrate form (NO_3) by nitrate bacteria such as *Nitrobacter,* through a process known as nitrification. Nitrates can then be absorbed directly by plants as primary nutrients and take part once again in organic synthesis within the plant.

The presence of ammonia nitrogen in ecosystems is a good measure of the relative balance between decomposition, bacterial action, and plant production. High levels of ammonia nitrogen in water usually indicate enrichment from fertilizer, animal wastes, or domestic sewage. Field measurements of ammonia nitrogen above the levels of 1 ppm generally indicate some major source of decomposition within the system in excess of that being utilized by bacterial action and plant growth. Field measurements of nitrites and nitrates are also good guides to the nutrient condition of lakes, streams, and estuaries. The same is true of dissolved phosphates, as we shall see later.

Nitrogen is returned to its atmospheric form by the actions of denitrifying bacteria such as *Pseudomonas, Thiobacillus,* and *Micrococcus denitrifcans.* Obviously, the cyclic flow of nitrogen throughout the ecosystem requires balances of bacterial action involving many species, so that appropriate levels of plant nutrients are maintained without excessive accumulation of decomposition products like ammonia. A 1969 report on environmental problems by a committee of the American Chemical Society pointed out that all life could be extinguished on earth by the extinction of perhaps just a dozen species of bacteria involved in the nitrogen cycle.

Nitrogen has an additional role in global ecology in the form of nitrogen oxides (NO_x) which enter the atmosphere from the combustion of fossil fuels, primarily petroleum products. Nitrogen oxides are air pollutants that contribute to photochemical smog, acid rain, and global warming.

THE PHOSPHORUS CYCLE

Phosphorus plays an essential role in almost every step of organic synthesis and has been described by Deevey (1970) as the "universal fuel of living organisms." Phosphorus in the form of ADP, ATP, and NADP is an energy source in photosynthesis and in many other metabolic pathways.

Phosphorus has its greatest reservoir in the lithosphere in the form of phosphate-bearing rocks. It is usually much less available to the biotic community than is nitrogen. In natural waters, it exists in the ratio of about 1 to 23 in relation to nitrogen (Odum, 1993). It is more likely than most other elements to limit ecosystem productivity.

In the protoplasm of plants and animals, phosphorus is broken down by cellular metabolism or the action of phosphatizing bacteria into dissolved phosphates, such as $CaHPO_4$ (Fig. 5.4). These dissolved phosphates may be utilized as primary nutrients in plants, or they may enter marine deposits and become fixed in relatively insoluble forms, such as $Ca_3(PO_4)2$, in phosphate rocks. Soluble phosphates usually reach plants through the metabolic actions of fungi known as mycorrhizae, which live in close symbiotic relationships with plant roots. Bone and guano de-

Figure 5.4   The phosphorus cycle (from Cook, 1984).

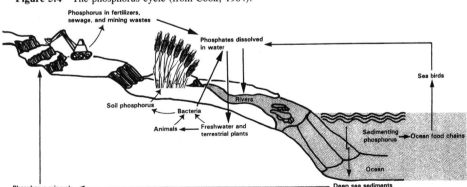

posits of birds may also lock up phosphates for long periods of time until they are dissolved by natural erosion or are artificially recovered in mining and crushing. The loss of phosphates to the oceans has generally been greater than the gain to land; soil erosion in good farmland in Illinois and Iowa has been estimated at 10 to 20 tons per acre per year (Odum, 1993). Fifty years of cultivation as practiced in the United States can readily reduce soil phosphate levels by one-third. The problem may be even greater in many countries such as India, where there is a shortage of fertilizers and the need to produce grain crops every year from the same parcel of land. India has interrupted the natural phosphorus cycle by burning animal wastes instead of using them as natural fertilizer. Cow dung is dried and used as a major fuel source in rural villages; this activity short-circuits both the nitrogen and phosphorus cycles and robs the soil of basic nutrients.

Although phosphorus is often a limiting factor in many ecosystems, a modern problem is excessive levels of phosphorus in lakes and rivers. Phosphates, or phosphorus anions, leach into surface waters from several pollution sources— domestic sewage, animal wastes, and phosphate detergents—stimulating plant growth and often producing undesirable plankton blooms. It has been estimated, for example, that Lake Erie formerly received over 15,000 tons of phosphorus yearly from detergents, wastewater, and agricultural runoff (Makarewicz and Bertram, 1991). The permissible phosphate loading for Lake Erie was estimated to be 9,000 pounds per day (1,600 tons/yr.). The excess phosphates created algal blooms, high bacterial counts, and the general problems associated with excessive eutrophication. These pollution discharges have now been reduced 84 percent in Lake Erie (Makarewicz and Bertram, 1991), and the quality of Lake Erie water has improved noticeably in recent years.

The main management problem associated with phosphorus is how to retain its levels in soils but reduce its levels in rivers, lakes, and estuaries; in other words, how to balance its loss from terrestrial ecosystems more closely with its return from the hydrosphere.

THE SULFUR CYCLE

Sulfur is a key element in proteins since it provides the linkage between polypeptide chains in protein molecules. Life on earth could not exist without sulfur, but it is less likely to be a limiting factor in ecosystem productivity than phosphorus. In nature, sulfur has a large reservoir in the soil and sediments as elemental sulfur and in various reduction and oxidation states including hydrogen sulfide (H_2S), sulfites (SO_2), and sulfates (SO_4). The oxidation of hydrogen sulfide is the basis of life in deep sea thermal vents.

Organic sulfur in plants and animals is reduced to hydrogen sulfide by bacterial action, and the H_2S is then oxidized to sulfates, such as ammonium sulfate, $(NH_4)_2SO_4$, by sulfur-oxidizing bacteria (Fig. 5.5). These sulfates are then taken up by plants as primary nutrients. Hydrogen sulfide can occasionally accumulate by rapid organic decomposition. In the Black Sea below 150 meters, concentrations of H_2S and H_2SO_4 can sometimes be so high as to exclude all forms of life

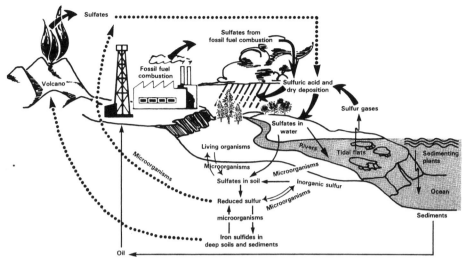

Figure 5.5 The sulfur cycle (from Cook, 1984).

other than sulfur bacteria (Brock, 1966). In polluted estuaries, large accumulations of algae, such as sea lettuce *(Ulva)* may undergo decomposition to produce obnoxious quantities of H_2S. In some coastal areas, this problem causes significant air pollution.

Sulfur is also a component of organic compounds in coal and petroleum; it is released in combustion as gaseous sulfur dioxide (SO_2), also an air pollutant. Sulfur dioxide forms dilute sulfuric acid in fog and rain and is a major component of acid rain.

SUMMARY

Biogeochemical cycles provide both a theoretical and practical basis for studying ecosystem function. They enable us to analyze the cyclic interactions of living organisms and their nonliving environments. Many principles of ecosystem dynamics are illustrated by the hydrologic, carbon, nitrogen, phosphorus, and sulfur cycles. These are representative of many such cycles involving biologically active elements and compounds.

In the hydrologic cycle, water passes through the atmosphere to the lithosphere and hydrosphere by precipitation, into the biosphere by absorption and photosynthesis, and back to the atmosphere by evapotranspiration. The quantity and quality of water is a key determinant of biological processes; there is no shortage of water on earth, but local disparities create many environmental problems.

In the carbon cycle, carbon dioxide in the atmosphere is incorporated into plant tissues through photosynthesis. It may then return to the atmosphere via respiration or decay, or pass into animals as organic carbon compounds. Carbon stored in coal or petroleum deposits is released as CO_2 during combustion. Atmospheric

CO_2 is increasing and promises to become a significant factor in global climate change.

The nitrogen cycle involves the microbial fixation of atmospheric nitrogen into ammonia and nitrates, compounds that serve as primary plant nutrients. Nitrogen-fixing bacteria exist primarily as free-living species in the soil and as symbiotic species in root nodules of legumes and some other plants. The decomposition of organic nitrogen involves other species of bacteria that return nitrogen to inorganic form. The nitrogen cycle is of special interest in soil fertility and agricultural production.

Phosphorus is much less abundant in the global ecosystem than is nitrogen, most of it existing in relatively insoluble form in phosphate rocks. Through slow dissolution, phosphates become available to plants as primary nutrients. Phosphates reach plants largely through the activities of mycorrhizal fungi. Through the decay and excretion of living organisms, phosphates are returned to soil by the action of phosphatizing bacteria. Phosphate levels are important regulators of plant growth in ecosystems. Low concentrations in soil or water may result in inadequate plant growth; high levels often result in excessive growth.

Sulfur in the forms of sulfates is also a basic plant nutrient and an essential component of proteins. In thermal-vent communities in the ocean, hydrogen sulfide is a primary sulfur and energy source in the absence of photosynthesis. Sulfur dioxide, a product of the combustion of coal and petroleum represents a major environmental factor in air pollution and acid rain. The decomposition of organic sulfur in plants and animals requires bacterial action to complete the cycle to produce sulfates, which are then available to plants.

In all the examples above, it is interesting to note the exchange of materials between the living and nonliving components of the global ecosystem. It is also noteworthy that all of these cycles involve the biological activities of either photosynthesis or bacterial metabolism.

REFERENCES

Brock, T. D. 1966. *Principles of Microbial Ecology*. Englewood Cliffs, NJ: Prentice-Hall.

Clark, W. C. 1989. Managing planet earth. *Scientific American* (special issue) 261: 46–54.

Cook, R.B. 1984. Interacting with the elements: Man and the biogeochemical cycles. *Environment* 26(7): 11–15, 38–40.

Deevey, E. S. 1970. Mineral cycles. *Scientific American* 223: 136–147.

Graves, W. (ed.). 1993. *Water: The power, promise and turmoil of North America's fresh water*. Washington, DC: National Geographic Society.

Makarewicz, J. C., and P. Bertram. 1991. Evidence for the restoration of the Lake Erie ecosystem. *BioScience* 41(4): 216–223.

McCormick, J. 1959. *Forest Ecology*. New York: Harper Bros.

Miller, G. T., Jr. 1982. *Living in the Environment*. 3rd ed. Belmont, CA: Wadsworth.

Odum, E. P. 1993. *Ecology and Our Endangered Life Support Systems*. 2nd ed. Sunderland, MA: Sinnauer Associates.

Raven, P., and G. B. Johnson. 1992. *Biology*. 3rd ed. St. Louis, MO: Mosby–Year Book.

Reichle, D. 1975. Advances in ecosystem analysis. *BioScience* 25(4): 257–264.

Chapter 6

ENERGY FLOW AND
TROPHIC
STRUCTURE

From a reductionist standpoint, ecosystems can be thought of as biochemical networks that funnel solar energy into and through living organisms. Whereas many elements and compounds cycle and recycle in biogeochemical interactions, energy does not return to its former state. Although it may move in various directions in an ecosystem, its movement is not cyclical. An excellent book on energy systems in ecology has been written by David Gates (1985).

The first and second laws of thermodynamics govern the passage of energy in ecosystems. Postulated by the work of Carnot and Joule in the 1930s, these physical laws are fundamental to ecosystem energetics. The first law of thermodynamics states that energy on earth is neither created nor destroyed; it is transformed from one type to another, for example, from light to chemical energy to heat to motion. The second law of thermodynamics states that no energy process occurs spontaneously unless it is a degradation or dissipation from a concentrated form to a dispersed form. Since some energy is dispersed at each transformation, no energy transformation is 100 percent efficient.

An important area of ecological research is energetic analysis, the study of the pathways and efficiencies of energy transfer in living systems. How does energy enter and pass through an ecosystem? How much energy is lost when solar energy is converted to chemical energy in plants through photosynthesis? How much is lost when plants are consumed by animals or people? How can we best utilize energy flow patterns to improve our own food supplies? These questions are basic to understanding the structure and function of ecological systems.

THE SOURCE OF ENERGY

The ultimate source of energy for most living organisms on earth is the sun, the only major exception being deep-sea thermal-vent organisms. For all other life forms, the sun supplies an incredible amount of energy. At sea level, the intensity of solar radiation averages 15,000 calories (g-cal) per square meter per minute. This totals 9 million calories per square meter per day, assuming 10 hours of sunshine, or more than 90 billion calories (90×10^9) per hectare per day. The total amount of solar energy striking the earth's surface each day is equivalent to the energy in 684 billion tons of coal (6.84×11^{11} tons), more than four times the coal reserves of the United States, enough to meet our national energy needs for 2,000 years. The solar energy striking the United States in 20 minutes would be sufficient to meet our country's energy needs for one year if it could be harnessed.

From a commercial standpoint, there are many problems associated with harnessing this energy. The problems of capturing, converting, storing, and transmitting solar energy raise a complicated set of technical, economic and political issues. Nonetheless, the potential is there. While solar energy is not always competitive at the present time with other energy sources for human use, in some places solar energy is one of the best options for supplying commercial energy needs.

From an ecological standpoint, of course, solar energy fuels life on earth. Only a relatively small portion of it is needed to do so. Most of the solar energy striking the earth's surface is reflected or absorbed, and thus scattered or transformed into heat. Of the sunlight falling on green plants, approximately 56 percent is reflected and 44 percent is absorbed (McNaughton and Wolf, 1973). The percentage absorbed varies with different species of plants and various environmental conditions. Of the amount absorbed by plants, most is spent in transpiration and other physical processes; less than 10 percent is utilized in photosynthesis. Ultimately, only about 1 percent of the total solar energy striking plants is converted to chemical energy through photosynthesis. Thus, the ecological efficiency of green plants is typically on the order of 1 percent.

FOOD CHAINS AND TROPHIC STRUCTURE

The transfer of energy from plants through a series of other organisms constitutes a food chain. The term *trophic level* refers to part of a food chain in which a group of organisms secures energy in the same general way. Green plants, algae, and autotrophic bacteria make up the basic trophic level of primary producers. All animals that obtain their energy directly from green plants, such as grasshoppers, meadow mice, cattle, and vegetarian human beings are in the same trophic level, that of herbivores or primary consumers. Sparrows eating grasshoppers and foxes eating meadow mice are in the trophic level of secondary consumers or primary carnivores. The hawk eating a bird that eats grasshoppers is in the fourth trophic level as a tertiary consumer or secondary carnivore.

The particular assemblage of trophic levels within an ecosystem is known as the *trophic structure*. Typically, ecosystems have three to six trophic levels through which energy and organic materials pass. In more vernacular terms, food chains usually have three to six links, or groups of organisms that derive their nutrition in these various ways.

An example of a short, practical food chain is alfalfa–cattle–humans in an agricultural ecosystem equivalent to grass–bison–wolves in a wilderness ecosystem. In aquatic systems, phytoplankton and other aquatic plants occupy the same trophic level as alfalfa or grass, and herbivorous animals, including various insect larvae, some crustacea, and a few herbivorous fish, occupy the same trophic level as cattle or bison. Menhaden and grass carp are two examples of herbivorous fish, sometimes described as "pasture fish."

The shorter the food chain, the greater the biomass that can be produced from a given amount of solar energy. The reason for this is that the majority of energy, normally about 90 percent, is lost in the form of heat or motion at each trophic level. A typical herbivore converts only about 10 percent of the plant energy it consumes into the chemical energy of its own body tissues, and a typical carnivore, in turn, converts only about 10 percent of the energy it consumes into its own body tissues. Thus a two- or three-link food chain produces more animal biomass than a five-link food chain. In practical terms, this means that a given prairie or alfalfa field can produce more pounds of bison or cattle than it can wolves or human beings. Similarly, a given pond or lake can produce more pounds of grass carp than trout.

These considerations help us to understand why the Arctic terrestrial environment and the Antarctic seas tend to be relatively productive during their short summer season. Typically, they have short, simple food chains. Figure 6.1 is a simplified diagram of an Arctic terrestrial food chain showing lichens as the primary producers; caribou, lemmings and snowshoe hares as the herbivores or primary consumers; and wolves, foxes, and human beings as carnivores or secondary consumers. Thus, although the growing season is short and conditions for most of

Figure 6.1 A simplified food chain for an Arctic terrestrial ecosystem. Arrows represent the flow of energy (from Southwick, 1976).

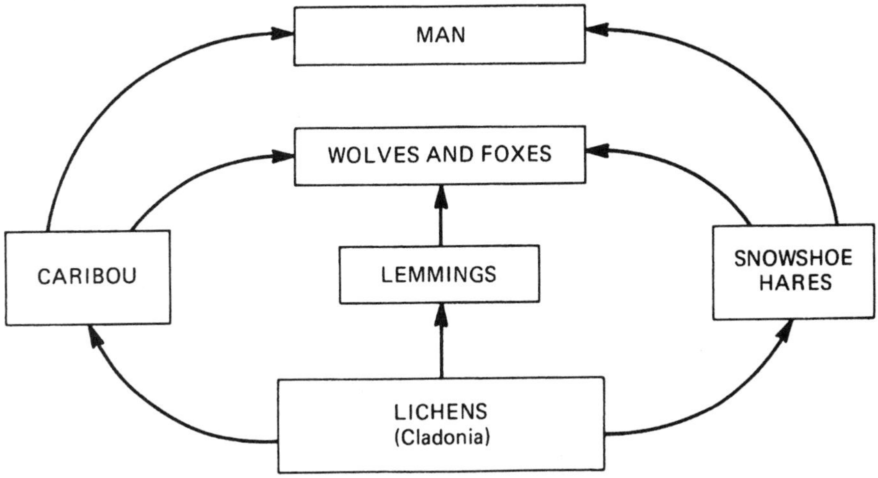

the year are harsh, the productivity of Arctic lands is high. This diagram is an oversimplification, of course; some other animals are present, many birds migrate in during the Arctic summer, and the human populations enhance their food supply by herding reindeer, a domesticated form of caribou. Both animals and human populations also supplement their food by feeding on marine organisms such as seals, whales, and fish. Nonetheless, the principle holds that Arctic food chains are relatively simple and direct.

Somewhat similar simplicities prevail in the Antarctic seas, which are remarkably productive considering the inhospitable climactic conditions. Figure 6.2 shows a simplified food chain beginning with phytoplankton, proceeding through shrimp or krill as the primary consumer or herbivore, and then on through fish, sea birds, squid, or whales to human consumers. These food chains show only three trophic levels from phytoplankton to baleen whales, from microscopic organisms to the largest animals on earth. Antarctic oceans are highly productive be-

Figure 6.2   A simplified food chain for an Antarctic aquatic ecosystem. Arrows represent the flow of energy (from Southwick, 1976).

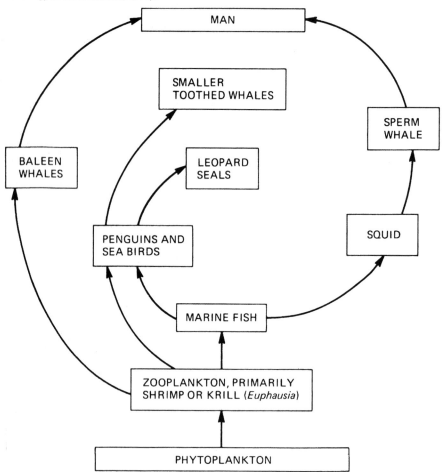

cause they have a 24-hour input of solar energy during the Antarctic summer (our winter), and they have a sharp turnover of water strata with an upwelling of nutrients from the bottom of the ocean to the surface at the Antarctic convergence. Both of these ecological factors favor the growth of plankton and produce huge populations of shrimp. Several nations, including Japan and Russia, have begun harvesting krill directly for human consumption, jumping ahead of whales in the food chain.

In human terms, the most direct and simple food chain is the vegetarian diet. In much of the developing world, people's diet is based on rice, wheat, beans, potatoes, or various vegetables. This type of diet has the advantage of providing the most calories and utilizing solar energy most efficiently. It has the disadvantage of potential malnutrition; that is, a diet dominated by rice may not have sufficient protein, vitamins, and minerals to support good growth and health. Many of the world's peoples cannot afford a diet well supplied in animal protein, however, and very often fruits and vegetables are cash crops, so a subsistence food such as rice or wheat supplies the basic caloric intake. A vegetarian diet can provide adequate nutrition if it contains a proper combination of legumes, fruits, and vegetables.

The food chains discussed above are all basically of the "grazing type," in which animal or human populations feed upon primary and secondary production. Another type of food chain in ecosystems is the "detritus" food chain; it proceeds from primary producers to decomposers, mainly fungi and bacteria. Some ecosystems may have a combination of detritus and grazing food chains; an example is when fungi grow on dead leaves or soil, and squirrels feed upon the fungi. Humans utilize the detritus food chain in consuming mushrooms. Figure 6.3 illustrates several types of grazing and detritus or decomposer food chains.

Trophic structures tend to become progressively more complex as we proceed from polar regions to temperate regions to the tropics, and also as we proceed from agriculture to natural ecosystems. Agriculture is, of course, intentionally simplified in order to funnel as much primary production as possible to humans. Natural systems become more complex in temperate and tropical zones because of greater biological diversity, a principle we will explore in more detail in Chapter 20. Figure 6.4 is a simplified diagram of some important food chains in a temperate pond in England. Even though this diagram is simplified, it remains complicated in comparison to Arctic and Antarctic food chains. In fact, it is more appropriate to call such trophic structures "food webs" rather than food chains. The interlocking nature of these relationships is so characteristic of ecology that a prominent ecologist, John Storer, chose the title, *The Web of Life*, to present some of the basic concepts of ecology in a small book (1953).

In most temperate food webs, the patterns of energy flow become so complicated that it is difficult or impossible to diagram all of the existing relationships. A typical temperate forest, for example, may have 40 or 50 species of insectivorous birds feeding on hundreds of species of insects. Numerous phyla and virtually countless species of invertebrates and microorganisms exist in the leaf litter and soil, representing several different trophic levels. No one has fully described all of these potential relationships, or even enumerated all the kinds of organisms in

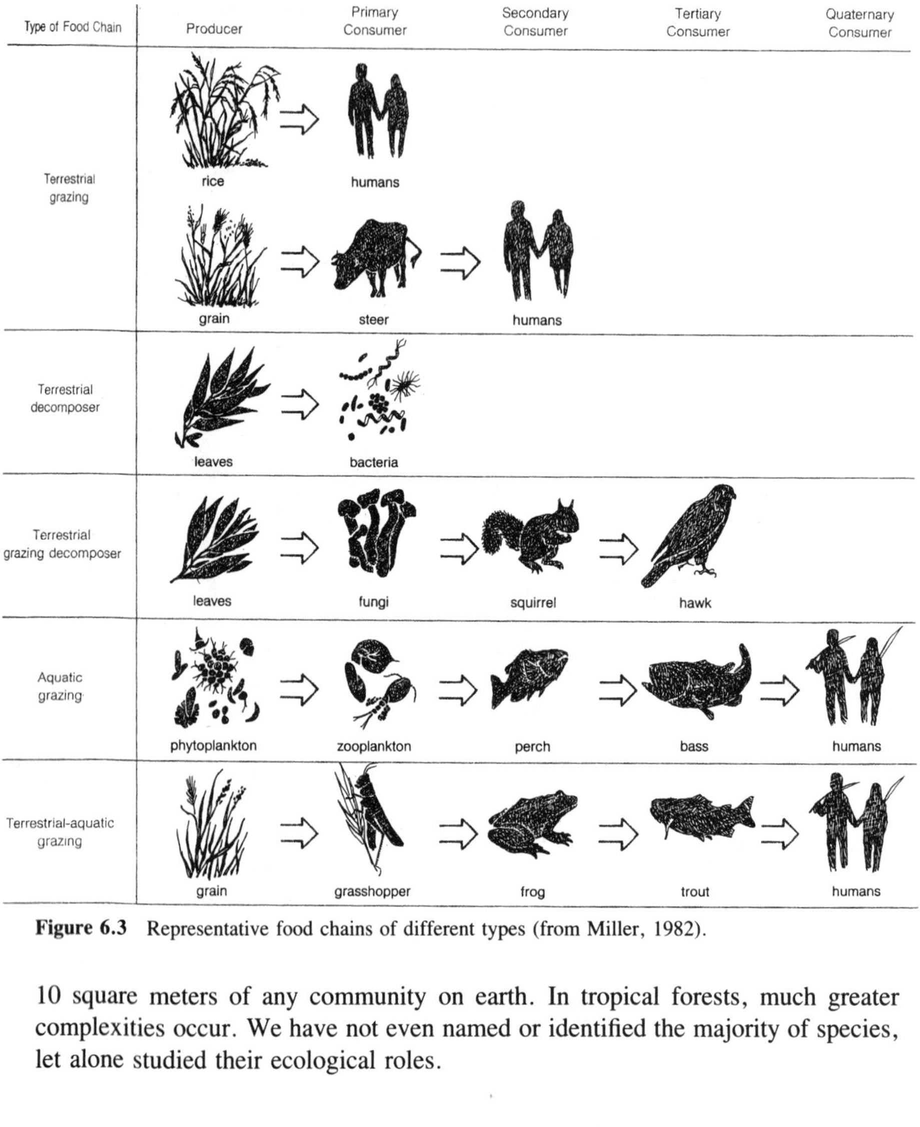

Type of Food Chain	Producer	Primary Consumer	Secondary Consumer	Tertiary Consumer	Quaternary Consumer
Terrestrial grazing	rice	humans			
	grain	steer	humans		
Terrestrial decomposer	leaves	bacteria			
Terrestrial grazing decomposer	leaves	fungi	squirrel	hawk	
Aquatic grazing	phytoplankton	zooplankton	perch	bass	humans
Terrestrial-aquatic grazing	grain	grasshopper	frog	trout	humans

Figure 6.3 Representative food chains of different types (from Miller, 1982).

10 square meters of any community on earth. In tropical forests, much greater complexities occur. We have not even named or identified the majority of species, let alone studied their ecological roles.

ECOLOGIC PYRAMIDS

Ecologic pyramids are a convenient way of diagramming certain numerical relationships within a trophic structure. They may be constructed on the basis of numbers of organisms, biomass, or energy flow.

Figure 6.5 represents a numbers pyramid for one acre of grassland. It is obviously not drawn to scale, but it helps to show the relative numbers of organisms of all species at each of four trophic levels: producers, herbivores, and the two levels of predators.

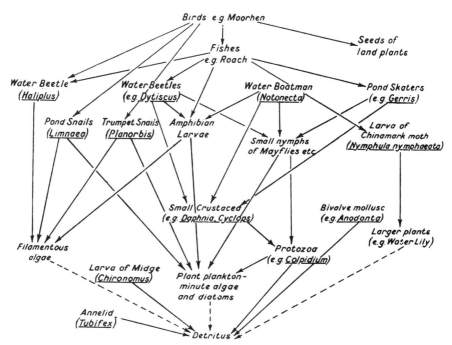

Figure 6.4 Food web for a temperate fresh-water ecosystem. Arrows represent the direction of consumption; dashed lines represent sources of nutrients (from Dowdeswell, 1961).

Figure 6.6 shows three types of pyramids for a hypothetical alfalfa–cattle–boy food chain based on 10 acres over the course of a year. These data help us to visualize both the numerical and energetic relationships within the system. They show it that requires almost 9 tons (17,850 pounds) of alfalfa in standing crop biomass, representing 20 million alfalfa plants, to support less than 5 calves totaling just over 1 ton in weight. These calves provide the beef protein for one boy for one year, if he were to live on a diet of hamburgers. In energetic terms, it requires almost 15 million calories of alfalfa to produce one million calories of beef, and this will add only 8,000 calories of human tissue to the growth of the boy. These figures dramatize the high ecological cost of a meat-eating diet.

The high energy cost of animal agriculture is shown even more dramatically in

Figure 6.5 A pyramid of numbers for one acre of grassland. The number of organisms in each trophic level are arranged with producers as a base and ascending levels of consumers: C-1, herbivorous invertebrates; C-2, first-order carnivores including spiders, ants, predatory beetles; C-3, second-order carnivores including birds, shrews, moles, etc. (from Odum, 1959).

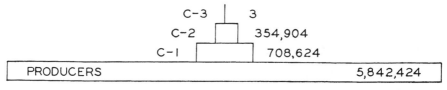

Figure 6.7. This shows that less than 1 percent of the gross primary production of green plants in a prairie ecosystem goes to the growth of beef cattle. Of 635 grams of carbon fixed per square meter per year by photosynthesis, fewer than 6 grams will become part of the calf. Of total gross primary production, 77 percent goes into plant respiration and decomposition, and twenty-three percent goes into animal consumption. Of the animal consumption, insects and meadow mice account for over one-fourth of the total, but on a well-managed farm, nearly three-fourths may be ingested by cattle. However, of that amount going into cattle (15.5 percent of the total original standing crop), the majority is either undigested or supplies energy for respiration; only 4 percent contributes to the actual growth of the calf. Under the best of conditions, this allows less than 1 percent of the original gross primary production for actual beef production.

Data such as these can contribute to agricultural management in several ways. They permit comparison of the efficiencies of different crops and analyses of how much energy is lost to insects, disease, rodents, birds, and other crop pests. They can also help in evaluating the cost effectiveness of fertilizers, pesticides, cultivation procedures, and watershed management.

Since modern agricultural researchers actually make many of these evaluations, at least in terms of net productivity and profit yield, there is an even greater need for such data on natural ecosystems. If we had ecologic pyramids and energy flow diagrams on estuaries, wetlands, forests, and natural grasslands, we would be in a better position to estimate their value in economic terms. When a developer wants to drain and fill parts of an estuary, such as San Francisco Bay, the ecologist is often hard-pressed to express the value of the estuary except in terms of esthetics and recreational value. These benefits are very important, of course, but it would also be helpful to know the ecologic function and energy budget of the estuary. What is its potential production of shellfish? How vital is it as a spawning and nursery ground for commercial and game fish? What role does it play in

Figure 6.6 Three types of ecological pyramids illustrated for a hypothetical alfalfa–calf–boy food chain computed on the basis of 10 acres for one year and plotted on a log scale (from Odum, 1971).

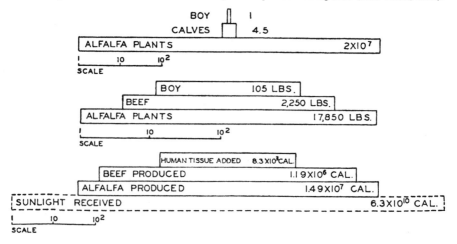

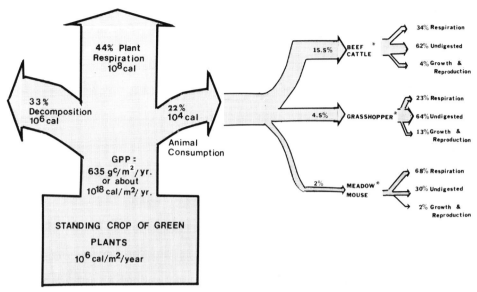

Figure 6.7 Energy flow in a simplified prairie ecosystem, based on the assumption that 70 percent of crop consumption is by cattle, 20 percent by insects, and 10 percent by meadow mice (from Southwick, 1976).

ecosystem balance in relationship to adjacent land forms? Does it provide essential habitat for the natural predators of mosquito larvae? How important is it as wintering grounds for waterfowl? Do its energy patterns offer productive opportunities for aquaculture? What is its role as a sink for carbon and a producer of oxygen? What is its role as a filter and purifier for pollutants? Questions like these involving trophic structure and energy flow should be explored before major development alters existing ecosystems. Ecological pyramids offer one way of organizing data to help answer some of these questions.

ECOLOGIC EFFICIENCIES

We know that the energetic efficiency of green plants is usually around 1 percent; that is, green plants incorporate only about 1 percent of the total solar energy striking their surfaces into organic compounds in photosynthesis. Approximately 99 percent of this surface energy is reflected, scattered, or dispersed in some other way.

With subsequent energy transformations through each trophic level, an additional 80 to 95 percent of the plant's chemical energy is lost through dispersion, heat loss, motion, and so forth. In other words, only 5 to 20 percent, averaging about 10 percent, of the energy in green plants is incorporated into herbivores, and only 5 to 20 percent of that in herbivores is incorporated into primary carnivores. Table 6.1 shows representative data on energy transfer at various trophic levels in three aquatic ecosystems. The ecologic efficiencies from herbivores to

TABLE 6.1
Efficiency of energy transfer at various trophic levels in aquatic ecosystems
(expressed as the percentage of energy reaching the trophic level that is converted
into organic material) (Odum, 1959).

Tropic Level	Cedar Bog Lake, Minnesota	Lake Mendota, Wisconsin	Silver Springs, Florida
Plants	0.10	0.40	1.20
Herbivores	13.3	8.7	16.0
Small carnivores	22.3	5.5	11.0
Large carnivores	absent	13.0	6.0

small carnivores in these aquatic systems fall within the range of 5.5 to 22.3 percent, meaning that each trophic level incorporates only this much of the energy it consumes into its own organic structure.

Engineers may point out that these efficiencies are relatively low, considerably less than that achieved by many machines. Machines vary in their efficiency (ratio of energy output to energy input) from 5 percent to nearly 100 percent. An automobile engine achieves an efficiency of approximately 25 percent in converting chemical energy to mechanical energy, and some machines that transmit mechanical energy only may have an efficiency of over 90 percent. In comparison, the 1 percent efficiencies of plants and the 5 to 20 percent efficiencies of animals seem low, but the comparisons are not entirely valid. Living organisms also achieve growth, reproduction, self-maintenance, and dispersal—processes that machines cannot accomplish—and all of these items enter into the energy budget.

Algal cultures have been utilized to achieve greater efficiencies in primary production, with hopes of helping to solve the world's food supply problems. In cultures of the single-celled algae *Chlorella,* with proper combinations of temperature, light, nutrients, and harvest rate, efficiencies of 20 percent and even 50 percent for certain short periods, were obtained. These methods have not proven feasible for mass production, however. In daylight, artificially managed algal cultures usually achieve no more than 2 to 8 percent efficiencies.

ENERGY AND HUMAN ACTIVITIES

Thus far we have looked at patterns of energy flow through natural and agricultural ecosystems. The other aspect of energy that is vitally important to global ecology is the human use of energy to fuel our domestic and commercial enterprises, our industry, transportation, communication, and all related human activities. These activities involve the production of heat, motion, and electricity by burning fossil fuels, fissioning the atom, and tapping a variety of alternative renewable sources such as hydroelectric power, wind, geothermal, solar, and tidal energy.

Our use and misuse of energy sources have tremendous impacts on the global environment. A wide variety of our pollution problems, including CO_2 buildup, urban smog, acid rain, particulate lead, radioactivity, and thermal pollution in rivers and lakes, are all related in one way or another to our use of energy. We will examine these problems individually in later chapters on different types of human impacts.

SUMMARY

Energy flows through ecosystems from sunlight to the chemical energy of plants and on to animals and eventually decomposers. It does not recycle or return to its former state. At each stage of energy transformation, the majority of energy is lost or dissipated as heat.

Approximately 1 percent of the solar energy striking the surface of green plants is converted into organic compounds within the plant. Ninety-nine percent, on the average, is reflected or converted to heat. As green plants are eaten by herbivores, typically 10 percent of the chemical energy in plants is converted into animal tissue, and 90 percent is lost in heat or motion. Similarly at the carnivore level of animals eating herbivores, approximately 10 percent of the prey species is converted into body tissues of the predator.

Food chains normally have three to six links, or trophic levels, through which energy passes. Short, simple food chains, like those found in Arctic and Antarctic ecosystems and in agriculture, yield the greatest quantity of plant or animal tissue per unit of original solar energy. Although such ecosystems may be more productive for human purposes, they are also potentially more unstable.

Trophic structures in temperate and tropical ecosystems become more complex as species diversity increases and complex food webs develop. In tropical ecosystems we have not yet identified or described most of the species, and we are a long way from understanding their ecological roles and energy pathways.

REFERENCES

Dowdeswell, W. H. 1961. *Animal Ecology.* New York: Harper and Row.
Gates, D. M. 1985. *Energy and Ecology.* Sunderland, MA: Sinauer Associates.
McNaughton, S. J., and L. L. Wolf. 1973. *General Ecology.* New York: Holt, Rinehart and Winston.
Miller, G. T., Jr. 1982. *Living in the Environment.* 3rd ed. Belmont, CA: Wadsworth.
Odum, E. P. 1959. *Fundamentals of Ecology.* 2nd ed. Philadelphia: W. B. Saunders.
Odum, E. P. 1971. *Fundamentals of Ecology.* 3rd ed. Philadelphia: W. B. Saunders.
Reichle, D. E. 1975. Advances in ecosystem ecology. *BioScience* 25(4): 257–264.
Ricklefs, R. E. 1973. *Ecology.* 1st ed. Portland, OR: Chiron Press.
Storer, J. H. 1953. *The Web of Life.* New York: New American Library of World Literature.

Chapter 7

ECOSYSTEM HOMEOSTASIS, SUCCESSION AND STABILITY

Ecosystem homeostasis is a technical term for the balance of nature. It involves more than the average person may attribute to that term, however. In its full sense, it refers not only to the relative balance of species, for example, the balance between predators and prey or hosts and parasites, but also to the balance between different biogeochemical cycles and energetic pathways within an ecosystem. A homeostatic condition within an ecosystem implies that all aspects of ecosystem function are in relative balance; there is a balance between production, consumption, and decomposition, as well as between species within the system. This does not mean that nature never changes, or that balances are precise. It does imply, however, a reasonable degree of regulatory function in natural systems.

HOMEOSTASIS

The concept of homeostasis in physiology was originally developed by the French scientist Claude Bernard in the nineteenth century and later expanded by the American physiologist Walter Cannon. It is the basis of our understanding of the regulation of body processes through nervous and endocrine control. This concept stimulated the discovery of vasomotor and metabolic responses in the regulation of heart rate, respiration, body temperature, and other basic functions. It also revealed the interplay of nervous and hormonal control in growth, reproduction, metabolism and behavior, and has contributed much to our understanding of health and disease.

In analogous ways, the concept of homeostasis at the ecosystem level helps us to understand regulatory processes within plant and animal communities. There are many differences, of course. An ecosystem has more loosely related parts than does an individual organism. Control mechanisms are less precise, and obviously less dependent upon direct neural or biochemical connections. Responses are much more subject to delay and variation.

Nonetheless, homeostasis in both an individual and an ecosystem depends upon the principle of feedback or cybernetics. In physiology, for example, we know that muscular activity increases carbon dioxide concentration and decreases oxygen levels in the blood. This stimulates a faster heart rate and breathing rate, which serve to expel carbon dioxide and bring in more oxygen. When CO_2 and O_2 levels return to normal, heart rate and breathing also return to normal. Similarly, when a large influx of carbohydrates, especially sugars such as glucose, are ingested, blood sugar levels increase. This normally triggers an increase of insulin secretion from the beta cells in the pancreatic islets of Langerhans, which metabolize blood glucose so that levels return to normal. Thus, the system remains in balance to meet the metabolic needs of the individual. If insulin production or function is deficient, blood glucose remains exceptionally high, creating a symptom of homeostatic imbalance that is typical of diabetes.

HOMEOSTASIS IN ECOSYSTEMS

In a balanced aquatic ecosystem, there is an analogous, though less accurately controlled, homeostasis involving carbon dioxide and oxygen. An increase in water temperature in the springtime increases metabolic rate and respiration in aquatic plants and animals, resulting in increased carbon dioxide and decreased oxygen levels. The higher level of free CO_2 plus increasing water temperature stimulate more rapid photosynthesis and plant growth. This reduces the CO_2 levels and adds more O_2 to the water; thus, both CO_2 and O_2 levels tend to return to normal. Many other factors influence CO_2 and O_2 levels in aquatic ecosystems, however, including decomposition rates, nutrient flow, and energy input. A combination of high nutrients, such as nitrates and phosphates, high temperatures, and many days of bright sunshine to provide high energy inputs often leads to excessive plankton blooms; that is, production exceeds consumption. Temporarily this will produce a high O_2 level and reduce CO_2 in the water. The situation changes rapidly, however, when phytoplankton blooms start to die. Then rapid decomposition quickly turns the tables, so that dissolved O_2 is rapidly depleted and CO_2 builds up quickly. This change may lead to fish kills through asphyxiation or lack of dissolved oxygen. One can imagine the complexity of ecosystems when one considers hundreds of species of plants and animals interacting over the common interfaces of many environmental factors, especially those involved in basic resources such as nutrients and energy.

An ecosystem can achieve a certain amount of self-regulation within limits, but if these limits are exceeded, as illustrated in the case above of plankton blooms and fish kills, the ecosystem may no longer be able to function normally and will thus show various patterns of change, injury, or breakdown. A major task of modern science is to to study ecosystems with the purpose of understanding natural homeostatic mechanisms and limitations. It is particularly important that we recognize the tolerance levels of different systems. How much disturbance can ecosystems of various types withstand? How much variation and perturbation is normal within the homeostatic abilities of the system, and at what point does serious injury or irreparable damage occur? These questions are being addressed by a major program of Long-term Ecological Research (LTER) within the National Science Foundation (NSF). The questions and the answers are of vital importance in the maintenance of viable environments for ourselves as well as for countless other organisms on earth.

It is obvious in many areas that we have exceeded ecosystem tolerance, through high pollution, excessive land scarring, and total displacement of some biotic communities. The concept of tolerance will emerge again when we discuss topics such as tropical deforestation, desertification, and pollution—cases in which we may never again be able to restore normal ecosystem structure and function, or, if we can, the cost may be exorbitant.

GLOBAL IMPLICATIONS OF ECOSYSTEM HOMEOSTASIS

In a very broad and massive way, the global ecosystem has also maintained certain homeostatic balances which are now changing due to human activities (Stern et al., 1992). Atmospheric carbon dioxide is probably the best example. Its increase in the twentieth century is well documented as a result of increased combustion and decreased CO_2 uptake due to deforestation. Although we can sketch the main outlines of the carbon cycle, we certainly do not understand all of its quantitative ramifications. What is the absorptive capacity of the oceans for carbon? How much carbon is released due to alteration of the biosphere? What are all of the climatic and ecologic implications of increasing atmospheric CO_2? Although carbon dioxide is probably the best-known example of developing imbalances in the global ecosystem, it is by no means the only one. Increasing methane, declining atmospheric ozone, and increasing acid precipitation are all indications of perturbations in global homeostasis. The existence of global homeostasis does not mean that the world will always be in a steady state; change is inevitable due to many processes including continental drift, volcanism, earthquakes, meteorite impacts, ecological succession, solar cycles, and so on. However, the relatively sudden onset (in a geological sense) of basic atmospheric changes in this century are topics of great interest and concern because of their widespread consequences. So also are the rapid changes in the biosphere, which we will discuss in several chapters in Part III, human impacts on the earth.

TROPHIC STRUCTURE AND ECOSYSTEM STABILITY

Ecosystems with simple trophic structures are usually more vulnerable to drastic ecologic change than ecosystems with complex trophic structures. In the Arctic terrestrial ecosystem discussed in Chapter 6, if the production of lichens becomes impaired, the entire system is impaired since lichens are at the base of virtually all food chains. Similarly, in Antarctic seas, if krill were eliminated by some ecologic accident, there would be a catastrophic decline of virtually all marine mammals, birds, and fish that depend upon krill for food.

There are, of course, a good many questions on what the term *stability* means. If interpreted to mean fairly rigid constancy, then stability is nonexistent in ecological systems. If the term means the ability of a system to bounce back from disturbances, as defined by Krebs (1972), then the concept of stability has practical significance. Stability also has significance if it means relative consistency in the numbers of species and their general population levels. There is little doubt that the best examples of unstable populations, of exaggerated swings in abundance and scarcity, tend to occur in the harsh and simplified ecosystems of cold or arid regions. Such classic instability occurs in the rise and fall of populations of lemmings, snowshoe hares, and desert locusts. These fluctuations are often accompanied by equally sharp fluctuations in vegetation and predator populations. If such prominent instabilities occur in the wet tropics, they have not yet been described extensively. So far as we know, they do not occur in tropical ecosystems unless the total system is destroyed, as in tropical deforestation. Then the question of stability has an entirely different meaning, one that is difficult to define and one that depends more on the system's ability to regenerate.

Apart from total alteration or destruction, the fact remains that temperate and tropical systems have less dependence on any one species. The temporary loss of one species does not necessarily endanger the entire ecosystem because alternate pathways for energy flow and nutrient cycling exist. There are exceptions, of course, for if a dominant species of grass in a temperate grassland community is eliminated, all herbivore species depending upon it would be adversely affected, but natural grasslands usually consist of many dozens of species of grasses and herbs. The original American prairies, for example, had a complex trophic structure, with over 100 species of plants and numerous herbivores and predators.

In general, there is ecologic strength and security in complex trophic structures that simplified systems do not have. It is no surprise that complex ecosystems are usually more stable and resistant to perturbation than simplified systems up to a point. The exceptions to this generalization lie mainly in the area of competition. Some studies have shown that species assemblages of gregarious insects have less stability than those of solitary insects. An example is the remarkable outbreaks of fire ants *(Solenopsis)* in South America and the southern United States. Several species of these highly social insects have shown extreme increases in number, creating a serious threat to other more stable insect populations and to wildlife, domestic animals and even human populations (Grisham, 1994). In the United States, fire ants are an invading species, expanding their geographic range and outcompeting native ant species, but dangerous outbreaks of fire ants are also occurring in Brazil, where they are native species. These are thought to be due to environmental disturbances such as deforestation and land clearing. In addition, multispecies assemblages tend to have a greater level of competition, leading to population instability, and long food chains (as distinct from food webs) can have inherent instabilities (Ricklefs, 1990). The relationships between diversity and stability are often quite complex and variable.

ECOSYSTEM DEVELOPMENT AND SUCCESSION

As mentioned in the previous section, ecosystems are not rigidly stable; they have a life history, so to speak, with periods of growth, development, and maturity. Gradual ecosystem change is known as *succession,* an orderly modification of the environment by changing plant and animal life. This process can be seen in an abandoned agricultural field, which passes through successive stages characterized by weeds and grasses, shrubs, small trees, and conifers, to become, finally, a deciduous forest (Fig. 7.1). The process may take 80 to 100 years, and may be set back or reversed by fire or other disturbances. The species that first invade are known as *pioneer species,* successive stages are known as *seres* or *seral stages,* and the final stage that results is known as the *mature* or *climax* stage. The species occurring in each stage are a function of climate, soil, previous land-use history, and surrounding communities. The climax community that ultimately results is primarily a function of climate, soils, and topography, and it is normally considered to be the most stable ecosystem type characteristic of a particular region. For example, the climax community of central Pennsylvania is a deciduous forest dominated by maples, oaks, and other broad-leaved trees, whereas the climax community of Alaska or northern Canada is a coniferous forest dominated by spruce, firs, and other evergreens.

Another commonly cited example of succession occurs in a newly formed lake, where vegetation develops within and around the lake, and phytoplankton and zooplankton develop within the water. The lake develops populations of crustacea, mollusks, aquatic insects, annelids, and other invertebrates, which then are rapidly

Figure 7.1 Primary succession. Here, exposed rock goes through a series of changes as one biotic community replaces another until a mature community is formed. During succession the plants of each community alter their habitat so that conditions become more suitable for other species (from Chiras, 1985).

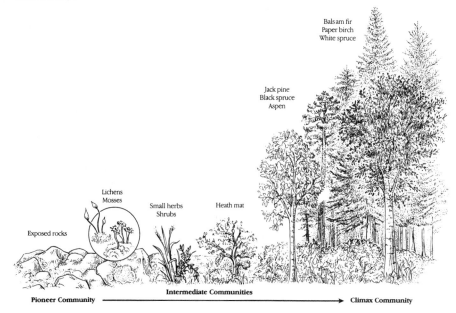

Balsam fir
Paper birch
White spruce

Jack pine
Black spruce
Aspen

Lichens
Mosses

Small herbs
Shrubs

Heath mat

Exposed rocks

Intermediate Communities

Pioneer Community ⟶ **Climax Community**

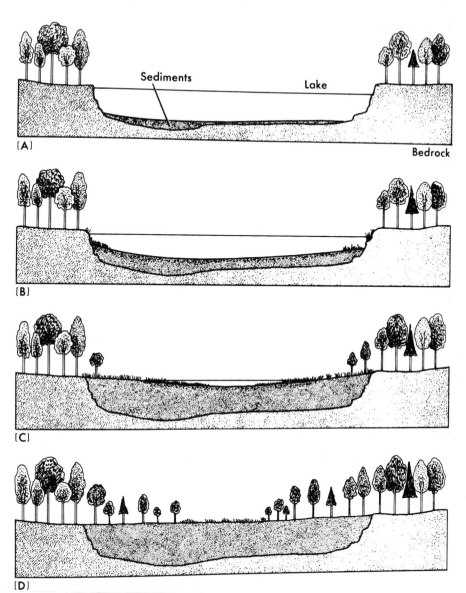

Figure 7.2 Successive stages in the natural aging and filling of a lake. Over many years sediments accumulate, the lake becomes shallow (A), and vegetation encroaches from the edges (B). The lake slowly becomes a marsh (C), a meadow, and finally a young forest (D) (from Whittaker, 1970).

joined by amphibia, fish, and waterfowl. The lake and its biological community gradually change as the bottom accumulates organic material and the waters become enriched with nutrients. If the lake is shallow, it will eventually succeed to a marsh and bog, but this process may take centuries or even millenia, and may not occur at all if the lake is sufficiently deep and has a scouring action. Figure 7.2 portrays a typical and successional pattern of a pond or small shallow lake.

The development of ecosystems is dramatically visible on newly formed volcanic islands, where windblown or waterborne plants and animals gain a foothold as pioneer species and begin the process of soil formation and biological development. When the volcanic island of Krakatoa, between Java and Sumatra, erupted violently in 1883, living organisms were extinguished. The island remained barren, burning, and lifeless. Within 3 years, 28 species of plants had arrived, increasing to 200 by 1950 (Elton, 1958). After 50 years, Krakatoa had developed a young forest with many species of animals, and within 100 years it had acquired at least 720 species of insects, 30 species of resident birds, a few reptiles, and some mammals (Sastrapradja, 1985). Similar successional stages are now being studied in the volcanically devastated landscape just east of Mount St. Helens in the state of Washington.

If not disturbed by fire, flooding, volcanic activity, or major human intervention, ecosystems tend to develop relatively stable climax communities which contain maximum biomass and diversity of living organisms in relation to the physical constraints of the environment. Table 7.1 lists some of the major characteristics of young and mature ecosystems (from Krebs, 1972). The early developmental stages of ecosystems typically have high levels of biological productivity in relation to biomass, but they exhibit low species diversity and less stability than mature or climax stages. Young ecosystems tend to have rapidly growing populations, with less specialization and fewer internal checks and balances than mature ecosystems. Young developmental stages are favored in agricultural practice, because they are most productive, but they must be managed carefully because they are also most vulnerable to sudden ecological change. They are sensitive to

Table 7.1

Characteristics of more mature and less mature ecosystems according to Margalef's theory (from Krebs, 1972).

	Ecosystem Condition	
Characteristic	Less Mature	More Mature
Structure		
Biomass	Small	Large
Species diversity	Low	High
Stratification	Less	More
Energy flow		
Food chains	Short	Long
Primary production per unit of biomass	High	Low
Individual populations		
Fluctuations	More pronounced	Less pronounced
Life cycles	Simple	Complex
Feeding relations	Generalized	Specialized
Size of individuals	Smaller	Larger
Life span of individuals	Short	Long
Population control mechanisms	Abiotic	Biotic
Exploitation by humans		
Potential yield	High	Low
Ability to withstand exploitation	Good	Poor

drought, floods, disease, and insect outbreaks. Whereas human agriculturalists have striven to maximize ecosystem productivity for human use, natural evolution has tended toward maximizing diversity.

THE CONSERVATION OF DIVERSITY

The eminent ecologist Charles Elton built a strong case for the conservation of biological diversity (Elton, 1958), and this still remains a priority in conservation (Soule and Kohm, 1991). Biodiversity was one of the major issues of the Earth Summit in Rio de Janeiro in June 1992. By conserving diverse ecosystems we help to ensure more natural, more complex, and more stable flora and fauna. In English hedgerows, for example, Elton noted a variety of beneficial species that could not survive in areas where field borders were stripped to barbed wire fences. Natural hedgerows included many insectivorous birds and mammals that provided natural insect control. Less dependence on massive insecticide application was necessary. Less pollution and less habitat destruction resulted.

In most of the world, agriculture, which is the applied management of food chains, has gone in the opposite direction, that of ecosystem simplification. Thus, we plow the prairie, replacing a hundred species of native prairie herbs and grasses with pure stands of wheat, corn, or alfalfa. Without question these plantings increase efficiency, productivity, and yield, but they also increase ecologic vulnerability. If we have a pure stand of one strain of wheat or corn, the possibility of a serious loss from a pathogen (such as wheat rust) or a herbivore (such as grasshoppers or corn borers) is amplified. Pesticides and fungicides have become a routine part of modern mechanized agriculture. Pesticide use is expanding worldwide, but many insect populations have become resistant, leading to greater costs and instabilities in agricultural management (Weber, 1994). Excessive pesticide use often destroys all semblance of ecological balance, eliminating beneficial species as well as harmful ones, so that long-term results are often detrimental (Pimentel and Lehman, 1993).

Agricultural practices also tend to simplify large-animal populations. In Africa, for example, humans have replaced complex ungulate faunas, which have included many species of gazelles, impalas, buffalo, hartebeest, wildebeest, zebras, eland, rhinos, elephants, giraffe, gerenuks, and dik-diks, with a single species of domestic cattle. This funnels more of the energy flow into a single manageable domestic species, but it also has a drastic effect on native grasslands, savannahs, and forests, and tends to produce exaggerated epidemics of diseases, such as rinderpest and foot-and-mouth disease. We clearly pay a price for oversimplifying ecosystems, thereby reducing their natural stability and resilience.

The question is often asked, Why do we need so many species? Of what practical value are so may different kinds of plants and animals that occur in natural ecosystems? Phrased in this way, such questions are not easy to answer. Much of the value comes from esthetic appreciation. The world would certainly continue to exist without peregrine falcons, Furbish louseworts, snow leopards, tigers, giant pandas, or grizzly bears, but we would all be poorer if they were lost. Aldo

Leopold pointed out (1949) that a mountain without a wolf or a bear may be safe for cows, but it is no longer the same. It has lost its mystique and its most venerable quality. Edwin O. Wilson (1992) of Harvard University has emphasized at least four reasons for the conservation of biodiversity. Aesthetics and the appreciation of nature, expressed by Aldo Leopold in the example above, is just one of these reasons. Others include ethical considerations, economic benefits from the discovery of new medicines and foods, and the significance of diversity in maintaining ecosystem functions. The entire issue of biodiversity is so important in global ecology that we will revisit it in Chapter 20 with further examples of its value. Here we will touch on some worldwide concerns related to this issue.

Paul and Anne Ehrlich (1981) have addressed the issue of biological diversity with a parable about the "rivet poppers." In this analogy a mechanic is seen taking rivets out of an airplane. There are more of them than necessary to hold the plane together, they add weight, and they can be sold separately for a good price if some are removed. The airline needs the money, especially for growth and expansion, so it is logical to remove extra rivets and save money at the same time. A time may come when removing $n + 1$ rivets will cause the crash of an airplane, especially in a storm or a time of stress, but no one knows when that time will be.

In a metaphorical sense, we probably can remove some species from an ecosystem without destroying the system, but since we do not know the role of each species or which may be essential to the entire system, we do not know when the removal of $n + 1$ species may result in the "crash" of the system.

The Ehrlichs' point is that the rivet poppers on Spaceship Earth include politicians, businesspeople, decision makers, in fact, all of us on this planet. We are responsible in one way or another for the depletion of forests, pollution of the air and waters, and loss of species. This does not mean that all human activities reduce diversity. There are many instances in which human activities enhance biotic diversity, either intentionally or inadvertently. Some successful species introductions, such as the introduction of the ring-necked pheasant, or Hungarian partridge, into the United States, have been favorable additions to biodiversity. Others, such as the introduction of the starling, have been undesirable additions, contributing to a loss of diversity in native species, in this case, the loss of hole-nesting birds such as the bluebird. Human activities have also enhanced diversity in managed areas such as parks, botanical gardens, domestic landscapes, and deliberate wildlife management projects. For example, providing nest boxes has enhanced wood duck populations, and constructing nest platforms in estuaries has increased osprey numbers.

In a similar way, some natural disturbances are beneficial to biodiversity. In tropical forests, windfalls of large mature trees enhance diversity by creating openings for secondary growth and forest regeneration. In prairie grasslands and ponderosa pine forests, intermittent fires also stimulate renewed successional growth and contribute to biodiversity. In the montane and boreal forests of North America, many large animals such as deer, elk and moose cannot survive browsing exclusively on climax species such as black spruce; they depend on natural fires to maintain a mixture of vegetation by creating forest openings and meadows. Many foresters feel that the underbrush and leaf litter in the ponderosa pine forests of western United States should be removed by light ground fires every few years to maintain soil fertility, prevent excessive accumulation of organic matter in the form of cellulose, and prevent more devastating fires that kill trees. In pine forests in the southeastern United States, very efficient fire suppression allows the dangerous accumulation of combustible materials, and a pine fungus disease has increased in the accumulated litter. In all of these forests, proper management should not exclude all fires, but should prevent totally devastating ones. These examples illustrate the need for understanding natural processes in ecosystems in order to main them properly.

The Florida Everglades provide another example of the importance of understanding natural ecosystem processes before the system can be maintained or properly managed. The unique plants and animals of the Everglades depend upon the proper balance of fresh and salt water flowing through its channels. A decline in fresh water occurred in the 1980s due to the diversion of rivers and ground water to the growing megalopolis of Miami. Large populations of birds have disappeared, and many characteristic features of the Everglades have been destroyed (Cohn, 1994; Culotta, 1995). In times of drought, many animals depend upon alligator wallows as a source of water. These pools hollowed out by alligators retain water longer than other parts of the swamp. It seems ironic, but is nevertheless ecologically true, that the survival of many fish depends upon their primary predator.

An expensive plan of environmental engineering is now being developed to restore the Everglades and ensure continued water supplies to the burgeoning Miami–Ft. Lauderdale metropolitan area. The U.S. Army Corps of Engineers is working on the restoration of the Kissimmee River, a fresh water source for Lake Okeechobee and urban, agricultural, and natural areas in South Florida. Culotta (1995) noted,

> With an estimated price tag of $370 million this is one of the most ambitious river restoration projects in U.S. history. It is, however, a mere drop in the watershed. . . . over the next 15 to 20 years, at a cost of $2 billion, the Corps and other state and federal agencies plan to replumb the entire Florida Everglades ecosystem.

We will note in Chapter 25 that curative measures in ecosystem health, as in human health, are far more expensive than preventive measures would have been in the first place.

Since so many of the world's ecosystems are altered by human activities, we can no longer apply a hands-off conservation policy to ensure their preservation. Programs of active management are sometimes necessary, and we can often apply modern principles of cropping and harvest to natural populations of fish, wildlife, and forests, to the mutual advantage of humans and the ecosystem. This approach was brilliantly documented by Aldo Leopold in his classic book *Game Management* (1933), and was also described by Watt in *Ecology and Resource Management* (1968) and Dasmann in *Environmental Conservation* (1984).

It is equally important to recognize that ecosystems have a certain level of "insult tolerance." If we continue to assault them with pollution or destruction, they will most certainly become irreversibly damaged. Yet the strength and resilience of some ecosystems is quite amazing, and we now have examples of ecosystems returning from badly damaged conditions to reasonable health and productivity (e.g., Lake Washington, the Thames River). Other ecosystems are sensitive and fragile, easily damaged, and not capable of sustaining massive insult. If we build concrete jungles around our lakes, rivers, and estuaries, and fill their waters with waste materials, we can no longer expect them to produce clams, oysters, gamefish, or support recreational pleasure. If we clearcut the tropical forest, we stress it beyond its ability to regenerate.

Much of the concern in global ecology today relates to irreversible changes in large-scale ecosystems: the conversion of forests to agriculture, the destruction of cropland by urban sprawl, the conversion of arid grasslands to deserts, the permanent changes in coastal zones, and so on.

THE CONTROVERSY OVER OLD-GROWTH FORESTS

A contentious issue in the United States, and elsewhere in the world, has been the selective logging of old-growth forests. This case illustrates that active management programs are not always enough to ensure conservation. In some cases total protection is urgently needed.

Old-growth forests, that is, primary uncut or virgin forests, have been decimated in the United States, with only a small percentage (less than 5 percent) of original forest still standing. There is pressure from business interests to cut these remaining forests, and the U.S. Forest Service has often been supportive. Those who want to log the forests maintain that selective logging opens up the forest, rejuvenates it, and stimulates new growth. The claim is made that without this rejuvenation the forests will die. Certainly selective logging will stimulate new growth, but it will also disturb a climax ecosystem in which the process of natural death and decay are essential to the survival of many organisms. The argument that the ecosystem will die is fallacious; it has survived for thousands of years, and will continue to do so short of any major environmental disturbance or major climate change. The point is that many organisms depend on undisturbed old-growth forests for survival, and if we disturb these few remaining forests, we will very likely cause the extinction of many species. These organisms may be small arthropods or microorganisms, or they may be more visible symbols of old-growth forests such as the spotted owl. In any case, the destruction of old growth will certainly mean a loss of biodiversity.

In upsetting the homeostatic conditions of ecological systems, we may be inflicting changes that could persist for centuries. This has happened in some of the desert regions of the Middle East and North Africa, as we will see in Chapter 13. It remains to be seen, however, whether we have learned from these historical lessons; thus the admonition of George Santayana that "Those who fail to learn from history are destined to repeat it," raises a worrisome prospect (Southwick, 1983). When scientific and historical facts are known, proper ecosystem management becomes a matter of ethics and value judgments. As pointed out by Odum

(1963), Leopold (1949), and Hutchinson (1948) 40 years ago, the careful management of ecosystems must be recognized by humankind as a moral responsibility.

SUMMARY

Ecosystem homeostasis refers to a relative balance of nutrient cycles, energy flow, and species composition in ecosystems. In balanced systems, feedback mechanisms exist to maintain relative consistency in biogeochemical patterns and the abundance and distribution of plants and animals.

This does not mean that ecosystems, however, are not static. They are dynamic, showing patterns of growth, development, and maturation. Young ecosystems tend to have low species diversity, low nutrient storage, low biomass, but high productivity. Mature ecosystems have higher species diversity, higher nutrient storage, and higher biomass, but lower productivity per unit biomass.

Ecosystems with simple trophic structure are often more vulnerable to change and perturbation than complex systems, though many specific exceptions to this principle can be found.

Human influence has tended to oversimplify ecosystems, especially in agricultural practice, where complex species assemblages of plants of animals are often reduced to monocultures, or single-species stands. Such simple systems have the advantage of funneling all ecosystem productivity into species of use to humans, such as rice, wheat, corn, or beef cattle, but they are also inherently unstable. We try to maintain the instability of agroecosystems with applications of fungicides, herbicides, insecticides, and fertilizer applications. The energetic costs of these efforts are very high.

Most ecosystems have a certain resilience or capability for restoration, though this capability varies greatly. Some ecological communities are very fragile, while others are surprisingly tough. Both history and current science can provide examples of irreversible damage, as well as remarkable recovery. Biodiversity is a vital aspect of ecosystem function and stability. Many human activities reduce biodiversity, threatening local and even global homeostasis. The need for greater scientific knowledge coupled with an ethical concern for ecosystem integrity is more apparent than ever.

REFERENCES

Chiras, D. 1985. *Environmental Science: A Framework for Decision Making.* Menlo Park, CA: Benjamin Cummings.

Cohn, J. P. 1994. *Restoring the Everglades.* BioScience 44: 579–583.

Culotta, E. 1995. Bringing back the Everglades. *Science* 268: 1688–1690.

Dasmann, R. 1984. *Environmental Conservation.* 5th ed. New York: Wiley.

Ehrlich, P., and A. Ehrlich. 1981. *Extinction: The Causes and Consequences of the Disappearance of Species.* New York: Random House.

Elton, C. 1958. *The Ecology of Invasions by Animals and Plants.* London: Methuen.

Grisham, J. 1994. Attack of the fire ant. *BioScience* 44: 587–590.

Hutchinson, G. E. 1948. Teleological mechanisms: Circular causal systems in ecology. *Annals of the New York Academy of Science* 50: 221–246.

Krebs, C. J. 1972. *Ecology: The Experimental Analysis of Abundance and Distribution.* New York: Harper and Row.

Leopold, A. 1933. *Game Management.* New York: Scribner's.

Leopold, A. 1949. *Sand County Almanac.* New York: Oxford University Press.

Odum, E. P. 1963. *Ecology.* New York: Holt, Rinehart, and Winston.

Pimentel, D., and H. Lehman (eds.). 1993. *Pesticide Question: Environment, Economics and Ethics.* New York: Chapman and Hall.

Ricklefs, R. E. 1990. *Ecology.* 3rd ed. New York: W. H. Freeman.

Sastrapradja, D. 1985. *Proceedings of the Symposium on 199 Years of Development of Krakatau and Its Surroundings.* Jakarta: Indonesian Institute of Sciences.

Soule, M. E., and K. A. Kohm. 1991. *Research Priorities for Conservation Biology.* Covelo, CA: Island Press.

Southwick, C. H. 1983. Environmental change and the Principle of Santayana. In *Environment and Population: Problems of Adaptation,* ed. J. B. Calhoun, pp. 93–96. New York: Praeger Scientific.

Stern, P. C., O. R. Young, and D. Druckman (eds.). 1992. *Global Environmental Change: Understanding the Human Dimensions.* Washington, DC: National Academy of Science.

Watt, K. E. F. 1968. *Ecology and Resource Management.* New York: McGraw-Hill.

Weber, P. 1994. Resistance to pesticides growing. In *Vital Signs,* ed. L. Brown, H. Kane, and D. M. Roodman, pp. 92–93. Washington, DC: World Watch Institute.

Whitaker, R. H. 1970. *Communities and Ecosystems.* New York: Macmillan.

Wilson, E. O. 1992. *The Diversity of Life.* Cambridge, MA: Harvard University Press.

Human Impacts on Planet Earth

Chapter 8

OUR GLOBAL CONDITION: A CLASH OF CONCEPTS

For many years, even centuries, there have been divergent views on human prospects in relation to the condition of our environment. Since the time of Plato (427–327 B.C.), many writings have described a utopian world where human beings would live in perfect harmony with each other and their environment. Sir Thomas More provided the Renaissance definition of Utopia in 1516, followed by others who wrote from a social, religious, economic, or environmental point of view (e.g., Francis Bacon, 1627; Louis-Sebastian Mercier, 1772; Edward Bellamy, 1888; H. G. Wells, 1905). These writings were predominantly philosophical and speculative, not scientific in the modern sense of the world, but they did express beliefs and attitudes about our future potential for the good life.

In contrast, other writers and philosophers worried deeply about the human future, focusing on the consequences of excessive population growth. Thomas Malthus (1798) foresaw the onslaught of Four Horsemen of the Apocalypse—poverty, disease, starvation, and war—unless human societies could exercise moral restraint and maintain their numbers in better balance with their economic resources. Others had anticipated his concerns—Machiavelli (1469–1527), Botero (1543–1617), Buffon (1707–1788), and Benjamin Franklin (1706–1790)—but Malthus brought the population and resource issue into its sharpest and most controversial focus.

In our own century, this dichotomy of attitudes was expanded with the dire predictions of Fairfield Osborn (*Our Plundered Planet*, 1948) and William Vogt (*Road to Survival*, 1948). Both saw human populations degrading our planet disastrously unless we reversed the trends of environmental deterioration. These pessimistic projections were followed by the enthusiastic optimism of many business and political leaders in the 1950s, who foresaw a brilliant future, the end of poverty, the conquest of disease, and abundant life (Sarnoff et al., 1956).

It would take a historical volume of considerable length to trace the rise and fall of environmental and social optimism throughout written history. We will return to some of these concepts in Chap 15 when we examine the role of scientific thought, but we can bring these views most clearly in focus now by reference to more recent studies and publications.

GLOBAL 2000

In 1982 the Council of Environmental Quality, an advisory branch of the White House, in the United States, published an extensive study of the global condition, entitled, *The Global 2000 Report to the President.* Twelve federal agencies of the U.S. government participated,[1] and more than 150 scientific experts contributed to the data analyses and writing. The final report, five years in the making (and actually completed in 1979 though not published until 1982), was 766 pages, making it one of the most extensive analyses of its type (Barney, Pickering, and Speth, 1982).

The main conclusions of Global 2000 were pessimistic, even alarming. They are summarized in the following quotation from page 1 of the report:

> If present trends continue, the world in 2000 will be more crowded, more polluted, less stable ecologically, and more vulnerable to disruption than the world we live in now. Serious stresses involving population, resources, and environment are clearly visible ahead. Despite greater material output, the world's people will be poorer in many ways than they are today. For hundreds of millions of the desperately poor, the outlook for food and other necessities of life will be no better. For many it will be worse. Barring revolutionary advances in technology, life for most people on earth will be more precarious in 2000 than it is now—unless the nations of the world act decisively to alter current trends.
>
> This, in essence, is the picture emerging from the U.S. government's projections of probable changes in world population, resources, and environment by the end of the century, as presented in the Global 2000 Study. They do not predict what will occur. Rather, they depict conditions that are likely to develop if there are no changes in public policies, institutions, or rates of technological advance, and if there are no wars or other major disruptions. A keener awareness of the nature of the current trends, however, may induce changes that will alter these trends and the projected outcome.

For human populations, Global 2000 projected growth from 4 billion in 1975 to 6.35 billion by the year 2000, an increase of more than 50 percent. The rate of growth will slow only marginally from 1.8 percent per year to 1.7 percent. In absolute numbers, this means that the increment of human beings on earth will still increase from 1.4 million people per week to nearly 2 million per week by 2000. As the report emphasizes, these figures are projections rather than predictions. Projections are mathematical estimates based on current trends; predictions

---

1. In addition to the CEQ, these were the Departments of Agriculture, Energy, and the Interior, the Agency for International Development, the Environmental Protection Agency, the Central Intelligence Agency, the Federal Emergency Management Agency, the National Science Foundation, the National Aeronautics and Space Administration, the National Oceanic and Atmospheric Administration, and the Office of Science and Technology Policy.

are statements that something will occur. Projections acknowledge the fact that trends can change or unforeseen events can occur; hence, they involve an element of statistical probability and tend to be less dogmatic than predictions.

World food supplies are projected to increase 90 percent from 1970 to 2000, indicating there will probably be some improvement in per capita food production on a global basis. Arable land, however, will increase only 4 percent by 2000, so most of the increase in food production will have to come from higher yields. The Green Revolution will have to continue its remarkable record of increasing yield, and will have to rely on genetic engineering in addition to selective breeding and better crop management. Despite the projected increase in global crop production, there will be many local and regional shortages of both food and water, as already seen in Ethiopia, Sudan, and Somalia where famine conditions prevailed from 1985 through 1993.

Global 2000 also projected significant losses of the world forests over the next 20 years, with a 50 percent per capita decline in growing stocks of timber. World forests are, in fact, disappearing at the rate of 18–20 million hectares per year, an area approximately one-half the size of California. Much of the loss is occurring in tropical forests, increasing the serious threat of major species extinctions in tropical ecosystems. Many biologists now estimate that as many as 20 percent of all species on earth will be lost if present trends continue.

Further evidence of environmental deterioration occurs in the worldwide loss of agricultural soils, and the extensive spread of deserts. In the 1980s world soil losses, from eroded soil washing into rivers, coastal areas, and oceans, exceeded 20 billion tons per year (Postel, 1989); more recently the loss has been estimated at 75 billion metric tons per year (Pimentel et al., 1995). Even the world's richest agricultural areas, such as the central United States, are experiencing serious depletion of soil reserves. More than 200 million tons of soil wash down the Mississippi alone every year, and underground aquifers of agricultural water in the western United States are dropping perceptively. On a worldwide basis, 6 million hectares are lost to desertification annually, an area about the size of Maine. World deserts, now occupying about 17 percent of the earth's land surface, were projected in the Global 2000 study to increase 20 percent by 2000. Many of Global 2000's projections are on target so far.

Global 2000 also projected serious declines in air quality and water quality. The former may have measurable impacts on climate, through the operation of the greenhouse effect, and the latter will increase competition for domestic water sources.

The overall message of Global 2000 was not one of despair but of great concern, and pessimism, if trends seen at that time continued. It attempted to focus attention on the major issues facing all nations and humankind in general, issues that were not receiving adequate political attention on the international scene. If the projections of Global 2000 continue to be borne out, the world will indeed be a harsher, sadder, and more competitive environment for many people, with the seeds of despair sprouting and pushing above the surface of everyday life.

OPPONENTS OF GLOBAL 2000

Almost as soon as *The Global 2000 Report* was published, critics responded. They objected to its negative tone, its disregard of progress, and its pessimistic focus on problems. Two of the most verbal critics, Julian Simon and the late Herman Kahn took up the debate right away, with a public discussion of world affairs at the AAAS meetings in Detroit in December of 1983. This was followed by a book and many articles presenting the other side of the coin. Simon and Kahn focused on human accomplishments, economic progress, and success stories in global affairs (Holden, 1983; Simon, 1984; Simon and Kahn, 1984).

In a series of conclusions diametrically opposed to those of Global 2000, Simon and Kahn summarized their findings as follows:

> If present trends continue, the world in 2000 will be less crowded (though more populated), less polluted, more stable ecologically, and less vulnerable to resource-supply disruption than the world we live in now. Stresses involving population, resources, and environment will be less in the future than now. . . . The world's people will be richer in most ways than they are today . . . The outlook for food and other necessities of life will be better [and] life for most people on earth will be less precarious economically than it is now.

Some of points listed by Simon and Kahn to support their optimistic projections were as follows:

- Life expectancies have been rising steadily in most of the world, a sign of demographic, scientific, and economic success.
- Birth rates have been falling in most nations. World population growth has declined from 2.2 percent yearly to 1.55 percent in the past 20 years.
- Food supplies have been improving dramatically since World War II. Famine death rates have declined.
- Trends in world forests are not worrisome, though local deforestation is troubling.
- There is no direct evidence for the rapid loss of species in the next 20 years.
- The availability of land will not constrain agriculture in coming decades.
- In the United States, higher-quality cropland is suffering less from erosion than in the past.
- Water does not pose a problem of physical scarcity, but it does require better management of existing supplies.
- The climate is not undergoing unusual or threatening changes.
- Mineral sources are becoming more abundant rather than less.
- World oil prices may well fall below recent levels.
- Nuclear energy gives every evidence of costing fewer lives per unit of energy produced than does coal or oil.
- Threats of air and water pollution have been vastly exaggerated.

In sum, Simon and Kahn foresaw a bright global future, with continuing improvement in the human condition and the environmental state. They have reaffirmed a strong belief in the power of science and technology to improve our lot, and they staunchly resisted the gloom-and-doom prognostications of Global 2000. The debate between ecologists and entrepreneurial economists has been reviewed and summarized in a recent book by Myers and Simon (1994).

The dispute between Simon and Kahn and the authors of the Global 2000 study is by no means the only one of its type. It is reminiscent, in fact, of a dispute 10 years earlier between the Club of Rome and the Sussex Group. In 1972, a report was issued under the auspices of the Club of Rome, a coalition of businesspeople concerned about global futures. The report, *The Limits to Growth,* (Meadows et al. 1972) was based on a major computer modeling of world trends in population, economics, and environment undertaken by a group of scientists and systems analysts at the Massachusetts Institute of Technology. The general conclusions of *The Limits to Growth* were similar to those of *The Global 2000 Report;* that is, projecting current trends, the computer models pointed to a number of dire consequences: overcrowding, increased pollution, environmental and economic deterioration, and increasing threats to the quality of life.

The importance of both the Club of Rome report and the Global 2000 study was their broad systems approach to global trends; both integrated data from many sources, including demography, economics, agriculture, industry, and environmental trends. In their analyses of interrelationships and feedback mechanisms, they represented pioneering efforts in global ecology.

Within a year, however, the *Limits to Growth* study was sternly refuted by a group of economists and physical and social scientists under the leadership of Christopher Freeman at the University of Sussex. Their report, *Models of Doom: A Critique of "The Limits to Growth,"* contested many of the methods and conclusions of the MIT study (Cole et al., 1973). Although they agreed in principle with the technique of mathematical model-building, they felt there were serious flaws in the attempt to model global processes in a unitary fashion, and they came to quite different conclusions about the future prospects of the world. They disagreed with much of the pessimism in the *Limits to Growth* study.

In many ways, the Global 2000–Simon and Kahn controversy seemed to be a replay of the previous decades, but the world scene is constantly changing, databases grow larger, new actors appear on stage, and the stakes become higher than ever before. The fact remains that we have abundant scientific literature projecting tragedy for the world unless we can effect major changes in current trends, and we also have abundant literature projecting a bright global future with improving conditions for most of the world's peoples. Brown (1991) discussed these two world views in terms of their disciplinary and cognitive backgrounds. He and his colleagues also produce two annual reports, *State of the World* and *Vital Signs,* that give up-to-date evaluations of global conditions.

The Meadows team has also reevaluated its "Limits to Growth" study after 20 years in a new book, *Beyond the Limits* (Meadows, Meadows, and Randers, 1992). They conclude that world populations have already exceeded the earth's limits in some ways, and that if present trends continue the twenty-first century will witness ecological and economic collapse. They also feel that this outcome is not inevitable and that there are ways of achieving sustainable development. They see the future as dependent on human choices and actions.

Environmental optimism emerged prominently in 1995 with observations that many environmental indicators have been improving. McKibben, in an article entitled, "An Explosion of Green" (1995), noted that forests in eastern United States have increased greatly since the 19th century. For example, Vermont was 35 percent forested in 1850, now it is 80 percent, and similar forest expansion has occurred in other parts of New England. Along with trees, there has been a return of many wildlife species: wild turkeys, beaver, white-tailed deer, black bear, coyotes, and even reported sightings of cougar. McKibben pointed out that all is not well with these forests, however. Many are intensely managed (the "working forest") to the detriment of other species, and many are in poor condition due to acid precipitation, insect outbreaks, or rampant development. Nonetheless, McKibben feels that we should recognize the overall favorable trends of reforestation and concentrate on solving specific problems.

A recent book by Easterbrook (1995) also emphasizes environmental success stories. Although Easterbrook acknowledges that the loss of rain forests is a tragedy and that global warming could trigger serious problems, he sees successful trends in many of our ecological relationships. He argues strongly for a new period of environmental optimism. Both McKibben and Easterbrook are environmental journalists who focus their views on the United States, not worldwide trends. Easterbrook's analyses have been evaluated from a broader scientific point of view by Raven (1995).

EVALUATION

Certain questions emerge as we review these conflicting studies. How can the divergence between optimism and pessimism be so sharp? How can writers and scientists come to totally different conclusions after looking at the same world? Who is right? Where do we find the closest proximity to truth?

Regarding the first questions, perhaps the important factors are the experience, training, and mindset of the scientists and writers. The *Global 2000 Report* was written primarily by environmental scientists, population experts, and biologists: demographers, agricultural scientists, soil scientists, hydrologists, climatologists, and ecologists. The Simon and Kahn studies, and many of the optimistic projections prior to their work, were done primarily by economists, businesspeople, technical experts in management, and journalists. The Meadows and Randers research was done primarily by computer modelers. Clearly, these authors look at the same situations with different points of view. One can view the United States and find tremendous progress in science, education, economics, and the well-being of our population. One can also see the alarming problems of United States in our urban ghettos, homeless populations, toxic waste dumps, social decline, budget deficits, and farm problems. The contrast is even more striking if we consider two countries as diverse as Switzerland and India. It is perfectly possible to tour the world by jet aircraft and air-conditioned taxi, staying in luxury hotels, and come away with the impression of great progress and prosperity. Likewise, one can tour the world visiting urban slums, refugee camps, exhausted deserts, areas of war and terrorism, and conclude that the visions of Malthus are here to haunt us today.

Our personal views are certainly influenced by individual circumstances. Although all Americans are not wealthy, and there are increasing numbers of U.S. citizens in the poverty range as defined in this country, collectively our nation is rich compared to most of the world. A person classified as poor in the United States, having $14,000 in annual income (in 1992) for a family of four, would qualify as middle-class or even relatively well-to-do in many countries of the world. We often forget that Americans, representing less than 5 percent of the world's population, use 25 to 30 percent of the world's resources to support ourselves. The enormous consumption of food, energy, and material goods in the United States, compared with the majority of the world's peoples, contributes to our self-delusion that all nations can have the good life if they only had free enterprise.

Diverse views on the state of the world were sharply apparent in world economic, political, and social affairs in the fall of 1990 and spring of 1991. Within a short span of just six months Iraq invaded Kuwait, and the world launched into a devastating war to liberate Kuwait, a war that resulted in an estimated 200,000 human casualties, millions of refugees, and unprecedented environmental damage to the Persian Gulf region. During the same period, the U.S. stock market soared to record levels, apparently unworried about the human, environmental and financial costs of the war. Obviously, the financial community, impressed with a highly successful short-term military victory, viewed the conflict in different light than those concerned with human health, environmental integrity, and refugee problems.

Diverse views were also present at the Rio Earth Summit in 1992. The United States stood alone as the only nation not willing to sign a pact to protect biodiversity and the U.S. was a reluctant signator of a global warming agreement. The U.S. government viewed these problems in a different way than most of the world's leaders.

With these potential biases in mind, it is no wonder that we can look at the world today and reach totally different conclusions. The role of science and education is to sort through the facts and fancies, with every possible attempt to achieve objectivity and reality. The special role of ecologists is to understand relationships, the interactions between the physical and biological worlds, between human activities and the state of the world, and between our attitudes and our actions.

In the following chapters, we examine the ways in which humans are impacting the earth, how we are changing its nature, and how it is influencing us. We will look at the health of the earth and the biosphere, ourselves included.

SUMMARY

Divergent views on the future of humankind and the fate of planet earth have occurred throughout history but have become sharply focused in recent years. *The Global 2000 Report* presented a pessimistic picture of future prospects: increased crowding, poverty and pollution, environmental deterioration, loss of forests, extinction of species, ecological instability, and an increasingly perilous existence for many of the world's peoples.

The computer modeling studies of Meadows and her colleagues (1972, 1992) also projected dangerous trends in global ecology which will lead to economic collapse if unchecked, but they also hold out some hope for major changes that can lead to sustainability.

In contrast, studies and publications by Simon and Kahn concluded that the world is entering a bright future: reduced population growth, better food supplies, better pollution control, less crowding, better health and greater longevity, and generally improving economic status for most of the world's peoples.

These different groups paint pictures of different worlds. The role of science, and especially ecology, is to understand the true relationships between human populations and environmental conditions, and to evaluate current trends in these relationships.

REFERENCES

Barney, G., T. R. Pickering, and G. Speth. 1982. *The Global 2000 Report to the President*. Washington, DC: U.S. Government Printing Office.

Brown, L. R. 1991. The new world order. In *State of the World, 1991*. ed. L. R. Brown et al. New York: W. W. Norton.

Brown, L. R., et al. (eds.). 1994. *State of the World, 1994*. New York: W. W. Norton.

Brown, L. R., H. Kane, and D. M. Roodman. 1994. *Vital Signs: The Trends That Are Shaping Our Future*. New York: W. W. Norton.

Cole, H. S. D., et al. 1973. *Models of Doom: A Critique of "The Limits to Growth."* New York: Universe Books.

Easterbrook, G. 1995. *A Moment on the Earth: The Coming Age of Environmental Optimism*. New York: Viking.

Holden, C. 1983. Simon and Kahn vs. Global 2000. *Science* 221: 341–343. Reprinted in Southwick, C. H. (ed.). 1985. *Global Ecology*. Sunderland, MA: Sinauer Associates.

Malthus, T. R. 1798. *An Essay on the Principle of Population*. Reprinted in *Everyman's Library*, 1914–1952. London: J. M. Dent; New York: E. P. Dutton.

McKibben, B. 1995. An explosion of green. *The Atlantic Monthly* 275(4): 61–83 (April 1995).

Meadows, D. H., et al. 1972. *The Limits to Growth*. New York: Universe Books.

Meadows, D. H., D. L. Meadows, and J. Randers. 1992. *Beyond the Limits*. Post Mills, VT: Chelsea Green.

Myers, N., and J. Simon. 1994. *Scarcity or Abundance: A Debate on the Environment*. New York: W. W. Norton.

Osborn, F. 1948. *Our Plundered Planet*. Boston: Little, Brown.

Pimentel, D., et al. 1995. Environmental and economic costs of soil erosion and conservation benefits. *Science* 267: 1117–1123.

Postel, S. 1989. Halting land degradation. In *State of the World, 1989*, ed. L. R. Brown et al. New York: W. W. Norton.

Raven, P. 1995. Review of *A Moment on the Earth: The Coming Age of Environmental Optimism* by G. Easterbrook. 1995. New York: Viking. *The Amicus Journal* 17(1): 42–45.

Sarnoff, D., et al. 1956. *The Fabulous Future*. New York: Dutton.

Simon, J. 1984. *Bright global future*. Bulletin of the Atomic Scientists. Reprinted in Southwick, C. H. (ed.). 1985. *Global Ecology*. Sunderland, MA: Sinauer Associates.

Simon, J., and H. Kahn (eds.) 1984. *The Resourceful Earth: A Response to Global 2000*. New York: Oxford University Press.

Vogt, W. 1948. *Road to Survival*. New York: William Sloane Associates.

Chapter 9

GLOBAL CHANGE
AND DEVELOPMENT

The term "global change" has many meanings. Often it is used to mean climate change, especially global warming and the atmospheric processes leading to global warming. In a broader sense, however, global change refers to all of the ecological changes occurring on planet earth. This involves much more than climate change per se; it involves all types of air and water pollution, changes in radiation, the release of toxic chemicals, land degradation, vegetational changes, loss of biodiversity, population growth, urbanization, and even constructional changes in the domestic environment if they are of worldwide importance collectively.

In this book, *global change* is used in this broader sense. It includes not only climate change but also the vast array of other ecological changes associated with or occurring at the same time as global climate change.

HUMAN MODIFICATION OF THE EARTH

We have known for a long time that human populations have been a major ecological force on the earth, modifying both physical and biological aspects of the world. Three important books emphasizing this human role are *Man and Nature: Physical Geography as Modified by Human Action*, written in 1864 by George Perkins Marsh (1965); *Man's Role in Changing the Face of the Earth*, edited by W. L. Thomas (1955–56); and *Earth as Transformed by Human Action*, edited by B. L. Turner (1990). These scholarly volumes offer clear documentation of human impacts on the geography and ecology of the earth. We will give specific examples from these books in later sections and relate them to more modern research.

Human impacts on the earth result from the interaction of at least three basic factors: numbers, activities, and duration. We may represent this relationship with the simple formula:

$$I_e = \int N \cdot A \cdot D$$

where I_e stands for impact on the environment; N, the numbers of people, or size of the human population; A, the activities of the people, which include land-use patterns, resource use, standards of living, and so on; and D, the duration of the human activities, whether they have occurred over decades, centuries, or millennia. This formulation is obviously an oversimplification because many subsidiary factors are involved such as livestock populations, machines and energy use, specific behavior patterns, conflict and warfare, and so on, but most factors fit into the general categories of numbers, activities, and duration.

For example, India and China, with a combined population of approximately 2 billion people, have a 7 times greater human population than the United States and Canada on 30 percent less land area. They have more than 10 times the livestock population. Furthermore their land has supported densely populated agricultural societies for thousands of years longer than North America has supported its population. Certainly their populations have had a much different impact on their environment than the indigenous populations of the United States and Canada.

In regard to activities, the present U.S. and Canadian populations have very different patterns of resource use and behavior than the people of India and China. In terms of diet, transportation, energy use, solid waste and pollution generation, mineral requirements, and standard of living, each individual in the United States and Canada produces far greater environmental impact than each individual in India and China. The United States alone, with less than 5 percent of the world's population, uses about 30 percent of the world's energy resources (Chiras, 1985). Thus, the United States and Canada differ from India and China in all three major categories *(N, A, and D)*, even though they share certain geographic features including latitude, climate, and topography.

Three of the most dramatic periods of environmental change due to human action have been (1) the rise of agriculture approximately 10,000 years ago, (2) the Industrial Revolution of the eighteenth and nineteenth centuries, and (3) the rapid population growth and worldwide agro-industrial expansion of the twentieth century, especially the past 50 years. These periods will be considered in subsequent chapters, but at this point we can highlight a few prominent aspects of global change since World War II.

A number of global trends show alarming patterns of deterioration. They emphasize the critical importance of establishing a better balance between economic and environmental relationships. These trends will also be discussed in subsequent chapters, but some of them can be mentioned here. The past 50 years have seen the following global changes.

1. World populations have increased 152 percent, from 2.3 billion in 1945 to 5.8 billion in 1996.

2. The annual net increase of people has increased 165 percent, from 35 million people per year to over 90 million per year.

3. Agricultural land use in tropical regions has doubled, much of it leading to unproductive conditions (Houghton, 1994).

4. Twenty to 40 percent of tropical forests have been clearcut, often converted to scarred wastelands.

5. Billions of tons of topsoil have been lost annually, and 17 percent of the world's vegetated land has been seriously degraded (Postel, 1994).

6. Ninety percent of the carbon held in vegetation and 25 percent of the carbon in soil has been released (Houghton, 1994).

7. Atmospheric carbon dioxide has increased approximately 17 percent, from around 300 parts per million to 350 ppm, raising the potential of unfavorable climate changes.

8. Emissions of other air pollutants have increased significantly despite improvements in air quality in many industrial cities. Sulfur dioxide emissions in the United States alone increased approximately 30 percent from 20 to 26 thousand tons, and nitrogen oxide emissions doubled from 10 to over 20 thousand metric tons per year.

9. Antarctic ozone levels in the upper atmosphere showed decreases as much as 50 percent; temperate regions showed a 4 to 6 percent decline, indicating a serious loss of this protective layer in the earth's upper atmosphere (Raven, Berg, and Johnson, 1995).

10. Marine fisheries' production of cod, herring, pilchard, and haddock declined an average of 75 percent since their peak years in the 1960s. More than a dozen other species of marine fish have shown seriously declining populations (Weber, 1994).

11. Coral reefs around the world have experienced extensive mortality from pollution, disease, overfishing, deforestation, dredging, and mining (Hughes, 1994).

12. 70 percent of the world's bird species have declined in numbers, and 1,000 species are now threatened with extinction (Youth, 1994).

This list could be extended to show many other changes in the earth's landscape, lakes, rivers, oceans, and biosphere. These changes indicate serious problems in the earth's systems that are endangering planetary health. Obviously, not all changes are detrimental or degrading; it is important to recognize favorable developments as well. There have certainly been many advancements in the human condition, in agriculture, food, industry, transportation, conservation, health, medicine, housing, education, communications, and living standards, but many of these favorable developments have been bought at the expense of future generations and the well-being of planet earth.

The relationships between population and environment are complex, and it is a mistake to oversimplify them. Some human impacts on planet earth are attributable to wealthy nations exploiting earth's resources to maintain high standards of living, such as multiple homes, multiple cars, and affluent lifestyles. Other human

Figure 9.1 Hong Kong's Kai Tak Airport, the third-busiest air cargo and the fourth-busiest air passenger terminal in the world, symbolizes growth and economic expansion, but it also represents a degree of crowding and noise that most Americans would not tolerate. Over Chinese objections, Hong Kong is moving ahead with plans to build a new airport on Lantau Island by the end of the century. (Photo by Steve Raymer © copyright National Geographic Society)

impacts are attributable to poor nations struggling to survive by overgrazing, over-cropping, and cutting all available vegetation for fuel.

Unfortunately, the gap between the rich and the poor has been widening. In 1960, the richest 20 percent of the world's people absorbed 70 percent of the global income; by 1989, their share climbed to 83 percent. In the same span of years, the poorest 20 percent had their share of global income drop from 2.3 percent to only 1.4 percent (Postel, 1994). The old adage "The rich get richer and the poor get poorer" was true for these years.

The extremes of wealth and poverty represent a continuum, of course, with many nations and certainly a majority of the world's peoples falling between the very rich and very poor. Nonetheless, this dichotomy between poverty and wealth illustrates the need for all peoples of the world to achieve a better economic and environmental balance and a sustainable mode of existence.

SUSTAINABLE DEVELOPMENT

For many nations and most of the world's peoples, the number one priority is economic development. Ideally, economic development should be sufficient to provide a reasonable standard of living and a secure future for a country's people, but this is an elusive goal. In 1975 Corita Kent stated the problem clearly:

As we grapple with the day's pressing problems, our view of the future is blurred and obscured. Yet the future threatens. Energy and ecology, unemployment and inflation, health and housing, war and peace. Even as we debate such issues, we exhaust the means to solve them. We burn our oil, deplete our lands, endanger our health, all to get us through the day. Inevitably, the situation grows more urgent.

"Sustainable development" has become a popular phrase among both econo-mists and ecologists. It has been defined as "a dynamic process designed to meet today's needs without compromising the ability of future generations to meet their own needs" (Corson, 1990). If sustainable development is our goal, human socie-ties must find ways of meeting current needs for food, jobs, housing, recreation, and the full range of human activities without destroying the productive capacity of the earth to provide these goods and benefits. We must foster agriculture with-out soil erosion, industry and transportation without excessive pollution, forestry and fisheries without depletion, and ultimately population stability without poverty and war.

In the past, the concept of sustainable development seemed utopian, a nice idea but not very practical. After all, there was always the frontier, providing unex-ploited resources and unlimited opportunities. The United States was founded on a frontier model and nourished by the dream of unlimited growth. President Frank-lin D. Roosevelt captured this philosophy in the 1930s when he said, "We shall expand indefinitely." He foresaw unlimited economic expansion with continuing growth in human prospects and well-being. In essence, this has always been the American Dream. The concept is an attractive one, and was a politically powerful idea as America emerged from the Great Depression, but today, in a more ecologi-cally aware world, it is naive to extend the concept to all realms of human activ-ity. We certainly know that growth and expansion cannot be indefinite in a finite

Figure 9.2   The world's insatiable appetite for petroleum produces environmental problems on land and sea. Here oil-field development along the Caspian Sea near Baku, Azerbaijan, has created an eco-logical mess that intensifies the country's slumping fishing and caviar industry. (Photo by George F. Mobley © copyright National Geographic Society)

system. Although the world needs development in many areas, ultimately the total sum of global development must be sustainable.

IMPLICATIONS OF SUSTAINABLE DEVELOPMENT

Sustainable development implies an equilibrium condition or steady state in terms of resource use, but it does not refer to a static condition. Both words are important in the phrase. *Sustainable* means that the process or activity can be maintained without exhaustion or collapse; *development* means that change and improvement can occur as a dynamic process.

Natural ecosystems provide a model. Natural ecosystems are sustainable in terms of resource utilization and energy use, but they also develop and change in the process of succession, as discussed in Chapter 7. Natural ecosystems are not static, but they are sustainable within the framework of the environmental conditions that support them.

For human populations, sustainability must ultimately mean dynamic equilibrium in numbers, energy use, food production, waste recycling, pollution control, industrial processes, and materials utilization. The earth cannot support an unlimited human population, an infinite number of motor vehicles, unchecked soil erosion, continued increase in air or water pollution, uncontrolled expansion of atmospheric carbon dioxide, or any other type of exponential growth.

While the notion of limited growth sounded un-American 40 or 50 years ago, it is more widely accepted now. Even many members of the business community, represented by the Club of Rome, a group of concerned business men, have supported and endorsed the model of limited growth in the book *The Limits to Growth* (Meadows et al., 1972). The idea of limiting growth has been perceived by others as unnecessarily negative, and, in fact, a major economic critique of the Meadows work is titled *Models of Doom: A Critique of "The Limits to Growth"* (Cole, et al., 1973). The controversy ignited by these two publications is simply another example of the economic-ecologic conflict that has been with us for many years.

The concept of limited growth has been incorporated in the concept of sustainable development. Although the latter concept recognizes limits to growth, it also acknowledges the positive aspects of development. Development can occur potentially in many aspects of human existence; it can take the form of better food and housing, better education and health care, expanded opportunities, and the spread of various intangibles associated with the quality of life. Certainly most of the world's peoples have tremendous growth opportunities in these areas. At the same time there must be reductions in population growth rates, pollution, environmental deterioration, conflict, and violence.

The concept of sustainable development was the theme for the United Nations Conference on Environment and Development (UNCED) held in Rio de Janeiro, Brazil, June 1–12, 1992. This conference was the world's largest and most visible environmental meeting since Stockholm in 1972. It attracted over 20,000 representatives from 180 countries including 118 heads of state. The conference addressed global issues such as atmospheric pollution; ozone depletion; climate

Figure 9.3 Urban crowding is an aspect of global change that increases pollution, strains sanitation systems, and adds immeasurably to the stresses and frustrations of modern life. This street scene in New Delhi, India, is typical of traffic conditions in many developing nations. (Photo by Steve Raymer © copyright National Geographic Society)

change; land resources including soil, forests, grasslands, and deserts; freshwater, coastal and marine resources; the conservation of biodiversity; management of toxic wastes; improvement in the quality of life and human health; and the problems of poverty and their relation to environmental degradation. The agenda of this meeting covered virtually the full range of global ecology, with the possible exception of the politics of war and peace. It also slighted and tried to avoid the fundamental issue of population growth. No nation seemed ready to confront this. Many also expressed skepticism that a conference with a broad environmental agenda can accomplish very much of a practical nature (Abelson, 1991), but there is no doubt that the conference focused the world's attention on environmental matters.

It is easy, of course, to state environmental platitudes; to do something about them is a completely different matter. If we are to make progress, we must first understand some of the scientific and social aspects of the problems we face in both environmental and human terms and then consider ways in which these problems can be solved. In the following chapters we will examine specific human impacts on the earth and their consequences in terms of environmental degradation. Sustainability is discussed more fully in Chapter 25.

SUMMARY

The phrase "global change" is sometimes used to refer to climate change, but it is more appropriately used in a broader sense to cover all the major ecological changes affecting the earth. Human populations have had a major impact on the global environment since the dawn of agriculture. This impact is accelerating dramatically because of increasing populations, human activities, and resource demands. These changes emphasize the need for sustainable development to meet human needs without further environmental destruction.

REFERENCES

Abelson, P. H. 1991. Sustainable future for planet earth. *Science* 253: 117.

Chiras, D. D. 1985. *Environmental Science: A Framework for Decision Making.* Menlo Park, CA: Benjamin/Cummings.

Cole. H. S. D., et al. 1973. *Models of Doom: A Critique of "The Limits to Growth."* New York: Universe Books.

Corson, W. H. 1990. *The Global Ecology Handbook.* Boston, MA: Beacon Press.

Houghton, R. A. 1994. The worldwide extent of land-use change. *BioScience* 44(5): 305–313.

Hughes, T. P. 1994. Large-scale degradation of a Caribbean coral reef. *Science* 265: 1547–1551.

Kent, C. 1975. *Group W.* Westinghouse Broadcasting Co.

Marsh, G. P. [1864] 1965. *Man and Nature: Physical Geography as Modified by Human Action,* ed. D. Lowenthal. Cambridge, MA: Harvard University Press, Belknap Press.

Meadows, D. H., et al. 1972. *The Limits to Growth.* New York: Universe Books.

Postel, S. 1994. Carrying capacity: Earth's bottom line. In *State of the World, 1994,* ed. L. R. Brown. New York: W. W. Norton.

Raven, P. H., L. R. Berg, and G. B. Johnson. 1995. *Environment.* Philadelphia: Saunders College Publishing.

Thomas, W. L. (ed.). 1955–1956. *Man's Role in Changing the Face of the Earth.* Chicago: University of Chicago Press.

Turner, B. L. (ed.). 1990. *Earth as Transformed by Human Action.* New York: Cambridge University Press.

Weber, P. 1994. *Net Loss: Fish, Jobs and the Marine Environment.* Worldwatch Paper 120.

Youth, H. 1994. Flying into trouble: The global decline of birds. *Worldwatch* 7(1): 10–19.

Chapter 10

LAND
DEGRADATION

The most basic needs of human and animal populations are food, water, habitat, and moderately clean air. The availability of these depends on a reasonably intact landscape. In the case of humans, landscapes may be artificial and distant from the sources of food and water but this does not eliminate the basic dependency.

Landscape is the ultimate resource in the production of food and the maintenance of water and habitat. Land provides 99 percent of human food, whereas aquatic systems provide less than 1 percent (Pimentel, et al., 1995). It is no surprise, therefore, that land was a primary concern of leading early ecologists (Leopold, 1949) or that landscape ecology is a current emphasis in modern ecology (Turner and Gardner, 1991).

A logical starting point for our consideration of human impacts upon the earth is land modification. Almost every major human activity since the dawn of agriculture has involved modification of landscape. Cultivating fields, deforesting hillsides, constructing villages, towns, cities, roads, canals, bridges, dams, harbors, airports—virtually everything we do on a significant scale modifies the natural environment. Much of this modification is beneficial to us, and some of it is even beneficial to nature at large, for example, when we restore a landscape devastated by a natural disaster such as a hurricane, earthquake, or volcanic eruption. Unfortunately, many of our current activities limit, damage, or destroy landscape in a significant way, and much of this represents a serious degradation of the land's ability to function.

THE LOSS OF FUNCTIONAL LANDSCAPE

The clearest evidence for land degradation comes from data on soil erosion, desertification, and deforestation. In recent years soil erosion, which represents a basic loss of the earth's productive capacity, has been estimated to be 24 to 25 billion tons of soil per year, equal to all the topsoil on all the wheat land of Australia (Corson, 1990; Raven, 1994). More current estimates of global soil loss are as high as 75 billion tons per year (Pimentel et al., 1995).

The United Nations estimated in 1993 that 70,000 square kilometers of farmland are abandoned every year in the world because the soil is lost. This is equal to 27,000 square miles, an area more than five times the size of Connecticut. During the decade of the 1980s, the amount of arable land per person in the world declined 19 percent due to soil loss (UN Economic and Social Council, 1993). This soil is lofted into the air in windstorms, washed down rivers in rainstorms, and ultimately ends up as sediment in rivers, lakes, seas, and oceans.

The sediment loads of some of the world's rivers are shown in Table 10.1. Much of this comes from the erosion of agricultural land. These figures are so large as to be virtually incomprehensible. What do they mean in terms of actual soil loss? Table 10.2 shows the relationship between cropping systems and soil erosion on typical American farmland in Missouri. Bare soil results in losses of 41 tons per acre per year, and continuous corn, nearly 20 tons per acre, whereas rotational systems reduce this loss to less than 3 tons per acre, and pasture cover of continuous bluegrass reduces soil erosion to only 0.3 tons per acre per year. These figures, especially the high loss rates under bare soil conditions, are alarming when we realize that the rates of natural soil formation are very slow, less than 0.5 tons per acre per year (Pimentel et al., 1995).

Soil is a highly complex mixture of organic and inorganic particles containing a diverse community of small invertebrate animals, including arthropods, annelids, and nematodes, and a rich array of microorganisms and fungi (Table 10.3). The entire complex takes hundreds and even thousands of years to form. In a temperate grassland or woodland, natural soil formation may occur at the rate of only 0.5 tons per acre per year. An inch of soil takes anywhere from 200 to 1,000

TABLE 10.1
Annual sediment load transported to the sea by major rivers, early 1980s (from Brown, 1989).

River System	Country/Region	Drainage Area (thousands of km²)	Annual Suspended Sediment Load (millions of tons)
Ganges-Brahmaputra	South Asia	1,480	3,000
Huang He (Yellow)	China	770	1,080
Amazon	South America	6,150	900
Chang Jiang (Yangtze)	China	1,940	478
Irrawaddy	Burma	430	265
Magdalena	Colombia	240	220
Mississippi	United States	3,270	210
Orinoco	Venezuela	990	210

Sources: D. E. Walling, Rainfall, runoff and erosion of the land: A global view, in K. J. Gregory, ed., *Energetics of Physical Environment* (New York: John Wiley & Sons, 1987). Ganges-Brahmaputra sediment load figure from Ocean Drilling Program, news release, Texas A&M University, College Station, September 4, 1987.

TABLE 10.2
Cropping systems and soil erosion.

Cropping System or Cultural Treatment	Average Annual Loss of Soil per Acre (tons)	Percentage of Total Rainfall Running Off Land
Bare soil, no crop	41.0	30
Continuous corn	19.7	29
Continuous wheat	10.1	23
Rotation: corn, wheat, clover	2.7	14
Continuous bluegrass	0.3	12

Note: Data are from average of 14 years of measurements of runoff and erosion at Missouri Experiment Station, Columbia (soil type: Shelby loam; length of slope: 90.75 feet; degree of slope: 3.68 percent).

years to form (Postel, 1989), and at the erosion rate of 40 tons per acre per year, an inch of topsoil can be lost in 10 years. Thus, under continuous corn cultivation or bare soil conditions, hundreds of years of soil formation may be lost in a few years. In hilly country and in urban construction, soil erosion is much greater, and may exceed 200 tons per acre (Brown, 1989; Goudie, 1994). The United States is losing topsoil 17 times faster than its formation rate. The state of Iowa is currently losing soil 30 times faster than the soil formation rate (Pimentel, 1994) and reported in 1981 that it had already lost one-half of its original topsoil.

The types of complex soil communities shown in Table 10.3 are important in soil fertility, and they do not develop overnight. They depend upon long-term ecological processes that are subject to short-term damage.

Much of the world's agriculture depends upon a mantle of topsoil that is only 6 to 8 inches thick (Brown and Wolfe, 1984). Agricultural practices common in the United States, which expose bare soil for several months of the year, are responsible for excessive losses of soil through both wind and water erosion. Stormy weather exacerbates soil erosion by producing strong winds, heavy rains,

TABLE 10.3
Soil communities: Typical abundance of microorganisms and invertebrate animals.

Organism	Abundance	Reference
Bacteria	1,000,000,000 per gram	Tepper, 1969
Actinomycetes	5,000,000 per gram	Tepper, 1969
Protozoa	500,000 per gram	Tepper, 1969
Algae	200,000 per gram	Tepper, 1969
Molds	20,000 per gram	Tepper, 1969
Nematodes	175,000 to 20,000,000 per square meter	Clarke, 1954
Molluscs (slugs and snails)	50,000 per acre	*Encyclopedia Britannica*
Myriapods (millipedes and centipedes)	1,000,000 per acre	*Encyclopedia Britannica*
Annelids (mainly earthworms)	1,000,000 per acre	*Encyclopedia Britannica*
Arthropods (e.g., insects, woodlice, spiders)	1,000,000 per acre	Clarke, 1954

and flash flooding. Under such conditions, years or decades of natural soil formation may be lost in a few days. Some agricultural experts have emphasized that
U.S. agriculture is not practiced on a sustainable basis, and that our basic agricultural productivity, one of our nation's great strengths, will decline in the twenty-
first century (Brown, 1989). Furthermore, according to the American Farmland
Trust (1994), the United States is losing 3,800 acres of farmland daily due to
development, primarily urban sprawl. For example, between 1970 and 1990 the
population of metropolitan Chicago increased by 4 percent, residential land area
increased by 46 percent, and commercial land development increased by 74 percent (*The Economist,* Oct. 15, 1994, p. 31).

The draft report of the U.S. Council on Environmental Quality, prepared for
the UN Conference on Environment and Development (UNCED) in 1992, a report
that was generally optimistic on the state of the U.S. environment, pointed out
that "40 percent of American cropland has eroded to the point of losing productivity" (Council of Environmental Quality, 1991).

The most dramatic example of the loss of agricultural productivity through
erosion in the United States was the Dust Bowl phenomenon of the 1930s. Years
of overgrazing and abusive farming techniques in the high plains region of Kansas, Oklahoma, northern Texas, and eastern Colorado, coupled with natural
drought conditions, created massive wind erosion, soil loss, and economic devastation. In 1936, choking dust storms stretched from Arizona and Utah as far east
as Illinois and Kentucky (Fig. 10.1). The human impact of this environmental
disaster has been vividly portrayed in John Steinbeck's novel *The Grapes of
Wrath.*

Both intentional and inadvertent soil conservation efforts in the United States
have made substantial progress in many areas. By removing or reducing grazing
pressure, allowing grass cover to return to devastated areas, and using more active
programs of contour plowing, strip farming, and crop rotation, some agricultural
lands have been restored. The U.S. Soil Bank program, which takes farmland out
of row-crop production for extended grass and alfalfa cover, has been very effective in reversing land degradation. The planting of legumes such as clover and
alfalfa has been effective in restoring soil fertility because of their role in nitrogen
fixation. One positive sign of soil conservation in the Great Plains region is the
decline in the estimated sediment load of the Mississippi River from 500 million
tons in 1952 (Gottschalk and Jones, 1955) to 300 million tons in 1982 (Brown
and Wolfe, 1984). This is still evidence of excessive erosion, but the figures at
least show an improvement since the 1950s.

EROSION IN DEVELOPING COUNTRIES

We have had the resources and the capital in the United States to do something
about soil erosion, but many of the world's poorer countries do not. Soil losses in
many developing countries throughout Asia, Africa, and Latin America are incredible. Land degradation is particularly bad in highlands, hilly country, and mountainous regions. The primary causes of severe land degradation are overgrazing,

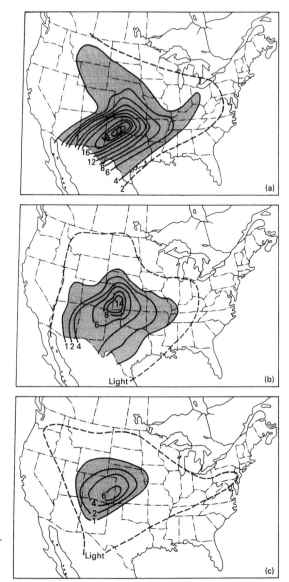

Figure 10.1 The concentration of dust storms (number of days per month) in the United States in 1936, illustrating the localization over the high plains of Texas, Colorado, Oklahoma, and Kansas: (a) March, (b) April, (c) May (from Goudie, 1994).

overcropping, deforestation and vegetational destruction. The underlying causes for this destruction are poverty and overpopulation, as people struggle to feed and shelter themselves. Figure 10.2 portrays typical severe erosion in Africa which destroys landscape, creates badlands, and takes land permanently out of a productive use, even for grazing.

Many countries suffering from this type of land degradation depend more and more upon lowlands and deltas for food production. Thus, countries such as India, Pakistan, China, Iraq, Vietnam, Egypt, Nigeria, and Brazil place more pressure

Figure 10.2 The removal of vegetation in southern Africa creates spectacular gully systems. The smelting of local iron ores in the early nineteenth century required the use of a great deal of firewood which may have contributed to this formation (from Goudie, 1994).

on their river basins and deltas to meet national food needs. But these areas have their own ecological problems, including excessive population pressures, intermittent flooding, and soil salination. Salination, or the buildup of salts and minerals in soil, is a major soil problem in many river basins. It results from flooding, and the overuse of irrigation, followed by rapid evaporation. Layers of salty minerals are deposited on the soil surface. In advanced cases, the land surface appears grayish or white, and the high salt concentrations are lethal to most types of plants.

The combined impacts of soil erosion and salination are depleting the earth's productive capacities at precisely the time when increasing populations are demanding more food. (We will look at this issue again in later chapters on desertization, deforestation, world agriculture and food prospects.) Only a small percentage of the earth's land surface is suitable for agriculture, and this amount is declining. Lester Brown (1991) has noted, "Each year, some 6 million hectares of land are so severely degraded that they lose their productive capacity, becoming wasteland." Worldwide, 80 percent of agricultural land shows moderate to severe erosion (Pimentel et al., 1995).

OTHER FORMS OF LAND DEGRADATION

The emphasis in the preceding sections has been on erosion, especially erosion related to agriculture and overgrazing. There are, however, other sources of erosion and many other forms of land degradation. Among the other sources of erosion are mining, especially surface mining, or strip-mining, logging, road building, pipeline and powerline construction, intense forest fires, chemical defoliation, and sanitary landfills. All of these activities contribute to the loss of soil by wind and water erosion and the increase of sedimentation in watercourses.

To this list we can add defunct industrial sites, some types of urban sprawl that create undesirable wastelands, abandoned housing developments, toxic waste dumps, land abandoned because of nuclear accidents as in the case of Chernobyl in the former Soviet Union, lands ravaged by war as in Kuwait, and so on. In the Chernobyl nuclear accident, over 100,000 people had to be moved from their home-sites, and eventually the radiation from this one site will affect millions of people. Much of the land around Chernobyl cannot be occupied because of radiation. Other lands became unsuitable for agriculture or habitation because of toxic runoff such as acid drainage from old coal mines or cyanide poisoning from gold mining.

Jacobson (1988) discussed the growing number of "environmental refugees," people fleeing areas unfit for human habitation. In a few cases this fleeing is due to natural events such as volcanic eruptions, as in the case of Mount Pinatubu in the Phillipines in 1991, or typhoons and tidal waves, as in the case of Bangladesh, also in 1991. But many of these environmental refugees are fleeing land despoiled by human activities. In Ethiopia, for example, over 1 million people had to aban-don once fertile farmlands that had turned into "stony deserts" from overgrazing and land misuse coupled with natural drought cycles. In 1991, the threat of starva-tion hung over 10 million people in Ethiopia, the Sudan, and Somalia. In the Sahel region of Africa in the mid-1980s hundreds of thousands of people migrated south to find water and food for themselves and their livestock. Some of the cities in Mauritania are now being engulfed by waves of eroding sand and dust, and homes must be abandoned.

In Latin America, millions of people occupy precarious homesites on hillsides on the edge of cities such as Rio de Janeiro, Brazil; Lima, Peru; La Paz, Bolivia; and Caracas, Venezuela, where they are subject to mudslides and total loss of their homes resulting from heavy rainstorms. Deforestation is a predisposing factor. In Bangladesh, hundreds of thousands of human deaths resulting from typhoons and tidal waves in recent years have been considered acts of God, but it is, after all, conditions of high population pressure and landscape deterioration in other areas that force people into marginal habitats in such high density.

In industrial countries such as the United States, one cause of people leaving their homesites for environmental reasons has been toxic chemical accumulations in soil and ground water. This has occurred in residential areas in western New York State and eastern Missouri, and it has certainly become a factor in the loca-tion of new housing developments.

In the history of environmental refugees, it is worth recalling two groups of people who had to flee their homelands in the Pacific because of atomic radiation. The native residents of Enewetok and Bikini in the Marshall Islands of the central Pacific had to leave in the 1950s when their islands were chosen for atomic bomb testing and were made intensely radioactive. Thirty years later in the early 1980s, the people of Enewetok were given the option to return, and many did, but it was soon discovered that the food they grew on the island was too radioactive for safe human consumption. Most of them had to evacuate their homeland again. For 40 years now, these people have been environmentally displaced, many of them to shantytowns on other islands. Justifying our actions as necessary for military research and development, we made some of these tropical islands uninhabitable.

Environmental migration can also occur in response to new environmental regulations. Certainly the loss of local industries because of law limiting air or water pollution has forced people to move. In the Pacific Northwest, some small towns dependent upon logging may lose their economic bases if the Endangered Species Act closes down logging to help preserve the spotted owl in old growth forests. People in these communities may consider themselves environmental refugees because of such environmental protection measures. Environmentalists maintain that we must protect the last remnants of old-growth forest before the logging industry destroys them, even if jobs are lost. The people who might lose their jobs angrily ask, "Are owls more important than people?" To this, the ecologist responds that we have already destroyed 98 percent of our old-growth forests, and the spotted owl is merely a representative species for an entire ecosystem. If we clearcut the remaining remnants of old-growth forests, we will certainly lose many species dependent upon this type of habitat. This topic will be considered later in the chapter on deforestation—it is mentioned here because clear-cutting represents a severe form of land degradation.

A similar situation prevails with global wetlands. In the lower 48 states, we have drained and filled approximately 50 percent of our nation's wetlands. In the late 1970s and early 1980s, we lost 2.6 million acres of wetland landscape in the name of development, ranging from agricultural expansion to commercial construction (Inkley, 1991). As a result, waterfowl populations are plunging to all-time lows. For example, pintail ducks in recent years have declined 62 percent below their long-term average populations, and both mallard and canvasback duck populations are at record low numbers in North America. Many other wetland species, such as amphibian populations, are also declining seriously as a result of landscape modification. We should consider these as the "canary in the mine," warning us of serious problems ahead.

In the 1970s, over 4,000 acres of land per day in the United States were modified from biologically productive landscape to urban, suburban, and commercial development. Although precise figures are not known for the present, the trend is still continuing, and it will certainly come to haunt future generations.

THE POSITIVE SIDE OF LAND MODIFICATION

It would be unfair to leave this topic without some mention of instances in which activities have greatly improved environmental conditions. There are many cases where human activities have improved environmental conditions for people, and a few where habitat has been enhanced for both people and wildlife. On the people-only side of the ledger, we can think of many of the world's gardens and parks, which have increased the beauty and diversity of rather dull landscapes. Europe has many more such places than the United States, including stately gardens surrounding magnificent works of architecture. These gardens often have a diversity of flora and wildlife that is more accessible to the enjoyment of many people than the original land on which they were built. Our own national parks provide a different opportunity for human recreation and wildlife protection, emphasizing more natural habitats. We haven't necessarily improved these environments, but we have tried to preserve them while making them more accessible to the public. Now there is concern that our most popular parks, such as Yosemite, cannot maintain natural values under the assault of too many people, and these problems are bringing forth difficult issues in land management and public policy.

Commercial ventures can also improve landscapes for human activities without entirely destroying natural features. Waikiki Beach in Honolulu is sometimes cited as an example. It is largely human-made, and now provides a beautiful beach setting with elegant hotels and restaurants for people's enjoyment. At the same time, inshore water quality has been maintained, offshore coral reefs have been protected to a reasonable extent, whales and dolphins swim close to the reefs, and on both land and water a moderate diversity of bird life is found. Obviously, much natural habitat has been lost, but for most people the existing Waikiki is more enjoyable than the thin spit of sand and marshland that originally existed. To note this is by no means to advocate converting all natural landscapes into resort centers, but simply to acknowledge that some human-induced changes can be made without excessive destruction of all natural habitat.

Other examples might include attractive parks, farms, estates, wildlife sanctuaries, and even cities where humans through planning and commitment have improved on nature. These will be fighting words to purists in the environmental arena but most people find such projects defensible. Even a staunch environmentalist, however, might acknowledge the benefits of environmental modification in harsher parts of the world. In the mid-Sahara desert, for example, with its incredible heat, lack of water, and blinding sandstorms, or in the deep interior of New Guinea, where humans must contend with mosquitos, leeches, and malaria, most of us would be very happy to see a comfortable rest house with electricity and running water. This is not to be taken as an argument against unspoiled wilderness—the need for wilderness is real, but we also need and can enjoy the right kinds of environmental modification.

SUMMARY

Human populations modify global environments in many ways, and have done so for centuries. Some modifications are necessary and beneficial for our own survival, but the majority of changes now represent serious degradation of landscape. This means that the functional capabilities of landscape are impaired, often permanently. Soil erosion occurs on a devastating scale in many parts of the world; in other areas, it represents a slow but insidious decline in life support and economic well-being. Other forms of land degradation include the buildup of toxic chemicals, the release or accumulation of radioactive elements, the destruction of land by covering it with asphalt or concrete, the creation of badlands from strip-mining and mine spoil, the spread of deserts, the loss of forests and wetlands, and the extension of urban and industrial blight. The landscape and the biosphere make up a living fabric spread over the surface of the earth. Thus, the degradation of land must be a concern of the highest priority in both ethical and practical terms. We are literally shredding and paving the surface of the earth at a rate that will come back to haunt us in the future.

REFERENCES

American Farmland Trust. 1994 (Dec.). News release. Washington, D.C.

Brown, L. R. 1989. Reexamining the world's food prospects. In *State of the World, 1989*, ed. L. R. Brown et al., pp. 41–58. New York: W. W. Norton.

Brown, L. R. 1991. The new world order. In *State of the World, 1991*, ed. L. R. Brown et al., pp. 3–20. New York: W. W. Norton.

Brown, L. R., and E. C. Wolfe. 1984. *Soil Erosion: Quiet Crisis in the World Economy*. Worldwatch Paper 60: 1–49. Reprinted in C. H. Southwick (ed.). 1985. *Global Ecology*, pp. 165–191. Sunderland, MA: Sinauer Associates.

Clarke, G. L. 1954. *Elements of Ecology*. New York: Wiley.

Corson, W. H. 1990. *The Global Ecology Handbook*. Boston: Beacon Press.

Council on Environmental Quality. 1991. U.S. National Report Prepared for Submission to the United Nations Conference on Environment and Development (draft). Washington, DC.

Gottschalk, L. C., and V. H. Jones. 1955. Valleys and hills, erosion and sedimentation. In *Yearbook of Agriculture: Water*, pp. 135–143. Washington, DC: U.S. Government Printing Office.

Goudie, A. 1994. *The Human Impact on the Natural Environment*. Cambridge, MA: MIT Press.

Inkley, D. B. 1991. *Comments on the Proposals for Late Migratory Bird Hunting Regulations, 1991–1992*. Washington, DC: National Wildlife Federation.

Jacobsen, J. L. 1988. Environmental refugees: A yardstick of habitability. Washington, DC: Worldwatch Paper 86.

Leopold, A. 1949. *Sand County Almanac*. New York: Oxford University Press.

Pimentel, D. 1994. Personal communication.

Pimentel, D., et al. 1995. Environmental and economic costs of soil erosion and conservation benefits. *Science* 267: 1117–1123.

Postel, S. 1989. Halting land degradation. In *State of the World, 1989*, ed. L. R. Brown et al. New York: W. W. Norton.

Raven, P. H. 1994. Defining biodiversity. *Nature Conservancy* 44(1): 10–15.

Tepper, B. 1969. Population growth of bacteria. *Biology of Populations*, ed. B. K. Sladen and F. B. Bang. New York: American Elsevier.

Turner, M. G., and R. H. Gardner (eds.). 1991. *Quantitative Methods in Landscape Ecology: The Analysis and Interpretation of Landscape Heterogeneity*. New York: Springer-Verlag.

UN Economic and Social Council. 1993. *Report of the Secretary General of the International Conference on Environment and Development*. New York: United Nations.

Chapter 11

DESERTIFICATION

$\mathbf{D}$esertification is a type of global change that refers to the conversion of arid or semiarid lands to deserts. Semiarid lands are usually grasslands, savannahs, or steppes. They become deserts through a progressive loss of rainfall and changes or loss of vegetation. These changes may be natural or induced by human action, or a result of various combinations of both natural and anthropogenic forces.

DEFINITIONS AND CATEGORIES OF DESERTS

Deserts are usually defined as dry regions with less than 25 cm (about 10 inches) of rainfall a year (ReVelle and ReVelle, 1992). These regions are also characterized by frequent atmospheric high pressures, low humidity, and high evaporation rates which greatly exceed precipitation. Deserts vary greatly in many physical and biological characteristics, from regions with virtually no rainfall to regions with seasonal monsoons that may cause temporary flash floods.

Deserts may be classified in various ways according to the amount of rainfall, snow, or fog they receive; their temperature regimes throughout diurnal, seasonal, or annual cycles; their topography, elevation, and latitude; their geological histories and continental positions; their flora and fauna; and the relative roles of natural and human factors in their formation.

So-called hyperarid deserts, such as the Atacama Desert of northern Chile, may average less than 2.5 cm rainfall per year (1 inch), and may have several years at a time of no rainfall at all. Such deserts, however may have plant and animal life that depends on coastal fog (Allan and Warren, 1993).

Thus, deserts may be moderate or extreme in their relative lack of precipitation;

tropical, temperate, or cold in temperature patterns; flat, undulating, or mountainous in topography; low or high in elevation and latitude; ancient or recent in geological terms; and natural, anthropogenic, or mixed in origin. Desert ecosystems and the richness of their biological communities vary tremendously according to any or all of these parameters.

NATURAL DESERTS

Natural deserts occur where naturally dry air currents have lost most of their moisture, and the regions are characterized by high atmospheric pressures, a lack of cloud cover, and abundant sunshine. Such regions often occur around 20° to 30° latitude north and south of the equator because air currents (known as Hadley cells) have lost most of their moisture over the tropics, which are within 23° of the equator. Most of the Sahara, Kalahari, Saudi Arabian, Rajasthan, and Australian deserts are examples of this type of desert.

The same conditions of dry air prevail in the rain shadows of mountains that intercept moisture-laden air. The term *rain shadow* refers to naturally arid regions on the leeward side of mountains. Mountains force moist air currents to rise and cool, causing precipitation in the form of rain, fog, or snow on the windward side of mountains. The air, now dry, then flows down the leeward slopes. The Mojave and Sonoran deserts of the southwestern United States and northern Mexico are examples of deserts in the rain shadows of the Sierra Nevada, San Gabriel Mountains, and Sierra Madre.

Natural deserts of relatively long geological histories often have rich communities of plant and animal life. The Sonoran Desert, for example, around Tucson, Arizona, is famous for mesquite, saguaro, cholla, and organpipe cacti, plus many other plants, and a diverse community of arthropods, reptiles, birds, and mammals. Among the mammals are pocket mice, grasshopper mice, kangaroo rats, desert cottontails, jackrabbits, peccaries, white-tailed deer, desert bighorn sheep, coyotes, coatimundis, and cougars. The Sonoran Desert has extremes of temperature and precipitation, with long, hot dry spells, interspersed by monsoonal rains and occasional snowfalls. The plants and animals of the Sonoran Desert have evolved fascinating adaptations for thermoregulation and water conservation to survive these conditions. When seasonal rains fall, the Sonoran Desert blooms in profusion, and living organisms seem to appear magically.

Natural deserts with more extreme conditions of dryness and temperature have much less abundant life. Much of the Sahara, for example, and large parts of the Saudi Arabian desert are barren, they are covered by rock and sand, lack plant life, and have daytime temperatures that may exceed 50° C (over 120° F), dropping to near 0° C (32° F) at night. These temperature extremes and the total lack of water in some areas produce large regions virtually devoid of resident plant and animal life. Migratory species may traverse such regions, but very few, if any, organisms live permanently under the most extreme desert conditions.

ANTHROPOGENIC DESERTS

Deserts resulting from human action in converting semiarid lands to true deserts usually present a different picture. In plant and animal life they resemble the most extreme natural deserts in their lack of biological communities. Former plant and animal life has disappeared and has been replaced by barren rock and sand.

The agents of anthropogenic desertification are usually overgrazing by domestic animals, overcropping in agriculture, cutting of brush, excessive withdrawal of ground water, salination, and sometimes frequently repeated use of fire. In other words, great numbers of cattle, sheep, or goats, combined with excessive use of water, plows, and fires can reduce grasslands to wastelands. This transformation is occurring extensively around the world, both on the margins of natural deserts and in vast tracts of natural grassland and savannah ecosystems (Allan and Warren, 1993).

A more complete definition of *desertification* is "a decrease in the productivity potential of land under arid, semi-arid and dry sub-humid climates, that may eventually lead to desert-like conditions" (Le Houérou, 1994). A related term, *desertization,* usually refers to an irreversible change of arid land into desert, that is, irreversible on a human time scale within 25 to 50 years.

Anthropogenic forces are not necessarily the sole cause of desertification. Human forces come into play most dramatically during drought cycles. Periods of reduced rainfall places greater stress on vegetation and water resources, plants cannot recover from grazing, crops fail, and ground water sources are exhausted. If irrigation is available, complete dependence on imported water coupled with high temperatures often leads to salination of soils.

We are now beginning to understand more about the relationships of rainfall and vegetative cover. If the vegetation is stripped from the land, and bare ground is exposed, surface temperatures increase, albedo (surface reflectivity) increases, hot thermal updrafts occur, humidity decreases, cloud cover is less likely to form, and rainfall declines. These changes are even more dramatic in large-scale deforestation. In both cases, the natural hydrological cycle is interrupted. Thus, the early stages of anthropogenic desertification are exacerbated by natural processes. Antienvironmentalists can say that desertification occurred because of a natural drought. Ecologists reply that human actions contributed to the drought.

THE EXTENT OF DESERTS AND DESERTIFICATION

In 1982 deserts occupied approximately 17 percent of the earth's land surface, but the Global 2000 report estimated that desertification was occurring at the rate of 6 million hectares per year, an area about the size of Maine. This spread of desert-like conditions included 3.2 million hectares of rangeland, 2.5 million hectares of rain-fed cropland, and 125 thousand hectares of irrigated croplands (Barney, Pickering, and Speth, 1982). The primary causes of desertification were identified as "overgrazing, destructive cropping practices, and use of woody plants for fuel."

The United Nations identified 2 billion hectares of land at high risk for desertification "as increasing numbers of people in the world's drier regions put more pressure on the land to meet their needs for livestock range, cropland, and fuelwood." The Global 2000 report projected that the world's desert areas would expand 20 percent by the year 2000.

The actual rate of desertification may, in fact, be greater than that projected in Global 2000. Some recent estimates of the extent of desert areas in the world give the figure of 26 percent of the earth's land surface (Smith, 1992). The exact figure depends on one's definition of desert, but Smith's use of the term is very conservative, and he defines true deserts as areas with rainfall below 12 cm per year (less than 5 inches). Allan and Warren (1993) state that almost 40 percent of the earth's land surface is occupied by deserts, but this includes extremely cold and frozen areas such as Antarctica.

Other scientists point out that 41 percent of the earth's land surfaces are "drylands" used for dryland farming, irrigated crop production, or grazing. Of these lands, 30 percent of the irrigated croplands are degraded, 47 percent of the dryland farms are degraded, and 73 percent of the rangelands are degraded (Kassas, 1994). According to the United Nations these deteriorating drylands provide life support for 900 million people who are facing economic hardship, social disruption, and, in some cases, famine.

Allan and Warren also point out that after oceans, deserts are among the most important elements in the global climate system. Deserts, like oceans, generate temperature and pressure conditions that have worldwide effects.

Anthropogenic desertification is a severe form of land degradation that seriously affects the biosphere and the ability of planet earth to support a growing human population. Figure 11.1 shows extensive areas with various degrees of desertification.

In the records of desertification, it is particularly interesting to review the recent history of two areas, northern Africa and the western United States. In Chapter 13, we broaden our historical scope and look at desertification in the Middle East during the dawn of agriculture.

DESERTIFICATION IN THE SAHEL

The Sahara is the world's largest desert. Stretching from the Atlantic Ocean to the Red Sea; a distance of more than 3,000 miles, it covers an area almost as large as the United States (3.5 million square miles for the Sahara, 3.6 million square miles for the United States). The southern edge of the Sahara is a semiarid savannah and shrubland known as the Sahel, extending from the nations of Mauritania and Senegal to the Sudan and Ethiopia in eastern Africa.

The Sahara has increased in size over the past 10,000 years, probably as a result of natural climate shifts, although human influences may have operated around its edges, especially in Egypt and the Sudan, where ancient populations flourished and then subsided to expand again in recent times.

In the last 50 years, the Sahel has been an area of active desertification.

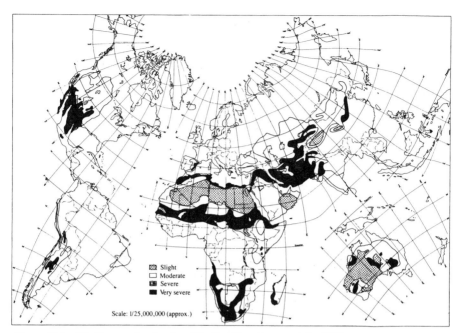

Figure 11.1   World areas showing desertification (from Barney et al., 1982).

Throughout the 1960s and 1970s the Sahara expanded southward a distance of 350 km in just 20 years (*World Bank Report,* cited in Brown, 1988). Burgeoning human and livestock populations denuded the landscape, and a severe drought cycle from 1968 to 1973 exacerbated the rate of land deterioration. Thousands of cattle died, famines occurred, and human malnutrition was extensive. The famines of Ethiopia, Sudan, and Somalia were intensified by this widespread desertification, though political and military factors also contributed to these human tragedies. Figure 11.2 shows the band of desertification stretching along the Sahel of Africa. This band of desertification was not uniform; it was very patchy and most severe in areas of human population concentration (Allan and Warren, 1993).

In the 1980s rains returned to the Sahel and vegetation returned in a very encouraging way (Tucker, Dregne, and Newcomb, 1991). The Sahara "retreated" northward again. This prompted some scientists to proclaim that the desertification of the Sahel was due to broad climate changes and not to anthropogenic forces, but this interpretation is an over simplification, as is attributing desertification solely to land abuse by humans.

Evidence that the two forces were interacting was provided in the 1970s by satellite photos of the Ekrafane ranch, a ranch of over 100,000 hectares in the Sahel in which grazing was strictly limited and cattle herds were rotated in different sectors of the ranch to prevent overgrazing. This ranch retained its vegetation and did not revert to desert during the worst of the drought that affected the area (Southwick, 1976). In space photos, the vegetation of Ekrafane stood out in stark contrast to the desertified areas surrounding it (Wade, 1974).

Other parts of the world in which desertification has been severe include eastern

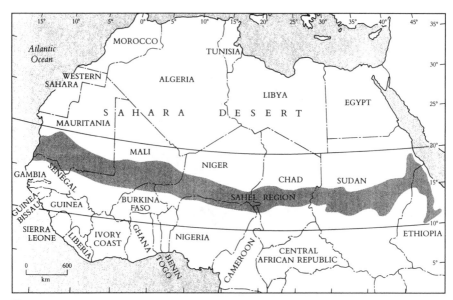

Figure 11.2   Desertification in the Sahel in the 1970s (from ReVelle and ReVelle, 1992).

Africa, especially Somalia, Sudan, and Ethiopia; southern Africa, especially Mozambique and parts of Zimbabwe and Botswana; western Asia and the Middle
East; Australia; and even parts of southern South America (Miller, 1992). Recent
estimates indicate that in addition to the 6 million hectares converted to desert
annually in the world, another 21 million hectares—an area of 81,000 square
miles, about the size of Kansas—are so degraded by soil loss every year that they
are no longer worth farming and provide only marginal grazing (Miller, 1992). In
Australia, a continent already composed of vast natural deserts, deforestation and
overgrazing have reduced tree cover from 14 percent of the land area to only 7
percent. More than 100 species of plants and 18 species of mammals in Australia
have become extinct in the last 200 years, and 4,000 other species of plants and
animals in Australia are on the endangered list (Caldicott, 1992).

DESERTIFICATION IN THE UNITED STATES

We tend to look upon desertification as a problem that occurs elsewhere in the
world—in Africa, Asia, or Australia—but we can find abundant evidence of it
within our own nation. The Dust Bowl of the 1930s in Texas, Oklahoma, Kansas,
and Colorado, was actually temporary desertification brought about by a combination of bad agricultural practices and extended drought. Because of the nation's
economic resilience, small population, and the availability of new lands to develop, we were able to recover from the Dust Bowl and reclaim many of its
damaged lands. We then developed extensive soil conservation programs and a
series of water management schemes that enabled midwestern agriculture to return
to productive states.

More permanent desertification has occurred in other western lands, especially in the American southwest. Grazing, fire, excessive irrigation, depletion of underground aquifers, and natural drought cycles have expanded the western deserts. The classic book *Deserts on the March* by Paul Sears (1935) of Yale University gives a vivid description of desertification in the United States in the early 1930s.

The World Resources Institute, in collaboration with the United Nations Environment Programme and the UN Development Programme (*World Resources, 1992–93*), estimates that between 1945 and the late 1980s, North and Central America, primarily in the west, had human-induced desertification to the extent of 145 million hectares, an area of over 55,000 square miles, equivalent in size to two-thirds the state of Utah. Two major threats to southwestern United States are high rates of population growth and excessive use of both ground and surface waters (Reisner, 1993). The water demands of cities such as Las Vegas, Phoenix, Tucson, Los Angeles, and San Diego place added stress on desert ecosystems. The overuse of water in western United States occurs to such an extent that the Colorado River supplying water to Utah, Nevada, Arizona, and southern California is used up before it reaches its natural destination in the Gulf of California. Other major river systems in Spain, Africa, the Middle East, India, and China are so overdrawn that they never reach the sea (Postel, 1995). The rivers feeding Lake Chad in central Africa have so declined in recent years that the lake has lost three-fourths of its surface area. The Aral Sea in Kazakhstan, once the world's fourth largest lake, has been shrinking in size since 1960 due to the diversion of its source rivers for irrigation. This excessive water use contributes to desertification in the watershed.

In the Near and Middle East, water is also in critically short supply. Israel, Jordon, Syria, and the Palestinians compete for sharply limited freshwater supplies, primarily from the Sea of Galilee, the Jordan River, and the mountain aquifer beneath the West Bank. The Jordan River, barely 20 feet wide and less than 6 feet deep at peak flow, stems from a watershed in the Golan Heights that Israel captured in the 1967 war. Israel's 5 million people currently have twice the water of Jordan's 4 million, and 8 times as much as 2.25 Palestinians (Schwarzbach, 1995). Israeli officials estimate that the region will have to find an additional 1 billion cubic meters of fresh water in the next 20 years. Myers has stated, "Were violence to break out again in the Middle East, it would not be over the region's most plentiful resource, oil, but over its scarcest, water" (quoted in Schwarzbach, 1995).

SUMMARY

Desertification is a form of land degradation in which semiarid grasslands, savannahs, and steppes lose vegetation and become arid deserts. Both natural and anthropogenic factors contribute to desert expansion. Most cases involve natural droughts interacting with overgrazing, overcropping, burning, excessive use of irrigation, and depletion of ground water sources.

Approximately 40 percent of the earth's land surfaces are classified as drylands, and are thus at risk of degradation and desertification. These lands provide life support for 900 million people.

True deserts occupied 17 percent of the earth's land surface in 1980 and increased about 6 million hectares annually throughout the 1980s. A 1992 study estimated that 26 percent of the earth's land surface was occupied by deserts with rainfall less than about 12 cm (about 5 inches) per year. The Sahel in Africa provided a dramatic case of desertification throughout the 1960s and 1970s as the Sahara Desert expanded southward. Recently, this advance has been reversed in some areas that have received more abundant rainfall.

Desertification has also occurred in Asia, Australia, and North and Central America in both modern and ancient times. The American southwest has experienced desert expansion, and the combination of rapid population growth plus heavy water use in this region increases the risk for continued desertification.

REFERENCES

Allan, T., and A. Warren (eds.). 1993. *Deserts: The Encroaching Wilderness*. New York: Oxford University Press.

Barney, G. O., T. R. Pickering, and G. Speth. 1982. *The Global 2000 Report to the President: Entering the Twenty-First Century*. New York: Penguin Books.

Brown, L. R., and C. Flavin. 1988. The Earth's Vital Signs. In *State of the World, 1988*. New York: W. W. Norton.

Caldicott, H. 1992. *If You Love this Planet: A Plan to Heal the Earth*. New York: W. W. Norton.

Kassas, M. 1994. *Desertification*. Environmental Education Dossiers, UNESCO, 7 (Mar. 1994). Barcelona, Spain: Centro UNESCO de Catalunya.

Le Houérou, H. N. 1994. *Desertification and Desertization: Climatic Fatality or Human Mismanagement*. Environmental Education Dosiers, UNESCO, 7 (Mar. 1994). Barcelona, Spain: Centro UNESCO de Catalunya.

Miller, G. T. 1992. Living in the Environment. 7th ed. Belmont, CA: Wadsworth.

Postel, S. 1995. Where have all the rivers gone? *Worldwatch* 8(3): 9–19.

Reisner, M. 1993. *Cadillac Desert: The American West and Its Disappearing Water*. New York: Penguin Books.

ReVelle, P., and C. ReVelle. 1992. *The Global Environment*. Boston: Jones and Bartlett.

Schwarzbach, D. A. 1995. Promised Land. (But what about the water?). *The Amicus Journal* 17(2): 35–39.

Sears, P. 1935. *Deserts on the March*. Norman: University of Oklahoma Press.

Smith, R. L. 1992. *Elements of Ecology*. 3rd ed. New York: HarperCollins.

Southwick, C. H. 1976. *Ecology and the Quality of our Environment*. New York: Van Nostrand and Reinhold.

Tucker, C. J., H. E. Dregne, and W. W. Newcomb. 1991. Expansion and contraction of the Sahara Desert from 1980 to 1990. *Science* 253: 299–301.

Wade, N. 1974. Sahelian drought: No victory for Western aid. *Science* 185: 234–237.

World Resources, 1992–93: A Guide to the Global Environment. New York: Oxford University Press.

Chapter 12

DEFORESTATION

Forests come in many forms and occupy many different environments: wet and dry, hot and cold, high and low. In general, forests are the natural biological communities on land areas with more than 75 cm of rain (about 30 inches) per year, and areas with appropriate temperatures and light. Forests do not grow above certain altitudes and latitudes because of low temperatures or low light during prolonged winters. In Colorado, at 40° north latitude, forests normally grow to about 3,400 meters (11,000 feet), but this varies with the direction and steepness of the slope, wind, soil and other factors.

THE EXTENT OF FOREST ECOSYSTEMS

About 10,000 years ago, forests and woodlands covered an estimated 6.2 billion hectares of the earth's land surface (Maini, 1990). This was equivalent to 23.9 million square miles of forest, covering 45.5 percent of the earth's land surface, exclusive of Antarctica.

By 1990 the earth's total forest lands had declined over 30 percent, to 4.3 billion hectares (16.6 million square miles), or only 28.7 percent of the earth's land surface (Maini, 1990). According to other estimates, the world's forest cover has declined 50 percent, from occupying 60 percent of the earth's land surface before people began clearing forests over 12,000 years ago to less than 30 percent of the earth's land surface today (*World Book,* 1992). Even many of our existing forests are in various states of deterioration from disease, pollution, overgrazing, or overcutting. One estimate indicates that only 12 percent of the earth's land surface consists of intact forests (Durning, 1993). In any case, we will soon have as much of the earth's land surface in deserts as in forests. Deserts are expanding, forests are shrinking.

THE ECOLOGICAL ROLE OF FORESTS

J. S. Maini of the Canadian Forest Service has referred to forests as the "heart and lungs of the world." He noted that most watersheds originate in forestlands, or lands once forested. Forests capture, retain, and slowly release water for the world's great rivers and lakes. Forests reduce soil erosion, maintain water quality, transpire great quantities of water vapor into the atmosphere, contribute to atmospheric humidity and cloud cover, mitigate certain types of air and water pollution, absorb carbon dioxide, release oxygen, and maintain rich assemblages of plant and animal life.

In short, forests play vital roles in all the biogeochemical cycles, especially the hydrologic and the carbon and oxygen cycles, and in creating and preserving biodiversity.

So what are we doing to forests? On a worldwide basis, we are cutting them down at a prodigious pace. If and when we replace them, we usually do so with a highly simplified stand of trees, often a monoculture. In this chapter we will first consider some different types of forests and then look more closely at human impacts on forests.

FOREST TYPES

Forests are classified in many ways, usually on the basis of climate and tree types. A common general classification begins with four climatic types: boreal, temperate, subtropical, and tropical. Within each of these broad categories, numerous specific types occur, some of which are shown in Table 12.1.

Each of these forest types occupies land with certain ecological qualities, and each maintains a characteristic biological community. Thus we associate pines, spruces, and firs with the northern coniferous forest, along with elk or caribou, beaver, showshoe hares, bears, lynx, ruffed grouse, snowy owls, and in some places wolves. In contrast, we associate palm trees, figs, teak, and mahogany, along with monkeys, tapir, paca, agouti, jaguars or leopards, parrots, toucans, or hornbills, with the tropical rain forest. Certain conspicuous plants and animals are indicator organisms; their presence usually indicates a characteristic biome, or broad-scale biological community with a typical assemblage of plant, animal, fungal, and microbial life forms.

TROPICAL FORESTS

Tropical forests have received the most publicity, in both the popular media and scientific literature. Some of the reasons for this are (1) tropical deforestation is extensive, severe, and dramatic; (2) ecologically, tropical forests are known for their exceptionally high biodiversity of plant and animal life; (3) tropical hard-

TABLE 12.1
A simplified classification of forest types.

Broad Category	Examples of Specific Types	Geographic Examples
Boreal forests	Taiga: spruce, fir, larch	Alaska Northern Canada Siberia Upper montane forests

Temperate forests	Coniferous:
	  pine, spruce, fir
	  hemlock, redwood
	  sequoia
		Pacific Northwest
		Eastern Canada
		Northern Europe
		Northern Japan
		Montane forests

	Deciduous:
	  oak, hickory, maple
	  beech, sycamore,
	  elm, birch, aspen
	  eucalyptus
		New England
		Central Europe
		Chile, Argentina
		Southeastern Australia

Subtropical forests	Coniferous:
	  pine, cypress,
	  pinion juniper
	Deciduous:
	  oak, hawthorn
	Chaparral:
	  scrub oak, cork oak
	  chamise
		Southeastern U.S.
		Southwestern U.S.
		Mexico
		Southcentral U.S.
		Mexico
		Southern California
		Mediterranean Basin

Tropical forests	Rain forests:
	  buttress trees,
	  lianas, stilt palms
	Monsoon forests:
	  similar
	Dry forests:
	  scrub oaks, palms
		Amazon Basin
		Malay Peninsula
		Borneo
		Southwest India
		West Bengal
		Western Costa Rica
		East Central Africa

woods are highly valued commercial products throughout the world, especially in prosperous nations; and (4) the developed and industrialized nations may find it convenient to emphasize the problems and shortcomings of less-developed countries. It is easier to see the faults in someone else than in ourselves, despite the fact that we are decimating some of our own temperate forests at rates faster than those occurring in the tropics.

The most famous tropical forests are tropical rain forests or monsoon forests. These constitute only about 58 percent of all tropical forests, whereas dry forests constitute 42 percent of tropical forests (Smith, 1992), but they do have the greatest biodiversity. A tropical rain forest may have hundreds of species of trees and thousands of species of animals per square kilometer, many times the numbers in temperate forests. All together, tropical rain forests contain several million species of plants and animals, perhaps one-half of all existing species.

Tropical rain and monsoon forests differ mainly in seasonality of rainfall. Precipitation occurs throughout the year in rain forests, whereas it is sharply seasonal in monsoon forests. Generally rain forests receive more rain, but exceptions occur.

There is no really sharp distinction between rain and monsoon forests; both usually receive more than 250 cm (100 inches) of rain per year.

Together, tropical rain and monsoon forests cover only 7 percent of the earth's land surface, and this percentage is diminishing yearly (Corson, 1990). These forests occur primarily in three major areas of the world: the Amazon Basin, equatorial West Africa, and tropical Asia (Fig. 12.1). Smaller areas of tropical wet forests occur in central America, southeastern Brazil, southwestern India, and southeastern Asia.

Estimates on the rates of tropical forest destruction vary widely. According to Myers (1992) tropical rain forests are being cut at the rate of 13,860,000 hectares per year, equivalent to 34 million acres or 53,000 square miles (Myers, 1992; Houghton, 1994). This is equal to the area of Pennsylvania and New Jersey combined. ReVelle and ReVelle (1992), however, place the rate of annual cutting at only 5 million acres per year, an area equivalent to the state of Massachusetts. The World Resources Institute (Hammond, 1994) gives the figure of 15,400,000 hectares of tropical forest destruction per year as the average between 1980 and 1990, and this figure is also given by Acharya (1995) based on FAO data. This is the equivalent of 38,038,000 acres per year (59,400 square miles), greater than the combined areas of Maryland, Delaware, New Jersey, Connecticut, Rhode Island, Massachusetts, New Hampshire, and Vermont. Most authors agree that 45 percent of the original tropical rain and monsoon forests have been destroyed, and the remaining forests are being lost at the rate of at least 1 percent per year.

The ecological consequences of this loss are severe. Probably the greatest long-term problem is the loss of biodiversity. This is a subtle problem but one with very serious long-term ethical, scientific, and economic ramifications. These issues will be considered in Chapter 20 on biodiversity.

Other problems arising from rapid tropical deforestation include erosion and general land degradation; increases in certain tropical diseases such as malaria, yellow fever, and trypanosomiasis; siltation of rivers and coastal areas; loss of

Figure 12.1   Tropical rain forests and the extent of tropical deforestation (from the Smithsonian Institution, 1988).

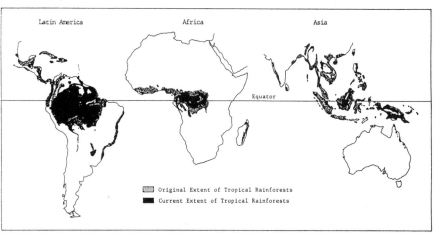

coral reefs; depletion of inshore fisheries; general economic decline; and long-term poverty. The potential also exists for serious climatic change, and possible global changes if virtually all the world's tropical forests are decimated. Portions of northeastern Brazil subject to extensive clear-cutting have already seen drastic declines in rainfall, an ominous indication of what could happen to other parts of the Amazon basin.

TEMPERATE DEFORESTATION

Although the environmental spotlight has been on the deforestation of tropical lands, the record in temperate areas is often no better and sometimes worse. Figure 12.2 shows the loss of natural forests in the United States from 1620 to 1920. The East and Midwest had the most extensive forest loss with the colonization and agriculturization of our country. Like the farmers in the Dust Bowl, however, our ancestors could move on to new frontiers. Along with a growing environmental awareness we also had the economic capacity to restore many parts of the environment that had been destroyed. Thus, extensive reforestation programs were undertaken, especially in the eastern half of the country. There is certainly more forest now in the eastern and midwestern United States than in 1920, but not as much as in 1620 before the European invasion. The recent expansion of forests in eastern United States has been documented by McKibben (1995).

In the western United States, however, our forestry record is dismal. Especially in the Pacific Northwest from California to Washington, we have devastated some of the most spectacular temperate forests in the world. Redwood and sequoia forests that once blanketed the coastal ranges have been reduced to sparse remnants amounting to only 4 percent of their original extent (Miller, 1992). Douglas fir, spruce, and hemlock forests have also been clearcut, leaving scarred and heavily eroded mountainsides. Old-growth forests with unique biological communities have been wiped out to preserve jobs and local economies in towns dependent on the logging industry. We have, in fact, destroyed a greater percentage of our original forests in the United States and Australia at a faster rate than have Brazil and other developing nations (Miller, 1992). Thus, we are not in a position to assume the moral high ground and preach to other nations about saving their forests.

In 1989 the U.S. Forest Service, charged with the management and conservation of forests representing 8.2 percent of the nation's land area, opened approximately 2.6 million acres to timber cutting. In the same year, the U.S. Forest Service reforested only 148,600 acres through tree planting programs (*Report of the U.S. Gen. Accounting Office,* 1990). Thus, over 17 times more acreage of Forest Service land was used for deforestation than for reforestation.

Other factors leading to the demise of American forests have been diseases, insect outbreaks, and pollution. In the eastern United States, Dutch elm disease, chestnut blight, gypsy moths, and acid rain have all caused considerable tree mortality and forest loss. In the western United States, ponderosa pine-bark beetle, bluestem fungus, spruce budworm, and urban smog from the Los Angeles basin

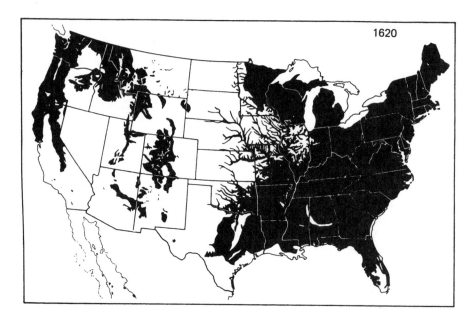

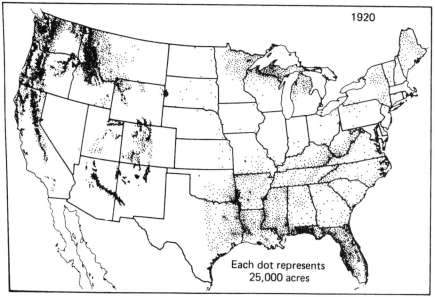

Figure 12.2 The loss of natural forests in the United States, 1620 to 1920. Substantial forest replantation and growth has occurred since 1920, especially in Eastern and Midwest United States (from Goudie, 1994).

Figure 12.3   Clear-cutting in the Pacific Northwest has produced a pockmarked landscape with many barren hills in the vicinity of Mount Rainier. Timber companies claim they plant many trees for each one felled, but the loss of old forests is permanent, and the consequences of erosional runoff affect other industries ranging from recreation to commercial fisheries. (Photo by James P. Blair © copyright National Geographic Society)

have been serious factors in forest decline. In some cases, disease, insects, and pollution operate synergistically to cause forest death.

CAUSES OF DEFORESTATION

Throughout the world there are three main driving forces of deforestation, all of them economic. In the developing world, and most of the tropics, forests are cut to provide agricultural land, both pasture and cropland. Both Brazil and Indonesia have extensive homesteading and translocation programs in which city dwellers are given economic incentives to move into forested areas and develop farms. These efforts are often tragic failures because thin tropical soils exposed after clearcutting forests are not good agricultural soils; they erode rapidly to impermeable hardpan, providing only one or two successful crop years and degrading quickly into unproductive tracts.

A second driving force in deforestation throughout the world in both tropic and temperate regions is the insatiable appetite of industrial nations such as the United States, Japan, and western Europe for lumber, pulpwood, and various forest products. Tropical hardwoods such as teak, rosewood, and mahogany have been used extravagantly by the wealthy nations. Temperate woods such as redwood, spruce, and fir have also been used extensively as construction materials in housing and furniture. Pulpwood is harvested in unbelievable quantities to meet the worldwide demand for newspapers, magazines, and packaging.

A third cause of deforestation is the acute need for fuel wood throughout much of the developing world. Wood remains a primary energy source in many countries in Asia, Africa, and Latin America. In Nepal, India, and much of Africa, villagers scrounge the countryside for shrubs, sticks, and even living trees to cook their daily meals. This is particularly true in the semiarid habitats of these continents, where a young sapling has a hazardous life. If not devoured by a cow, goat, or camel, it may be cut down by a local villager and end up providing the energy to cook that evening's rice or gruel. Firewood and brush supplies 52 percent of all energy use in sub-Saharan Africa (Hammond, 1994). Approximately 25 percent of the world's population (1.5 billion out of 5.8 billion) depend upon firewood for their primary source of fuel.

THE ROLE OF FORESTS IN GLOBAL ECOLOGY

The global importance of forests is a vital topic about which we have very little hard data. Without doubt, the world's forests play key roles in biogeochemical cycles, in soil formation, in climate patterns, and in biodiversity. They are essential in the global carbon budget, but their precise role in absorbing atmospheric carbon dioxide and maintaining global air quality is not fully understood. The destruction of tropical forests alone is estimated to add 1.1 to 3.6 billion tons of carbon to the atmosphere, about 30 percent of global carbon emissions (Acharya, 1995). Forests certainly provide many other significant ecosystem services on a global scale such as the moderation and stabilization of weather, the maintenance and renewal of water resources, the mitigation of pollution, and so on.

WHAT CAN BE DONE?

We have not been discussing possible solutions to these environmental problems up to this point with the philosophy that it is first important to recognize the extent and causes of these problems. We might begin here, however, to think of a few rather obvious needs and potential solutions.

It is obvious that we must have greater awareness of these problems and their consequences on the part of the general public and especially our political and business leaders. This is happening but rather slowly. Certainly the Earth Summit in Brazil was a step in the right direction. Even before the Earth Summit, Brazil was making serious efforts to reduce its rate of deforestation. In many developing nations, there seems to be a greater awareness of the problem than in the United States.

With greater public and political awareness, the next major steps must involve broad-scale conservation programs, reforestation projects, and more intelligent efforts at sustainable forest use. These activities will require multiple approaches, including recycling of forest products, enhanced planting, and economic development that does not require forest destruction.

New jobs must be found in small communities totally dependent on logging. Environmental restoration should be viewed as a source of jobs that are more important than those associated with environmental exploitation. Recreation and appropriate ecotourism must be recognized as legitimate reasons for conservation as well as sources of income.

There is increased awareness now that intact tropical forests can be far more valuable in the sustained harvest of natural products, such as foods, medicinals, fibers, and unique organisms for biotechnology, than if they are decimated in one fell swoop by timber cutting (Anderson, 1992; de Onis, 1992). Illar Muul has written extensively and persuasively on "integrated conservation strategies" for tropical forests. These are strategies that involve the multiple uses of forests ranging from selective logging to natural history tourism (Muul, 1994). Selective logging can be done by aerial means using airships or balloons to remove mature trees without building roads or skidding logs through the forest over drag trails that kill hundreds of small saplings for each tree removed. Muul has demonstrated in Malaysia that natural history tourism can benefit national economies and provide local jobs. He also advocates proper use of forest products including medicinals, food products, and animals to yield better long-term profits than logging. For example, Muul has shown that clear-cut logging in Malaysia can produce a profit of $4,800 per hectare in the first year, but this cannot be repeated for 60 years; hence the yield per year from clear-cutting is only $80 per hectare. On the other hand, nature tourism can produce $5,000 per hectare per year in limited areas, and the sustainable harvest of plant and animal products, $1,000 per hectare per year. In other words, an integrated conservation approach in tropical forests can potentially yield over $6,000 per hectare per year, year after year without destroying the forest, while clear-cutting can yield only $80 per hectare per year on a 60-year cycle. The U.S.-based organization Conservation International is demonstrating similar benefits of forest conservation in Latin America.

The implementation of such integrated programs requires substantial social and economic restructuring. It requires not only public recognition of the values of intact forests but also changes in the "property rights to forests, pricing of forest products, and political power over the disposition of forests" (Durning, 1993). With such programs, we will be able to use forests productively on a long-term basis, instead of destroying them for a quick one-time financial reward. These topics will be explored more fully in Chapter 25 on sustainable development.

SUMMARY

Forests originally covered an estimated 45 to 60 percent of the earth's land surface before the advent of agricultural and industrial revolutions. Now they have been reduced 30 to 50 percent according to different estimates, to cover approximately 30 percent of the earth's land surface. Intact forests may constitute as little as 12 percent of the earth's land area. If present trends continue, the earth will soon have virtually as much desert as forest landscape.

Forests perform many vital functions in all the biogeochemical cycles, in the maintenance of air and water quality, in soil formation, in the development and preservation of biodiversity, and in the moderation of weather and climate.

Forests are decimated through a variety of human economic activities sparked by the need for more agricultural land, more pasture, more timber, paper, and wood products, and more fuel wood. Both rich and poor nations contribute to extensive deforestation.

The destruction of forests leads to a tragic series of ecological and economic problems, which have been particularly acute in many countries of Asia, Africa, and Latin America. The industrialized countries of Europe and North America are also experiencing substantial forest loss, and do not have an enviable record in managing this problem.

Solutions must be found to counter the loss of forests. These solutions will require greater efforts in conservation, recycling of forest products, reforestation, and the sustainable use of intact forests. Demonstration projects in Southeast Asia and Latin America have shown that integrated conservation strategies in tropical forests can produce far greater economic yields through sustainable activities than by clear-cutting.

REFERENCES

Acharya, A. 1995. Tropical forests vanishing, pp 116–117 in *Vital Signs,* ed. L. R. Brown, N. Lenssen, H. Kane. New York: W. W. Norton.

Anderson, A. (ed.) 1992. *Alternatives to Deforestation: Steps Toward Sustainable Use of the Amazon Rain Forest.* Irvington, NY: Columbia University Press.

Corson, W. H. 1990. *The Global Ecology Handbook.* Boston: Beacon Press.

de Onis, J. 1992. *The Green Cathedral: Sustainable Development of Amazonia.* New York: Oxford University Press.

Durning, A. T. 1993. *Saving the Forests: What Will it Take?* Worldwatch Paper 117. Washington, DC: Worldwatch Institute.

Goudie, A. 1994. *The Human Impact on the Natural Environment.* 4th ed. Cambridge, MA: MIT Press.

Hammond, A. L. (ed.). 1994. *World Resources, 1994–95: A Guide to the Global Environment.* New York: Oxford University Press.

Houghton, R. A. (ed.). 1994. The worldwide extent of land use change. *BioScience* 44: 305–313.

Maini, J. S. 1990. Forests: Barometers of environment and economy. In *Planet Under Stress,* ed. C. Mungall and D. J. McLaren, New York: Oxford University Press.

McKibben, B. 1995. An explosion of green. *The Atlantic Monthly* 275(4): 61–83 (April 1995).

Miller, G. T. 1992. *Living in the Environment.* 7th ed. Belmont, CA: Wadsworth.

Muul, I. 1994. *Tropical Forests: Integrated Conservation Strategies and the Concept of Critical Mass.* UN Man and Biosphere Digest 15. Paris: UNESCO.

Myers, N. 1992. *The Primary Source.* New York: W. W. Norton.

ReVelle, P., and C. ReVelle. 1992. *The Global Environment: Securing a Sustainable Future.* Boston: Jones and Bartlett.

Smith, R. L. 1992. *Elements of Ecology.* 3rd ed. New York: Harper Collins.

U.S. Gen. Acct. Office (GAO). 1990. Forest Service timber harvesting, planting, assistance programs, and tax provisions. Washington, DC.

World Book Encyclopedia. 1992. Chicago: Scott Fetzer.

Chapter 13

HISTORICAL ASPECTS OF ENVIRONMENTAL DESTRUCTION

When we view the problems of deforestation, desertification, land-cover change, and soil erosion that are occurring today, it is logical to ask if these are solely products of modern societies. Are they entirely the consequences of mechanized agriculture, modern industry, and recent population growth? Or have these problems of environmental degradation occurred in the ancient past long before the industrial revolution? In other words, are modern societies the only culprits, or do these problems have ancient origins?

This chapter reviews some aspects of human-environmental relationships in several broad stages: those of early hunter-gatherer societies, in early agricultural societies and the first great civilizations, and in modern industrial and agrarian societies of the last few centuries. Our intention is to gain some perspective on ecological aspects of the human condition, to learn from history rather than arbitrarily assume that environmental destruction is a feature of the modern world.

THE ECOLOGICAL RELATIONSHIPS OF HUNTER-GATHERERS

Most current evidence indicates that early hunter-gatherer societies were successful practicing ecologists. Such societies survived in competitive biotic communities in which human beings were neither the strongest, swiftest, or hardiest members. Humans did have several distinct advantages, however. They had close, effective social organizations, exceptional manipulative abilities, and emerging intelligence. Thus, our ancestral societies developed tools and fire as early as 2,500,000 years ago, and they accumulated and transmitted knowledge much faster than nonhuman primates.

Most of this knowledge was ecological, relating to the environment and its most effective use. It included detailed knowledge of food and water resources,

like that which can still be seen in some indigenous peoples of today. The Kalahari San people, for example, can find water in barren desert where other people
would die of thirst. The Australian aborigines can locate grubs and lizards in the
Australian deserts far better than a modern biologist. The ecologist Elton (1933)
pointed out, "The Arawak of the South American equatorial forest knows where
to find every kind of animal and catch it, and also the names of the trees and the
uses to which they can be put." Bates (1960) found that native people on the
Micronesian atoll of Ifaluk had detailed knowledge of how the plants on the islands could be used for food, medicine, construction, and ornament. They also
possessed detailed knowledge of the reefs and sea around them. In all societies of
hunter-gatherers, and among Paleolithic peoples for more than 2 million years,
survival depended on knowledge of the environment.

There is increasing evidence that early societies were ecologically well adapted
to their environments. The popular conception that they barely clung to existence
through a precarious and difficult struggle is probably misleading. Certainly living
conditions were not comfortable or convenient, but modern studies of both ancient
and current hunter-gatherers have shown that they frequently had ample resources
(Lee and DeVore, 1968). Although population estimates of early human societies
are difficult to make with any certainty, most experts feel that the total human
population of the world in Paleolithic times was less than 5 million people, less
than the population of New York City (Goudie, 1994). Their demands on resources were negligible. Present-day hunter-gatherers are declining in number, but
this decline is a result of modern forces: a deteriorating environment and competition from agricultural peoples. In some cases their environment has been totally
destroyed by forest clear-cutting, as in parts of Borneo and the Philippines.

Several studies on hunter-gatherer people have shown that if environments were
intact, malnutrition was rare and starvation infrequent, indicating a relatively
sound ecologic balance in their way of life (Dunn, 1968; Neel, 1970). Many of
these same peoples, however, have high infant mortality rates, primarily due to
infectious diseases, and high "social mortality" (e.g., infanticide, geronticide,
warfare). Skeletal remains also show evidence of high injury rates and certain
chronic diseases such as arthritis. These factors along with other social customs
reduced life expectancies sharply and limited populations.

Practical knowledge of their environment did not necessarily mean that early
humans were good conservationists. They knew enough to survive and even pros-
per, but they exploited their environment at every opportunity. They were persistent foragers and relentless hunters whose primary goal was survival. They were
nomadic "in part because prolonged habitation in any one area depleted game and
firewood and accumulated wastes to the extent that the region was no longer habitable" (Guthrie, 1971). Some scholars have proposed that overzealous hunting by
early humans contributed to the widespread extinction of animals in the Pleistocene epoch[1] (Martin and Wright, 1967). This is a controversial hypothesis, because major environmental changes driven by climatic shifts were occurring at the

---

1. The Pleistocene epoch, also known as the Ice Age, was the period from 10,000 to over 1,000,000 years ago in
which great populations for large mammals, including woolly mammoths, mastodons, giant pigs, royal bison, camels,
horses, giant armadillos, and great ground sloths roamed the high temperate regions of North America and Europe.

same time. There is some evidence, however, that early societies were not always conservative in their hunting practices or environmental protection. For example, the Plains Indians of North America were known to have killed more bison than they could use by driving them over cliffs. These deaths, however, did not come close to the numbers of bison slaughtered by white settlers and frontiersmen.

The facts are that the hunter-gatherers existed on earth in small numbers for millions of years without extensive environmental damage. As pointed out by Lee and DeVore (1968), "Cultural man has been on earth for some 2,000,000 years; for over 99 percent of this period he has lived as a hunter-gatherer. Only in the last 10,000 years has man begun to domesticate plants and animals, to use metals, and to harness energy other than the human body." The civilizations of agricultural and industrial people have a long way to go before they can match the longevity on earth of hunter-gatherer societies. Of all human activities, agriculture has had the greatest impact, not only in stimulating rapid population growth and the development of elaborate civilizations, but also in altering the earth's land surface.

THE DOMESTICATION OF PLANTS AND ANIMALS

With the rise of civilizations and their elaborate division of labor, more niches for the "nonecologist" became available. The weaver, potter, and toolmaker did not require the same broad ecological knowledge as the hunter and gatherer. The most significant ecological achievements of sedentary societies were the domestication of plants and animals for greater productivity and control over the means of subsistence. Thus, the development of food production around 8,000 B.C. in the Near East and later in most other parts of the world altered the entire pattern of human existence. Permanent villages became established, intergroup cooperation and trade routes developed, and a demand arose for a new type of ecologic knowledge associated with the husbandry of plants and animals. Economics displaced ecology as the vital key to survival success. This period of history has been dramatically portrayed by Dr. Jacob Bronowski in the PBS video series "The Ascent of Man."

The beginning of recorded history, about 3,000 B.C., showed civilizations in Egypt and Mesopotamia with cities and a high degree of vocational specialization. The Bronze Age, beginning shortly after 3,000 B.C., and the Iron Age, starting about 1,000 B.C., accelerated agriculture, exploration, and conquest but did little to expand human concern for the environment. Environmental exploitation now took place on a much broader scale. Major land changes occurred. Forests were cut, fields cleared, pastures grazed and plowed, canals excavated and the landscape was carved to fit the new economic demands of man. A general increase in aridity occurred through much of the world then occupied by civilized people, and some of the great land barrens of the modern world developed during this time (Marsh, 1864; Sauer, 1938; Sears, 1935; Thomas, 1955–1956).

Two major theories have been advanced to explain the development of these great land barrens, especially those in the Middle East and North Africa. The theories differ in their views on the role of humans. One views humans as the victim of climatic change, and the other views humans as the perpetrator of climatic change. The first theory says that societies had to adapt to desert conditions imposed on them by great continental forces of climate and geography, planetary forces over which they had no control; the second, that human societies themselves were the major desert-making force, the real perpetrators of environmental change.

The humans-as-victims view has prevailed in most academic circles until recent years, possibly because it seemed more acceptable to the human ego. Some scientific research supports this concept (Kerr, 1985; Tucker, Dregne, and Newcomb, 1991); evidence of considerable synchrony in waves of desertification globally suggests that broad climatic shifts did occur. Increasing evidence from archaeology and ecology, however, suggests that the role of human beings as a desert-making force has been underestimated. The most accurate picture shows a combination of human population pressures on semiarid lands and natural drought cycles to be the true cause of desertification (Goudie, 1994; Stewart and Tiessen, 1990). This view does not assert that desertification is entirely anthropogenic, but it does conclude that humans were influential in accelerating and intensifying desert formation and were often the key factor (Allan and Warren, 1993).

The geographer Carl Sauer believed that there were three great periods of habitat destruction in human history. The first period beginning about seven thousand years ago in the Near East, when the great herds of pastoral societies and the early successes of agriculture caused extensive land scarring, erosion, and irreversible aridity in the Middle East, Asia, and Africa. The second occurred during the latter days of Rome and the disorderly period following, when many of the Mediterranean lands were despoiled. The third period of destruction accompanied the transatlantic expansion of European civilizations into the New World, when disastrous land exploitation occurred throughout the Americas. Sauer documented these views in an important book called *Land and Life,* first published in 1938 and reprinted in a more recent anthology (Leighly, 1967).

To these three periods of environmental destruction outlined by Sauer, we should add a fourth period, that of the 20th century when burgeoning populations, exponential growth, and new forms of pollution are placing increasing pressures on landscape.

Does each major technological advance carry with it the possibility of destruction? Keeping in mind the effects of the domestication of hoofed animals, the expansion of early civilizations, the invasion of the New World, the development of gunpowder, and the rise of modern medicine, the internal combustion engine, the nuclear age, and the destructive potentials of high-tech military firepower, one is tempted to answer yes.

THE RISE OF AGRICULTURE

Agricultural societies set into motion several major forces that had significant geographic and meteorologic consequences: deforestation, overgrazing, intensive burning, overcropping, land scarring, and extravagant use of irrigation. The geographic consequences included increased erosion, soil loss, declining water tables, and salination. Meteorologic and climatic consequences included reduced atmospheric humidity and cloud cover, increased heat reflectivity, and lower amounts of rainfall. Ecological studies have shown that forests and grasslands help to maintain the levels of precipitation necessary for their own existence, and that deforestation results not only in an immediate lowering of ground water levels but also lower levels of rainfall over the long term. Forests recycle moisture back into their immediate atmosphere by transpiration from whence it falls again as rain. Transpiration return from an acre of forest may reach 2,500 gallons of water per day (McCormick, 1959). This sustains a natural system of water reuse, through the hydrologic cycle, as described in Chapter 5. If the forest is removed over a large area, this natural reuse cycle is broken, and water is lost through rapid runoff. Both ground water levels and atmospheric moisture decline. Grasslands perform the same function to a lesser degree, with smaller amounts of available moisture. It is interesting to examine Middle Eastern history in light of this principle of hydrologic recycling.

CIVILIZATIONS OF THE MIDDLE EAST

The span of history from 5,000 B.C. to 200 A.D., which we know primarily as the period of great civilizations—Sumerian, Babylonian, Assyrian, Phoenician, Egyptian, Grecian, and Roman—was also a period of unprecedented environmental disturbance. We tend to concentrate our attention on the superb achievements of these civilizations in literature, art, government, and science, while we virtually forget their incompetence in land management. These golden civilizations prospered at the expense of their environments. They left a landscape that has never recovered.

As specific examples of the destructive nature of these civilizations, we can cite critical landscape changes in many areas of the Middle East, North Africa, and Mediterranean lands. As late as 4,000 B.C., the headwaters of the Tigris and Euphrates rivers were covered with forests and grasslands (Saggs, 1962). These forests and grasslands flourished under favorable precipitation conditions which occurred in the Holocene period from 9,000 B.C. to nearly 4,000 B.C. In fact, most of the area now occupied by Iran and Iraq was productive and well watered (Sauer, 1938). Domestic cattle appeared in the seventh millennium B.C., probably around 6,300 B.C. (Perkins, 1969). Herds of domestic cattle, sheep, and goats found very favorable pasture in these virgin grasslands. Their great success provided a major stimulus for the developing civilizations of Mesopotamia, but it also provided the first significant onslaught on these grasslands and their adjacent forests. The herders prospered, utilizing the stored capital of thousands of years, and the upsurge of prosperity was followed by more people and larger herds. Sheep graze grasses down to their roots, exposing soil to erosion. Goats climb stunted trees eating leaves and twigs. Forests were cut to provide additional pasture, and more land was exposed to the increasing livestock populations. This period was also characterized by a universal shift to drier conditions.

Urban communities developed around 4,000 B.C., and further deforestation occurred to provide timbers for developing cities. Elaborate agricultural practices converted the cleared land to food production, and irrigation canals extended the reach of the rivers. At first, the erosion and silt load of the canals was manageable, and the alluvial soil along the rivers was remarkably fertile. With passing generations, however, siltation increased and it became necessary to occupy great numbers of slaves and laborers with the job of keeping irrigation channels free of silt (Dasmann, 1968). After 3,000 B.C., several meters of silt clogged the irrigation canals of the Tigris and Euphrates. All of these changes—loss of vegetation and soil, lowered water tables, declining agricultural productivity, diminishing rainfall, and the added economic burden of siltation in the irrigation canals—were significant factors in the fall of the great Babylonian empire (Saggs, 1962). Successive waves of invaders—Kassites, Elamites, Assyrians, and eventually Persians—conquered this tragic land and increased the devastation.

At the same time that the early Mesopotamian empires were flourishing in the Middle East, there was a thriving civilization in the Indus Valley of present-day Pakistan. A prosperous and advanced culture existed at the site of Mohenjo-Daro in the province of Sind (Wallbank, 1958). There is evidence here of an urban civilization with well-planned streets, dwellings, public buildings, and a well-engineered drainage system. A sophisticated governmental structure apparently existed. The people were skillful in the use of bronze, copper, silver, and lead. They made beautifully painted pottery, delicate jewelry, and carefully woven cotton textiles. This prosperous civilization came to an abrupt end about 1,500 B.C. No one knows the reason for its collapse, but there was again a loss in the means of subsistence. The Sind today is mostly desert, except in those portions bordering the Indus River and its related irrigation canals.

Human populations, over a period of 3,000 years, from the sixth to the third millennium B.C., had become a major force in environmental change. Powerful civilizations and glorious cities were intimately dependent upon the land that gave them birth. When the pastoral and agricultural riches of the land were destroyed, these civilizations could no longer be supported, decline began, and they became more vulnerable to invading armies.

RELEVANCE OF ENVIRONMENTAL HISTORY
TO MODERN PROBLEMS

To what extent are the ancient environmental events of the Middle East relevant to modern problems? The difficult social, political, and economic problems of the Middle East today are certainly related to conflicts over land and resources. Problems arising from shortages of space, arable land, resources and water are clearly exacerbated by cultural and religious differences, but land and resources remain the focal points of conflict. Would there be as much tension and dispute over the Gaza Strip and the West Bank if land and space were sufficiently abundant to accommodate everyone who had a historical claim to it? Would Iraq have invaded Kuwait if it had all the resources and access to the Persian Gulf that it wanted?

We don't know the answers to these questions, but it is obvious that political conflicts often arise when competing groups insist that certain land and resources are theirs. In the Middle East, different groups invoke long-standing historical rights to these limited landscapes. Perhaps, after all, some of the modern conflicts in the Middle East have their roots in environmental degradation that occurred 5,000 to 10,000 years ago, even before modern religions developed. Perhaps these conflicts first arose when the shift occurred from a rich and prosperous landscape to harsh desert environment. The cultural and religious factors involved are complicated, but the fact remains that different social groups are claiming ancient rights to landscape and resources that are in short supply. The close connection between limited resources and social conflict is clearly an area we need to study as the present world population expands and our own global environment deteriorates.

NORTH AFRICA AND THE MEDITERRANEAN

In North Africa, familiar patterns of ancient land degradation have been traced. The Sahara, although a natural desert, within written history was not as extensive as it is now. In arid parts of Egypt and the Sudan, Davidson (1959) found abundant evidence of productivity within historical times:

> Even as late as the third millennium large numbers of cattle are known to have found grazing in lower Nubia (. . . now part of Sudan) where, as Arkell says, "desert conditions are so severe today that the owner of an ox-driven water-wheel has difficulty in keeping one or two beasts alive throughout the year." And anyone who has traveled in these dusty latitudes will have noticed how the wilderness of sand and rock that lies to the west of the Nile, far out upon the empty plains, is scored with ancient wadi beds which must once have carried a steady seasonal flow of water, but are now as dry as the desert air.

Throughout Saharan Africa there were campsites and herding communities as late as 1800 B.C. in areas now too arid to support life (Phillipson, 1993). It would be extravagant and incorrect to claim that human activities produced the Sahara Desert, for it was largely the result of great climatic shifts over the course of geologic history. But the evidence is clear that humans extended Saharan conditions and forced once productive and well-watered areas to become barren wastelands. The Sahara is still actively extending its arid conditions, in some areas at the rate of many miles per year (Wade, 1974). In recent centuries, over 250 million acres (390,000 square miles) in Africa have been converted from agricultural and pastoral production to desert, primarily through the agency of humans and their livestock. The most recent developments in this process will be discussed later in this chapter.

According to Sauer (1938), the second great period of human destruction of the landscape occurred during the latter days of Rome. Phoenicia, Greece, and

Rome all prospered from the riches of the Mediterranean lands. One of the best accounts of this Roman exploitation was provided long ago by Marsh (1864):

> The Roman Empire, at the period of its greatest expansion, comprised the regions of the earth most distinguished by a happy combination of physical advantages. The provinces bordering on the principal and secondary basins of the Mediterranean enjoyed a healthfulness and equability of climate, a fertility of soil, a variety of vegetable and mineral products, and natural facilities for the transportation and distribution of exchangeable commodities, which have not been possessed in an equal degree by any territory of like extent in the Old World or the New. The abundance of the land and of the waters adequately supplied every material want . . . the luxurious harvests of cereals that waved on every field from the shores of the Rhine to the banks of the Nile, the vines that festooned the hillsides of Syria, of Italy, and of Greece.

Even much of North Africa was rich and productive; the higher mountains had extensive cedar forests which provided further wealth for Rome (Fig. 13.1). But Marsh hastened to add a more recent description of these lands, as he observed them in the mid-nineteenth century during his tenure as United States ambassador to Italy:

> If we compare the present physical conditions of the countries of which I am speaking, with the descriptions the ancient historians and geographers have given of their fertility and general capability of ministering to human uses, we shall find that more than one half of their whole extent—including the provinces most celebrated for the profusion and variety of their spontaneous and their cultivated products, and for the wealth and social advancement of their inhabitants—is either deserted by civilized man and surrendered to hopeless desolation, or at least greatly reduced in both productiveness and population. Vast forests have disappeared from mountain spurs and ridges; the vegetable

Figure 13.1 The remains of a once-forested hillside in North Africa. Pollen deposits indicate that this range of hills was covered by coniferous forest before the time of the Roman expansion. Now only the inorganic skeletons of a former ecosystem remain. The vegetation and soil have been irretrievably lost (photo by C. H. Southwick).

earth accumulated beneath the trees by the decay of leaves and fallen trunks, the soil of the alpine pastures which skirted and indented the woods, and the mould of the upland fields, are washed away; meadows, once fertilized by irrigation are waste and unproductive. . . .

Besides the direct testimony of history to the ancient fertility of the regions to which I refer—Northern Africa, and the greater Arabian peninsula, Syria, Mesopotamia, Armenia, and many other provinces of Asia Minor, Greece, Sicily, and parts of even Italy and Spain—the multitude and extent of yet remaining architectural ruins, and of decayed works of internal improvement, show that at former epochs a dense population inhabited those now lonely districts.

It appears, then, that the fairest and fruitfulest provinces of the Roman Empire, precisely that portion of terrestrial surface, in short, which, about the commencement of the Christian era, was endowed with the greatest superiority of soil, climate and position, which had been carried to the highest pitch of physical improvement, and which thus combined the natural and artificial conditions best fitting it for the habitation and enjoyment of a dense and highly refined and cultivated population, is now completely exhausted of its fertility, or so diminished in productiveness, as, with the exception of a few favored oases which have escaped the general ruin, to be no longer capable of affording sustenance to civilized man.

The decay of these once flourishing countries is partly due, no doubt, to that class of geological causes, whose action we can neither resist nor guide, and partly also to the direct violence of hostile human force; but it is, in far greater proportion, the result of man's ignorant disregard of the laws of nature.

ENVIRONMENTAL DESTRUCTION IN OTHER PARTS OF THE WORLD

The phenomenon of ecological deterioration has been discussed at length by Paul Sears (1935), Fairfield Osborn (1948), and Carl Sauer (1967). They have shown that the environmental destructiveness of human populations has not been limited to the Middle East, Africa, and the Mediterranean, but has, in fact, occurred throughout China, India, sub-Saharan Africa, and the New World. In India, for example, the original vegetation of Rajasthan and the southern borders of the Gangetic basin, areas that are now arid and semidesert, was deciduous forest and far richer in natural soil moisture than at the present time (Champion, 1936).

The Great Thar Desert of western India has increased its size by 60,000 sq. mi in the last 100 years (Ehrlich and Ehrlich, 1972). Dr. A. Krishnan of the Central Arid Zone Research Institute of Jodhpur, India, said in 1973 that archeological evidence and recent carbon[14]-dating studies had established that the Rajasthan Desert was largely caused by deforestation, overgrazing, and deterioration of the soil (*Calcutta Statesman,* Sept. 15, 1973, p. 6). This conclusion also held out some hope: since the desert was primarily human-caused, it could be reclaimed, Dr. Krishnan felt, by adequate soil stabilization and large-scale establishment of grasses and trees. Such reclamation, of course, would require high capital expenditures and relief from human and livestock population pressures, both of which are virtually impossible to achieve over large areas. Ironically, the Rajasthan Desert is now threatened by rapid human population growth, stimulated in part by a large irrigation project in western India. One consequence of this project will be increased salination and further loss of natural vegetation.

In North America, the present desolate shifting-sand areas bordering the Colorado River were rich grassland pastures until major grazing began in the seventeenth century (Sauer, 1967). Mexico was extensively forested over vast areas
now occupied by dry and rocky plains (Osborn, 1948). Of all these destructive
changes wrought by humans, Sears (1935) wrote:

> Whenever we turn, to Asia, Europe, or Africa, we shall find the same story repeated
> with an almost mechanical regularity. The net productiveness of the land has been de-
> creased. Fertility has been consumed and soil destroyed at a rate far in excess of the
> capacity of either man or nature to replace.

Fairfield Osborn, in *Our Plundered Planet* (1948), supported this view and
pointed out that destructive land-use practices have not been confined to ancient
peoples and former centuries; they have occurred very recently as well. In Australia, for example, forest destruction, overgrazing, and unlimited burning, have
greatly accelerated wind erosion, extending the Australian deserts within the last
100 years.

The most dramatic desert formation in recent years has been the southern
expansion of the Sahara Desert into the countries of Mauritania, Mali, Burkina
Faso, Niger, Chad, the Sudan, and Ethiopia, in the region of central Africa known
as the Sahel. Although drought conditions throughout the late 1960s and early
1970s triggered this great southern movement of the desert, several scientific studies have shown that human activities have contributed to desert formation in the
Sahel (Wade, 1974). Edward Fei, chief of the AID Special Task Force on the
Sahel, said, "The desertification is man-caused, exacerbated by many years of
lower rainfall." The French hydrologist Marcel Roche stated, "The phenomenon
of desertification . . . is perhaps due to the process of human and animal occupa-
tion, certainly not to climatic changes." These scientists concluded that the primary agents of desert expansion into the Sahel are overgrazing, excessive cutting
of trees for firewood, concentrations of large populations in limited areas, and
excessive wind and water erosion during storms (Fig. 13.2). Well-intentioned efforts, funded by foreign aid, to drill wells into deep aquifers to provide local water
sources exacerbated desertification by concentrating nomadic peoples and their
livestock in virtual refugee camps where environmental damage was intensified.

Dramatic evidence of the effect of overgrazing was provided on the Ekrafane
ranch in the Sahel where a 250,000 acre range was divided into five sectors and
cattle were allowed to graze on only one sector a year. Wade (1974) noted, "Although the ranch was started only 5 years ago, at the same time as the drought
began, the simple protection afforded the land was enough to make the difference
between pasture and desert." The Ekrafane ranch maintained green vegetation that
was visible from an earth satellite photo showing it and the surrounding desert.

Recent evidence shows that some vegetation has returned to parts of the Sahel
with the advent of rains. Rainfall and vegetation decreased from 1980 to 1984,
and the southern edge of the Sahara moved 240 km southward in this period. In
1985 and 1986, precipitation increased and the edge of the desert moved northward again 143 km (Tucker, Dregne, and Newcomb, 1991). These studies show

Figure 13.2 Desertification in the Sahel. Several years of drought in sub-Saharan Africa in the 1970s, coupled with heavy grazing pressure and excessive concentrations of livestock and people around water sources, greatly extended the Sahara Desert with a tragic loss of human and animal life (UPI photo from Southwick, 1976).

the dynamic state of the desertification process and how it is influenced by annual changes in weather. The research also illustrates the interaction of weather and human factors in desertification, and it emphasizes the need for continuing studies to discover whether the Sahara is expanding or contracting over the long term.

The American Dust Bowl of the 1920s and 1930s is another well-documented story of land misuse (Fig. 13.3). A once fertile grassland changed into a dusty pit of ecologic and human tragedy. Although the United States had the land resources and capital reserves to recover from the Dust Bowl, a similar situation could occur again in the Great Plains if agricultural pressure becomes too intense at a time of unfavorably dry weather (Stone, 1993).

Throughout much of present-day Latin America and in many tropical or subtropical regions of the world, excessive use of slash-and-burn agriculture is actively practiced. Forests are cut and burned on excessively short cycles, and the cleared areas are planted in bananas, manioc, or some other starchy tropical plant. This practice denudes forest tracts rapidly, permits only one or two years of productive agriculture, and leaves a wake of destruction (Fig. 13.4). After a few years, the subsistence farmer must repeat the process in a new area. A forest that may have taken 1,000 years to develop can be destroyed in a few years. In the

Figure 13.3 Devastated farmlands in the American Dust Bowl, converted from productive grassland to virtual desert through human land abuse (photo from U.S. Soil Conservation Service).

tropics, where natural soils are thin and very susceptible to rapid leaching and erosion, the entire ecosystem may be irreparably damaged.

Not all slash-and-burn agriculture is harmful to the environment. As originally practiced by low-density human populations, such "shifting agriculture," or "swidden agriculture," opened up small areas of forest for cultivation for one or two years and then allowed them to revert to the natural successional stages of secondary forest growth. The process was then repeated elsewhere, and each site was not revisited for 15 to 25 years, an interval of time that allowed the soil and the forest to recover. Shifting agriculture, when properly applied, has the ecological advantages of recycling nutrients locked up in biomass and even enhancing biodiversity. The problem with modern-day slash-and-burn agriculture is that it is now practiced on large areas that are revisited yearly and never allowed to recover. This produces an environmental wasteland, an economic disaster, and a human tragedy. The remnants of such destroyed landscapes are found throughout many areas of Latin America, Africa, and Asia.

Warfare, of course, has also had devastating and tragic effects on the environment. In Southeast Asia, the widespread use of herbicides, the cratering effects of bombs, shells, and rockets, and the physical disturbances of heavy vehicles churning across the landscape during the Vietnam War produced extensive ecological damage, some of which may persist for decades or even centuries. These effects will be discussed in Chapter 24.

Figure 13.4 Views of slash-and-burn agriculture in the Amazon basin. The tropical forest is cut and burned to clear open space for marginal agricultural planting. One or two productive years may be obtained before the thin tropical soils are depleted and baked into an impervious hardpan (photos by C. H. Southwick).

SUMMARY

The advent of agriculture about 9,000 years ago led to unprecedented population growth and prosperity. Great civilizations arose in the Middle East from the newly found wealth of agricultural development, wealth based on the natural capital of well-watered forests, grasslands, and soils. The key innovation of this period was the domestication of plants and animals. Over several millenia, however, these rich landscapes deteriorated into harsh desert environments through the combined action of land abuse and recurrent droughts.

Scientific evidence from archaeology, ecology, geography, and history shows that many environments throughout the world have been rendered barren and inhospitable by excessive pressure from the axes, plows, hoofed animals, fires, and military campaigns of human societies. Human destruction of the natural environment has interacted with natural drought cycles to convert forests and grasslands into desert habitats. The disappearance of great civilizations has been linked to environmental collapse, that is, the destruction of natural resources on which they depended. In the decade of the 1990s, there is little evidence that we recognize this aspect of our own history or realize its applicability to our present predicaments. In country after country around the world, including the United States, the conservation movement is losing ground to economic concerns. In both temperate and tropical lands, deforestation is occurring at increasing rates. In Asia, overgrazing and energy development schemes are invading national parks and damaging what remains of natural environments. In the Himalayan mountains, the landscape is being carved to within an inch of its life to support populations that won't stop growing. In Africa and Latin America, excessive slash-and-burn agriculture is encroaching on natural areas that have existed in ecological balance for thousands of years.

Much of this environmental destruction is done in the name of economic necessity. The pressures of population growth and economic development place a greater burden than ever on the finite blanket of life the covers this planet. That blanket is becoming torn and shredded in thousands of places. Like a wound or burn on the skin, its ability to heal depends on the extent of the lesion and the health of the surrounding tissue. A point can be reached where the healing capacity is lost.

The philosopher George Santayana has wisely said that those who fail to learn from history are condemned to repeat it (Southwick, 1983). Humanity will ultimately pay a heavy price for our collective failure to learn from the lessons of past environmental mistakes. The many instances of regional environmental deterioration, if not addressed soon, will coalesce into global deterioration.

REFERENCES

Allan, T., and A. Warren. 1993. *Deserts: The Enroaching Wilderness*. New York: Oxford University Press.

Bates, M. 1960. *The Forest and the Sea: A Look at the Economy of Nature and the Ecology of Man*. New York: Mentor Books.

Champion, H. 1936. *A Preliminary Survey of the Forest Types of India and Burma.* Indian Forest Records (Silviculture Series) 1 (1).

Dasmann, R. 1968. *Environmental Conservation.* New York: Wiley.

Davidson, B. 1959. *The Lost Cities of Africa.* Boston: Little, Brown.

Dunn, F. 1968. Health and disease in hunter gatherers. In *Man the Hunter,* ed. R. Lee and I. DeVore. Chicago: Aldine.

Ehrlich, P., and A. Ehrlich. 1972. *Population, Resources, Environment: Issues in Human Ecology.* 2nd ed. San Francisco: W. H. Freeman.

Elton, C. 1933. *The Ecology of Animals.* London: Methuen.

Goudie, A. 1994. *The Human Impact on the Natural Environment.* 4th ed. Cambridge, MA: MIT Press.

Guthrie, D. 1971. Primitive man's relationship to nature. *BioScience* 21: 721–723.

Kerr, R. 1985. Fifteen years of African drought. *Science* 227: 1453–1454.

Lee, R., and I. DeVore. (eds.). 1968. *Man the Hunter.* Chicago: Aldine.

Leighly, J. (ed.) 1967. *Land and Life: A Selection from the Writings of Carl Ortwin Sauer.* Berkeley: University of California Press.

McCormick, J. 1959. *Forest Ecology.* New York: Harper Bros.

Marsh, G. P. [1864] 1965. *Man and Nature: Physical Geography As Modified by Human Action,* ed. D. Lowenthal. Cambridge, MA: Harvard University Press, Belknap Press.

Martin, P., and H. Wright. 1967. *Pleistocene Extinctions.* Proc. 7th Cong. Internat. Assoc. for Quaternary Research 6. New Haven, CT: Yale University Press.

Neel, J. 1970. Lessons from a "primitive" people. *Science* 170: 815–822.

Osborn, F. 1948. *Our Plundered Planet.* Boston: Little, Brown.

Perkins, D. 1969. Fauna of Catal Huyuk: Evidence of early cattle domestication in Anatolia. *Science* 164: 177–179.

Phillipson, D. 1993. *African Archaeology.* New York: Cambridge University Press.

Saggs, H. 1962. *The Greatness That Was Babylon.* New York: Hawthorn Books.

Sauer, C. 1938. Theme of plant and animal destruction in economic history. *Journal of Farm Economics* 20: 765–774.

Sauer, C. 1967. *Land and Life: A Selection from the Writings of Carl Ortwin Sauer,* ed. J. Leighly. Berkeley: University of California Press.

Sears, P. 1935. *Deserts on the March.* Norman: University of Oklahoma Press.

Southwick, C. 1976. *Ecology and the Quality of Our Environment.* 2nd ed. Boston: Willard Grant.

Southwick, C. 1983. Environmental change and the principle of Santayana. In *Environment and Population: Problems of Adaptation,* ed. J. Calhoun. New York: Praeger Scientific.

Stewart, J., and H. Tiessen. 1990. Grasslands into deserts? In *Planet Under Stress,* ed. C. Mungall and D. McLaren. New York: Oxford University Press.

Stone, R. 1993. FAO sounds soil-loss siren. *Science* 261: 423.

Thomas, W. (ed.). 1955–56. *Man's Role in Changing the Face of the Earth.* Chicago: University of Chicago Press.

Tucker, C., H. Dregne, and W. Newcomb. 1991. Expansion and contraction of the Sahara Desert from 1980 to 1990. *Science* 253: 299–301.

Wade, N. 1974. Sahelian drought: No victory for Western aid. *Science* 185: 234–237.

Wallbank, W. 1958. *A Short History of India and Pakistan.* New York: Scott, Foresman.

Chapter 14

POPULATION
ECOLOGY

The problems of environmental deterioration that we have been considering—deforestation, desertification, erosion and soil loss—are all closely linked to population growth. Desertification of the Middle East followed the origins of agriculture and the expansive growth of human populations. Deforestation of both temperate and tropical nations has been and is currently associated with the need to provide more resources for growing populations. It is also obvious that air and water pollution, and a broad range of secondary environmental problems, are products of population growth. Population growth is clearly one of the driving forces of global change.

Numbers per se are not the only factors; consumption rates per individual figure prominently in the equations of environmental impact. For example, the United States with less than 5 percent of the world's population, uses 25 percent of the world's nonrenewable energy resources, 33 percent of the world's mineral resources, and produces about 33 percent of the world's pollution and trash (Miller, 1992). A baby born in the United States will damage the earth's environment from 20 to 100 times more in its lifetime than a baby born in a less developed nation (Ehrlich and Ehrlich, 1990).

Thus, the environmental impacts of population growth are a result of both numbers and resource use. With the human population of the world increasing at the rate of about 1.7 million people per week, questions of where this is taking us are absolutely critical.

Two broad topics to consider in population ecology are: (1) basic biological principles of population growth, regulation and fluctuation in species varying from microorganisms to elephants, and (2) patterns of human population growth. These subjects are considered in this chapter and the following one.

It may seem irrelevant to discuss the population growth of organisms as varied as yeast cells, insects, rodents, deer, and elephants in a book presuming to focus on the human aspects of global ecology, but it is often valuable in science to use a comparative approach. Human populations do not exist in the world as separate and distinct entities. We can learn a great deal from comparative biology and evolutionary ecology. While we cannot extrapolate directly from populations of lower organisms to human populations, we can discern patterns and principles that can guide our thinking, research, and planning.

BASIC POPULATION BIOLOGY

Virtually all organisms have the potential for substantial population growth. A single female insect or fish may produce hundreds or thousands of eggs at a time. A female house mouse has the potential to ovulate 10 to 12 ova every five days, but only a few of these become fertilized and develop into living young. If fertilization occurs at its maximal rate in house mice, one pair of house mice can produce over 3,000 descendants in a single year. This great reproductive output is due to the fact that gestation in the house mouse *(Mus musculus)* is only 21 days, offspring are weaned at 21 days of age, and young can mature sexually within 21 days after weaning. Thus, at 42 days of age, young mice are ready to breed. Furthermore, the female house mouse has postpartum estrus, which means that she comes into estrus, or breeding readiness, within 24 hours after the birth of her litter. At this time, she ovulates a new set of eggs and is ready to mate again. Thus, the female mouse has the physiological capacity to support both pregnancy and lactation at the same time.

Most mammals, of course, do not possess such great powers of reproduction, and some of the largest animals have only slow growth potentials. Elephants and whales, for example, have long gestation periods (12 to 18 months for whales, 20 to 22 months for elephants), and they usually give birth to only one young at a time. With life spans of 30 to 60 years, a female may still give birth to a dozen or more young in her lifetime, and both elephants and whales are capable of substantial population growth if they are protected and given the right ecological conditions.

The human female normally ovulates 300 to 400 ova in her lifetime, any one of which can potentially develop into a child. If conception occurs at the maximum rate, however, a human female can give birth to one baby every 12 to 15 months, or a total of perhaps 25 in her lifetime if only single births occur. This still means a reproductive potential 8 to 10 times greater than the average family size in most countries.

Many plants, animals, and microorganisms have the reproductive potential for geometric or exponential growth in favorable habitats. Geometric growth refers to a type of population increase represented by a series of numbers having a common multiple, such as 2, 4, 8, 16, 32, 64, 128, where each number is twice the former. Geometric growth has dramatic, even awesome, consequences if it continues very long. We generally fail to recognize these consequences (see Box 14.1). Geometric or exponential growth cannot continue indefinitely, and the pattern by which it comes to an end constitutes a population growth curve.

BOX 14.1
The Realities of Geometric Population Growth

The results of geometric growth in population numbers are so incredible that we tend to forget the simple arithmetic involved. Geometric growth simply means numbers that are increasing by a constant multiple, say, for simplicity's sake, a factor of two. This growth would cause a constant doubling in a given time period. Meadows et al. (1992) illustrated the power of such geometric growth with three stories.

If we take a piece of paper and fold it in half, we have doubled its thickness. Fold it again, and it becomes four times its original thickness. If we could continue folding it 40 times, how thick would it be? Several inches? A foot, 10 feet, 100 feet? The fact is, we cannot fold a paper 40 times, but if we could, it would form a stack high enough to reach from the earth to the moon.

A Persian legend tells of a courtier who presented the king with a beautiful chessboard and asked the king to give him one grain of rice for the first square, two for the second, four for the third, and so on, doubling for each of the 64 squares. By the 21st square, more than a million grains of rice were required, and by the 64th square more than all the rice in the world.

A third story comes from a French riddle for children illustrating the suddenness by which geometric growth reaches a limit. If we own a large pond with one lily pad that doubles in size each day as it grows, and we know the lily pads will cover the pond and choke off all other life in 30 days if not checked, when should we try to control the growth of the lily plant? If we decide not to worry until the pond is half covered, we assume we will have plenty of time to take action. On what day will it be half covered? The 29th day. Thus we will have only one day to act! (from Meadows, et al. 1992)

Many factors can terminate population growth, producing either a decline in reproduction or an increase in mortality. These factors include space limitations, food shortages, disease, competition, predation, dispersal or emigration, changes in social behavior such as aggression, changes in genetic composition, and any combination of the above.

PATTERNS OF POPULATION CHANGE

Populations are dynamic entities. Never static, and rarely stable, they usually exhibit some form of growth, decline, or fluctuation. Population change can be represented as follows:

$$P_c = \frac{dN}{dt} = (B - D) + (I - E)$$

where P_c is population change, dN is the change in number of individuals, dt is the change in time, B is births, D is deaths, I is immigration, or movement into the population, and E is emigration, or movement out of the population. Another way of expressing this is $dN/dt = (B + I) - (D + E)$. Stated simply, the population at a given moment is the difference between factors increasing the population (births and in-migration) and factors decreasing the population (deaths and out-migration). These factors are all influenced by ecological and behavioral qualities of the population and by the community and ecosystem in which the population lives.

The dynamic interplay of these factors produces the attributes or characteristics of a population: size, spatial distribution, density, sex and age structure, birth rate, death rate, dispersal rate, rate of increase or decrease, and so on.

GROWTH PATTERNS

Populations in new environments or populations finding new resources often increase in number, typically showing one of four basic growth patterns (Fig. 14.1): exponential (J-shaped), irruptive (Malthusian), logistic (S-shaped), or domed (capped).

All patterns of growth especially the J-shaped pattern, are temporary, and the models differ primarily in the patterns that terminate growth. The exponential model makes no commitment as to how growth will end.

Irruptive, or Malthusian, growth comes to an abrupt end as a result of a sud-

Figure 14.1 Some basic patterns of population growth. (from Southwick, 1976)

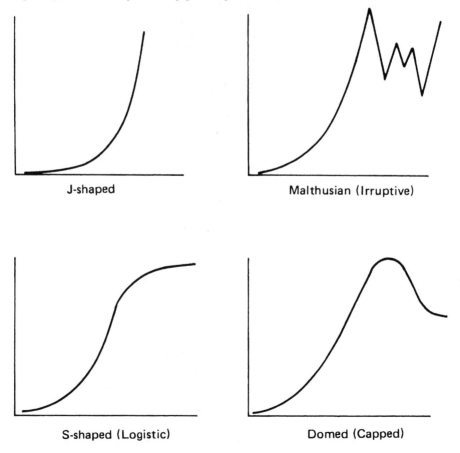

J-shaped

Malthusian (Irruptive)

S-shaped (Logistic)

Domed (Capped)

denly increased mortality, a sudden decline in reproduction, or both. The points at which these sudden changes occur are often unpredictable.

Logistic growth comes to a gradual reduction of growth in an orderly and more predictable fashion. The population reaches an upper level, usually designated the *asymptote,* or *carrying capacity,* often represented by the letter K. As the population approaches K, various factors come into play, sometimes called *environmental resistance,* which depress reproduction and increase mortality. The population ceases to grow when these two attributes are in balance. Logistic growth tends to involve rapid and continual negative feedback; that is, as the population grows beyond an inflection point, various forces act to reduce the rate of growth.

The exponential and irruptive patterns of growth are sometimes represented by the simplified formulation

$$\frac{dN}{dt} = rN$$

where N is the population number and r is reproductive potential or the *intrinsic rate of natural increase*. The outcome of this type of growth is unpredictable.

The logistic pattern is often represented by,

$$\frac{dN}{dt} = rN\left(\frac{K-N}{K}\right)$$

where K is the asymptote or carrying capacity. This formulation indicates that as N approaches K, the numerator approaches zero, and K divided into zero is itself zero. Since any number multiplied by zero is zero, the growth of the population stops.

The domed, or capped, pattern shows the tendency of some populations to overshoot the carrying capacity of the environment and subsequently fall back to a lower level. The difference between this and the Malthusian pattern is that the decline to a lower level is gradual, not catastrophic. In both domed and Malthusian growth, feedback mechanisms limiting population growth are delayed, usually until a threshold point is reached causing a more drastic reaction.

A central question in ecology is, Which of these patterns is most typical of populations of living organisms? And secondly, What kinds of factors contribute to one or the other? These questions are also central to many areas of human affairs: anthropology, conservation, economics, history, politics and international relations, to mention just a few. They are extremely important to our own future.

MALTHUSIAN GROWTH

Toward the end of the eighteenth century, a professor of political economy in England, Thomas Malthus, noted the tendency of populations, including human populations, to increase rapidly. In 1798, he published his famous *Essay on the*

Principle of Population in which he proposed the idea that populations tend to increase faster than their means of subsistence. He theorized that populations increase geometrically, while their food supplies increase only arithmetically. An arithmetic increase is a series of numbers having a common difference, for example, 2, 4, 6, 8, 10, 12. In a geometric series, each number has a common multiple. Thus, a geometric increase based on a multiple of 2 would lead to 64 in the same six-generation series (2, 4, 8, 16, 32, 64).

Malthus thought that once populations outstrip their food supplies they are decimated by starvation, poverty, disease, and warfare, which he called the "Four Horsemen of the Apocalypse."

Malthus was referring to a print by the artist Albrecht Dürer in a 1498 A.D. edition of the Book of Revelation. It showed a terrible scene of war, starvation, disease, and suffering as the fate of humankind. It was based on the original apocalyptic writings of the Biblical Book of Revelation, presumed to have been written by the Apostle John, although the authorship is still a matter of controversy. Some scholars attribute the first apocalyptic visions to Jewish authorship in the Old Testament Book of Daniel as early as 200 B.C. In any case, the apocalyptic writings seemed to be an effort to rationalize the existence of a righteous God with the sufferings and depravity of human beings on earth. Malthus was primarily concerned with human populations, and he used the apocalyptic vision to argue for moral restraint as the only solution to human population pressures.

Malthus's concepts produced a storm of protest from both the church and the scientific community. The clergy saw his ideas as anti-God and as a threat to natural procreation. The scientific community viewed them as an insult to the capabilities of science and technology. There was increasing optimism throughout the nineteenth century that science could grow more food, conquer disease, and provide more satisfactorily for expanding populations. The prevailing beliefs in the Baconian Creed (science shall conquer all problems), and the Abrahamic concept of land (go ye forth and subdue all nature) were clearly anti-Malthusian. Thus, the ideas of Malthus fell into general disrepute and were considered little more than historical oddities.

An interesting footnote on history, however, is that the ideas of Malthus strongly impressed Charles Darwin, who also became convinced that populations tended to grow expansively, leading to an inevitable struggle for existence, natural selection, and survival of the fittest. Darwin's concepts of evolution, developed in his classic book *Origin of Species* in 1859, relied to a considerable extent on Malthus's ideas about population. Although *Origin of Species* was also a controversial work, the ideas of Darwin were accepted more readily by scientists than those of Malthus. Throughout the late nineteenth and first half of the twentieth century, Malthusian doctrine was generally condemned by scientists and philosophers.

In more recent years, however, the concepts of Malthus have become more respected. His name is reappearing in the literature of economics, sociology, and biology as more of a population scholar rather than an historical heretic. Modern societies often find themselves enmeshed in many of the problems Malthus predicted. Although the world does not have such massive famines as in former years, malnutrition and starvation are still prevalent. Poverty, war, and epidemic disease are still common on the global scene. Even with the "green revolution," which produced bumper-crop yields of rice and wheat in India and Indonesia in the 1960s and 1970s, starvation conditions still prevailed in parts of Bangladesh, Cambodia, Ethiopia, Sudan, and Somalia. In the early 1970s many nations in the Sahel experienced famine conditions, forcing mass migrations and causing mass mortality in both people and livestock.

Disease has also continued to be a serious problem for modern societies. Some infectious diseases have shown signs of substantial resurgence, especially in poorer and more crowded nations. Malaria, nearly conquered in the 1960s, increased tremendously around the world in the 1980s. Cholera epidemics occurred throughout Asia and Africa 20 and 30 years ago, subsided, and then broke out in even greater fury in South America in 1991. Peru alone had over 332,000 reported cases in that one year. Trypanosomiasis, schistosomiasis, influenza, and tuberculosis have all increased in recent years despite the advantages of modern medicine to control them. And the most serious epidemic of all, AIDS, which was first recognized clinically in 1981, is in 1995 still spreading throughout most of the world. All our best scientific promises for a cure or a preventative vaccine for AIDS are still many years away.

In reference to global epidemics, sometimes called pandemics, Nobel Laureate Dr. Joshua Lederberg stated in 1969:

> There is a considerable amount of self-delusion . . . that the antibiotics will take care of any bacteriological infection . . . that the plague has been conquered by medicine; that virus infections will somehow be taken care of. But, . . . when you see a pandemic like Hong Kong flu, you have a foretaste of what really can happen. That was a world-wide epidemic. The attack rate was something like 20–30 percent of the world's population. . . . It was not a particularly lethal one, but it is only a minor accident that it was not. Such events are undoubtedly going to occur in the future that will be very much nastier.

Dr. Lederberg was not referring to AIDS, and did not even know about the HIV viruses at the time of his 1969 statement. If the HIV viruses had evolved to be mosquito-borne, which they have not, the AIDS epidemic would now be an absolute Malthusian nightmare in much of the world. We are at least aware now of emerging viruses and antibiotic-resistant bacteria and protozoa. These medical and public health problems have been discussed in a book by Garrett (1994), and will be considered further in Chap. 22.

The possibility of infectious disease acting as a major controling factor on human populations still exists, despite advances in medical science and public health. The worldwide realities of war, starvation, disease, and poverty led Spengler (1969) to conclude, "Malthus' fears may at last be irremediably confirmed."

If the ideas of Malthus could be condensed into a single graph, it would show a curve similar to that in Figure 14.2. This curve shows a geometric or exponential increase in the early stages of population growth followed by a series of catastrophic events. The upper limits of population growth are characterized by sud-

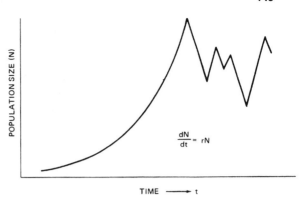

Figure 14.2 Population growth
pattern typical of Malthusian, or
irruptive, growth. The equation,
$dN/dt = rN$, represents only the geometric or exponential form of growth
up to the point of decline. N = population size, t = time, d = change in,
and r = intrinsic rate of increase
(from Southwick, 1976).

den, often drastic, mortality through the mechanisms of starvation, disease, and
violence.

LOGISTIC GROWTH

The first major scientific challenge to the Malthusian theory of population growth
appeared in the 1830s in the work of Pierre Verhulst. In 1838 Verhulst proposed
that populations normally grow in an orderly fashion describing an S-shaped
curve. His theory became known as the logistic theory of population growth (Fig.
14.3). The logistic theory asserted that populations have a slow initial growth rate,
which increases exponentially until the rate reaches a maximum, known as the
point of inflection, and then progressively lessens as the population approaches
the upper limit of its growth. The upper limit, or asymptote, is approached gradually and in an orderly, predictable fashion. The logistic and Malthusian population
curves do not differ in their early stages—both show a slow start followed by
a period of rapid growth—but they differ fundamentally in the later stages as
growth ceases.

Following the work of Verhulst, the logistic theory received little attention until

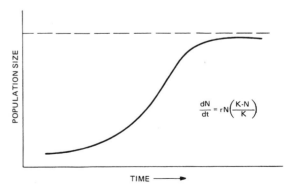

Figure 14.3 Population growth
typical of logistic, or S-shaped,
growth. K = asymptote of the population, or carrying capacity of the
environment (from Southwick,
1976).

it was independently derived and popularized in the 1920s by Raymond Pearl. Pearl applied the logistic curve to experimental populations of yeast and fruit flies (Figs. 14.4 and 14.5) and other small organisms in the laboratory. His work triggered a burst of research activity resulting in the application of the logistic curve to the population growth of protozoa, pond snails, thrips, ants, bees, and eventually vertebrate animals and human populations (Allee, et al. 1949). Figures 14.6 and 14.7 show the logistic curves for water fleas and worker ants.

With an abundance of evidence from many diverse sources, the logistic theory attained wide acceptance. It was proclaimed as a law of population growth by some biologists and was used to predict future population levels of both animal and human populations (Pearl, 1925, 1930). Pearl's investigations led him to conclude (1925):

> It has been shown in what has preceded that populations of living things, from the simplest, represented by yeast, to the highest, represented by man, grow in accordance with the logistic curve. One can now feel more certain that this curve is a first and tolerably close approximation to a real law of growth for human populations. . . . We can, for example, upon a more adequate scientific basis than mere guess-work, predict future populations, or estimate past populations, outside the range of known census counts.

Some biologists accepted Pearl's interpretations rather freely, as indicated in this statement by Clarke (1954): "Populations of a wide variety of organisms, ranging from bacteria to whales, have been found to follow the logistic curve in their growth form. . . . The growth of man's population follows a similar pattern whether examined in individual regions or in the world as a whole." Clarke presented a graph of population growth in the United States fitted with a logistic curve predicting an upper asymptote of 184 million by the year 2100 A.D. With our population in 1995 over 260 million and growing, we can clearly see the dangers of oversimplifying ecological problems.

Other biologists and mathematicians have criticized the logistic theory, pointing out that it does not accurately describe population growth in either laboratory or natural populations. Sang (1950) did a careful reexamination of the *Drosophila* studies and concluded: "The data we have summarized show that only in very exceptional circumstances would logistic growth occur in a *Drosophila* culture. . . . The ecological situation is too complex to be adequately described by the Pearl-Verhulst Law." Sang discussed general aspects of the logistic theory:

> This acceptance of a plausible formula, in spite of its inaccuracy, may be justified, provided this attitude does not inhibit further research, particularly into fundamental conceptions. In the case reviewed, it seems that wide acceptance of the logistic law has led to just this kind of inhibition of further work, and that many interesting ecological and physiological processes have been ignored as a result.

There is much inertia on the subject of human population research and planning, perhaps because of the deep and widespread acceptance of the logistic "law" of population growth.

The study of vertebrate populations in both natural and confined habitats has

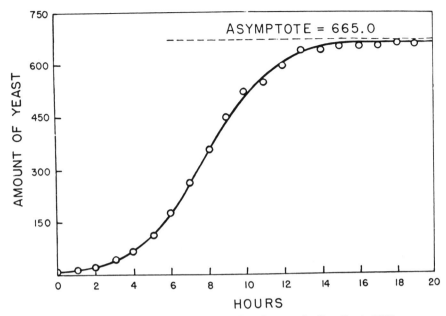

Figure 14.4   Logistic growth of a laboratory population of yeast cells (from Pearl, 1925).

Figure 14.5   Logistic growth of a laboratory population of fruit flies, *Drosophila* (from Pearl, 1925).

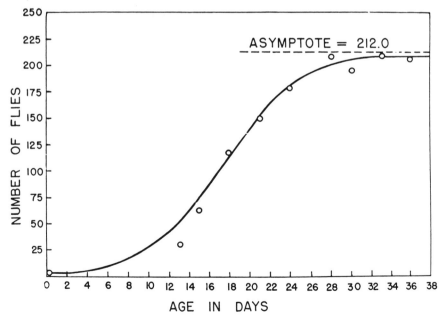

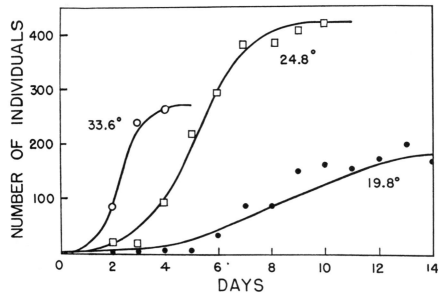

Figure 14.6 Logistic growth of three laboratory populations of water fleas, *Moina,* at different temperatures (from Allee et al., 1949).

produced some population growth curves deviating considerably from logistic growth and tending toward irruptive, or Malthusian, growth. The sheep population of Australia showed logistic-like growth from 1840 to 1890, but its upper level occurred abruptly and was followed by several sharp declines (Fig. 14.8).

Mule deer in the Kaibab National Forest of northern Arizona grew from an estimated population of 4,000 in 1906 after an extensive predator-control campaign and the reduction of domestic livestock in the forest to an estimated population of 100,000 by 1924. Then in the winters of 1924 and 1925, the deer died by the thousands after the high population had destroyed its food supply. The growth and rapid decline of this population has been described as a classic Malthusian curve (Fig. 14.9).

Some authors have refuted this Kaibab deer story, calling it a "myth which is propagated in textbooks" (Burk, 1973), and others have pointed out that the census data on Kaibab deer were inadequate (Caughley, 1970). Although these authors have quite properly prompted biologists to look more critically at the data, most of the original reports confirm a tremendous buildup of deer numbers between 1906 and 1924, with one field observer counting 1,700 deer in one meadow in the summer of 1924 (Allen, 1954). Official predator-control records show that in the 25 years after 1906 the known predator kill in the Kaibab included 4,889 coyotes, 781 mountain lions, 30 wolves, and 545 bobcats, plus an uncounted number of eagles. Many field biologists visiting the Kaibab from 1918 to 1924 reported a dangerously expanding herd and a seriously declining food supply. Many of them predicted impending disaster for the forest and the deer. In the winter of 1924, the lower level of the forest "looked as though a swarm of locusts had swept through it, leaving the range torn, gray, stripped and dying" (Rasmussen, 1941). In many areas, 80 to 90 percent of the forage was gone, and deer died by thousands, some estimates suggesting that 60 percent of the herd perished by starvation in two winters.

Although the precise reasons for the rapid population growth of the mule deer are not fully known, it is fairly certain that it did occur through a combination of factors: a young forest at an early and luxuriant stage of successional growth, protection from hunting, reduction in predation, and less competition from livestock. The population crash was apparently a combined result of environmental deterioration and starvation.

Another example of Malthusian growth and decline in a large animal population is the history of reindeer *(Rangifer tarandus)* on St. Matthew's Island in the

Figure 14.7 Logistic growth of worker ants, *Atta sexdens,* within a nest (from Allee et al., 1949).

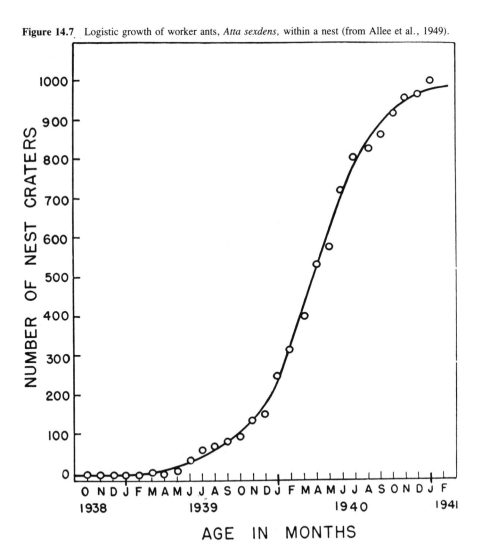

AGE IN MONTHS

Bering Sea. In 1944, 29 reindeer were introduced to the island, and these gave rise to a population of 6,000 by 1963. In the mid-1960s, a massive die-off reduced the population to below 50 animals (Caughley, 1970).

Many other examples of irruptive, or Malthusian, Growth in animal populations can be given. Among vertebrates, several species of small mammals are prone to population irruptions: house mice, Norway rats, bandicoot rats, meadow voles, lemmings, snowshoe hares, European rabbits, and prairie dogs, to mention just a few. Some bird populations can also increase in Malthusian proportions; house sparrows, starlings, and quelea, or African finches, are examples. Irruptive population patterns are also very common in many insects, especially agricultural and forest pests such as locusts, grasshoppers, gypsy moths, miller moths, spruce budworms, corn borers, bean beetles, rice borers, and spittle bugs. In addition, both zooplankton and phytoplankton are prone to irruptive population growth. This is often stimulated by phosphorus and nitrogen pollution (i.e., excessive enrichment of lakes and streams) but may also occur naturally in response to weather or seasonal changes.

These numerous examples of Malthusian growth should not obscure the fact

Figure 14.8 Growth of the sheep population of south Australia, 1840–90. Annual rainfall in inches on lower chart (from Davidson, 1938).

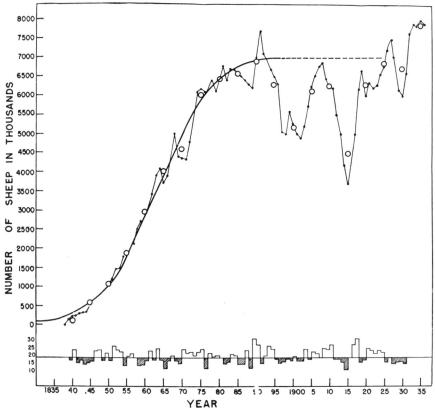

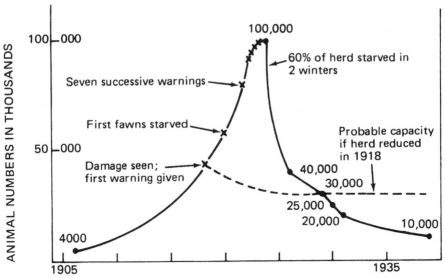

Figure 14.9 Population estimates of the Kaibab mule deer eruption and crash, based on Leopold's interpretation of field notes and descriptions of Rasmussen (from Leopold, 1943).

that most plants and animals live in reasonable balance with their environments and are not generally subject to sudden surges of growth or catastrophic declines. We do not have great outbreaks of robins, vireos, or warblers. The majority of animals showing irruptive population patterns tend to be herbivores or primary consumers, and they often respond to sudden increases in primary production with rapid population growth.

REPRODUCTIVE PATTERNS: "R" AND "K" SELECTION

Various species of plants and animals have evolved with relatively different emphases on reproductive output and survival. In all cases, the "goal" (or, more properly, the result) of natural selection has been continuation of the species. In some species, known as "r"-selected or "r"-patterned species, maximum energy is put into reproduction. High reproductive output may take the form of a large number of eggs (as in insects or fish), large clutch size (as in birds), or large litter size (as in mammals). Relatively less energy goes into parental care or into ensuring the survival of eggs or young produced. Many insects and fish, for example, produce prodigious numbers of eggs, but provide no care for them; varying numbers survive on their own, depending on environmental conditions. Organisms with "r"-selected patterns also reproduce frequently, mature quickly, have short life spans, and are capable of very rapid, explosive population growth. The term "r" refers to the intrinsic rate of increase, implied, in this case, to be high.

Organisms with "K" selection have evolved to put less energy into reproduction per se and proportionately more into the survival of those few young that are produced. These are animals with small litter sizes or single young, close and effective parental care, and long life spans (Wilson, 1975). Elephants are an example. They have only one young at a time, a long gestation period, a long period of very protective parental care, and a long life span. "K"-patterned species are more likely to have logistic patterns of population growth.

These two patterns are often called "reproductive strategies," and this term is embedded in the biological literature. The term, however, is inappropriate in the sense that *strategy* normally refers to a conscious plan, as we might think of a political, economic, military, or financial strategy. Certainly animals do not consciously think about whether to have few young or many young. (The term has even been used to refer to the reproductive patterns of plants, whether they produce few seeds or many seeds.) For this reason, it is more appropriate to refer to "r" and "K" as patterns or forms of selection rather than strategies.

In any case, the terms "r" and "K" are relative. Sometimes closely related species may differ enough in their reproductive patterns to warrant calling one an "r" species and the other a "K" species. For example, house mice are typically "r"-patterned in their reproductive performance. They have large litters of 8 to 10, frequent litters often 6 to 7 times per year, rapid maturation, a short period of parental care by the mother only, a short life span, rapid population growth, and frequently show Malthusian growth patterns.

On the other hand, grasshopper mice *(Onychromys torridus)* are more typically "K"-patterned, with small litters of 2 or 3, usually only 2 or 3 litters per year, a longer period of maturation, a longer period of parental care by both mother and father, a longer life span, and slower population growth that is more typically logistic.

These relative patterns have considerable significance in conservation biology in which the goal is to preserve species in nature. Many rare and endangered species are "K"-selected. Thus, their survival and population growth must be very carefully monitored and protected. Some examples are mountain gorillas and other primates, grizzly bears, giant pandas, California condors, whooping cranes, humpback whales, and African and Asian rhinos.

DENSITY-DEPENDENT AND DENSITY-INDEPENDENT FACTORS

In either logistic or Malthusian patterns, and with either "K"- or "r"-selected populations, growth may be limited by two broad types of factors: (1) density-dependent factors, in which the intensity of action varies according to population density, and (2) density-independent factors, in which the intensity of action is not necessarily related to population density.

Most factors regulating population are at least density-related. For example, a mortality factor that kills 10 percent of the population at low density and 70 percent of the population at high density is said to be "density-dependent." Normally, ecological factors influencing both reproduction and mortality, such as food supplies, habitat, disease, and predation, are density-dependent, or at least related to density in some way.

Density-independent factors are those that operate at the same intensity regardless of population density. A hurricane, flood, cold wave, or forest fire might kill 95 percent of the population regardless of its density. It is debatable whether any true density-independent factors exist, short of natural catastrophes, such as a severe drought or a volcanic eruption, which might eliminate an entire population. But even an environmental disaster can be survived by a few individuals who find unusually favorable places of refuge or have unique access to very limited resources. If the number of protective sites or resources for survival are sharply limited, a higher percentage of a low-density population than a high-density population can be accommodated. Thus, the effects of even a major calamity could be density-dependent. Certainly the great majority of factors regulating populations are related in some way to density or population size, and virtually all factors that operate in a systematic rather than a stochastic way are density-dependent.

Density-dependent factors may regulate populations by either decreasing reproduction (birth rate or natality) or increasing mortality (death rate). The balance between these factors often determines whether Malthusian or logistic patterns will prevail. In the logistic pattern, increasing "environmental resistance" produces a gradual decline in reproduction and/or a gradual increase in mortality. In the Malthusian pattern, a lag or threshold effect is common, (Chapter 4) whereby sudden mortality often occurs in response to increasing density, and the population effect may be drastic. This pattern occurs, for example, in vole populations when high density triggers a fatal epizootic of tuberculosis, in lemmings when high density triggers mass emigration and dispersal mortality, and in human populations when high density coupled with crop failure produces famine.

These examples illustrate the importance of studying the type of factors operating on populations, and the various ways in which these factors are related to population size and density. The principle of density-dependence applies to physical factors such as weather, biological factors such as food supply and disease, and behavioral factors such as sexual and parental behavior, territorialism, aggression, dispersal, and other responses to crowding.

SUMMARY

Most living organisms have the reproductive potential to support substantial population growth. Typically, population growth begins slowly, but if ecological conditions are favorable the potential exists for geometric or exponential growth. Such growth is inevitably temporary, and regulated in one of three patterns: irruptive, or Malthusian, characterized by sudden, even catastrophic, declines in growth; logistic, in which an asymptote is attained gradually and more predictably; or domed, in which the population overshoots its ultimate level but returns to an asymptote in a gradual manner.

Thomas Malthus felt that irruptive growth was typical of most populations, including human populations, unless behavioral restraints were present. On the other hand, logistic growth was proclaimed a "law of population growth" by Raymond Pearl and other scientists on the basis of limited laboratory and statistical studies. Regional and national human populations may show either logistic or Malthusian patterns; we do not know yet which pattern the world population will show. Hopefully, it will be logistic. Presumably, we have the opportunity and ability to choose our pattern of growth. We do not need to simply let things go, drifting toward some unknown, unwanted, and potentially tragic ending. We should have the means to plan our own future.

In natural populations of animals, many examples of both irruptive and logistic population growth can be found, depending on ecological conditions and reproductive characteristics of the species.

Some species, designated "r"-selected, have evolved high reproductive output, and often show irruptive growth. Other species, designated "K"-selected, have evolved lower reproductive potentials but more elaborate systems of parental care to ensure survival. These species more typically show logistic growth. Many exceptions to these generalizations exist depending on ecological conditions. In fact, some of the logistic patterns in laboratory studies actually occurred in confined populations of r-selected species.

REFERENCES

Allee, W., et al. 1949. *Principles of Animal Ecology*. Philadelphia: Saunders.

Allen, D. 1954. *Our Wildlife Legacy*. New York: Funk and Wagnalls.

Burk, C. 1973. The Kaibab deer incident: A long-persisting myth. *BioScience* 23: 113–114.

Caughley, G. 1970. Eruption of ungulate populations, with emphasis on Himalayan Thar in New Zealand. *Ecology* 51: 53–72.

Clarke, G. 1954. *Elements of Ecology*. New York: Wiley.

Davidson, J. 1938. On the ecology of the growth of the sheep population in South Australia. *Trans. Roy. Soc. of South Australia* 62:141–148.

Ehrlich, P., and Ehrlich, A. 1990. *The Population Explosion*. New York: Doubleday.

Garrett, L. 1994. *The Coming Plague: Newly Emerging Diseases in a World Out of Balance*. New York: Farrar, Straus and Giroux.

Lederberg, J. 1969. "News and Comments" (House Foreign Affairs Subcommittee hearings on December 2, 1969). *Science* 166: 1490.

Leopold, A. 1943. Deer irruptions. *Wisconsin Conservation Bulletin* 321: 3–11.

Malthus, T. 1798. *An Essay on the Principles of Population*. London: Johnson. [1952 edition: Everyman's Library, New York: E. P. Dutton]

Meadows, D., D. Meadows, and J. Randers. 1992. Beyond the Limits: Confronting Global Collapse and Envisioning a Sustainable Future. Post Mills, VT: Chelsea Green.

Miller, G. 1992. *Living in the Environment*. 7th ed. Belmont, CA: Wadsworth.

Pearl, R. 1925. *The Biology of Population Growth*. New York: Knopf.

Pearl, R. 1930. *Introduction to Medical Biometry and Statistics*. Philadelphia: Saunders.

Rasmussen, D. 1941. Biotic communities of the Kaibab Plateau, Arizona. *Ecological Monographs* 2: 229–275.

Sang, J. 1950. Population growth in Drosophila cultures. *Biological Reviews* 25: 188–219.

Southwick, C. 1976. *Ecology and the Quality of Our Environment*. 2nd Ed. Boston, MA: Willard Grant.

Spengler, J. 1969. Population problem: In search of a solution. Science, 166: 1234–1238.

Wilson, E. 1975. *Sociobiology: The New Synthesis*. Cambridge, MA: Harvard University Press, Belknap Press.

Chapter 15

HUMAN
POPULATIONS

On a global basis, the human population has shown a J-shaped pattern of growth over the past two thousand years (Fig. 15.1). The world's population, which was less than 500 million people for most of human history, now stands at 5.8 billion. This population is increasing at the annual rate of 1.5 percent, which means a net increase of almost 87 million people per year, or about 1.7 million per week (World Population Data Sheet, 1995).

Many statistics can be assembled around these stark facts. The world's population early in the agricultural revolution (about 8,000 B.C.) was probably no more than 10 million (ReVelle and ReVelle, 1992). By the birth of Christ, there may have been 200 million people in the world. Human beings took more than 3 million years to reach a population of 1 billion people, sometime in the eighteenth century. The second billion came in only 130 years, the third billion in 30 years, the fourth billion in 15 years, the fifth billion in 12 years, and the sixth billion will be achieved in about 11 years, before the year 2,000. Current projections indicate 8 billion by the year 2020 (World Population Data Sheet, 1995).

At the beginning of the present century, the world's population was only 1.7 billion. The number of human beings has more than tripled in the last 90 years. No one can predict what the human population will be at the end of the twenty-first century. Even by the mid-twenty-first century, the world's population could be 12 to 14 billion.

In global terms the big question is, How and at what levels will this J-shaped growth be limited? Malthusian theory would predict human misery and catastrophe; logistic theory would predict a gradual resolution and adaptation to some higher level, perhaps 8 to 10 billion; the domed model of growth would predict an overpopulation of 8 to 10 billion with readjustment to lower levels.

Most ecologists consider human population growth to be one of the greatest

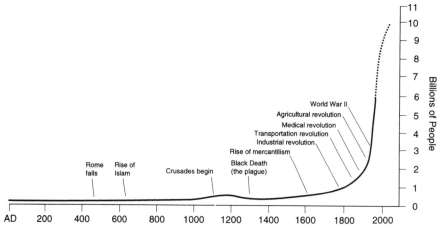

Figure 15.1 The J-shaped curve of human population growth in the world over the past 2000 years (from Corson, 1990).

problems in global ecology and a major driving force of environmental degradation. They see excessive consumption as an equally important cause of pollution and environmental deterioration. Most agree that the two factors work hand in hand to threaten the world's ecological integrity.

Not all scholars agree with these basic conclusions, however. Julian Simon, a prominent economist, insists that there is no population or environmental crisis (1992). He asserts that population growth, economic growth, and a resource-rich world coupled with modern technology will produce greater prosperity and better health for increasing numbers of people (see Chapter 8). This view does not seem very realistic in today's world, however, and few scientists accept his interpretation of world affairs.

In contrast to the rosy picture that Simon paints, we can offer these sobering facts (Miller, 1992):

- One out of five people in the world, including one out of three children under the age of five, is hungry or malnourished.
- One out of five people has inadequate housing, and an estimated 150 million people are homeless.
- 17 million people are refugees, stateless, landless, and often homeless.
- One out of three people have poor health care and not enough fuel to cook food or keep warm.
- One out of four adults cannot read or write.
- Over a billion people are seriously ill with preventable infectious diseases, including malaria, tuberculosis, schistosomiasis, trypanosomiasis, and filariasis.
- Every day at least 30,000 children under the age of five die in poor countries of conditions that could be prevented at a cost of about $5 per child per year.

It is true, as Dr. Simon states, that the world is making progress in that more people are living longer and fuller lives than ever before, but it is also true that more people are living in misery and poverty than ever before.

Some scientists have calculated that an optimal human population on earth in terms of reasonable living standards is no more than 2 billion people (Pimentel et al., 1994). They estimate that between 1 and 2 billion people could be supported in relative prosperity. This estimate is based on the availability of agricultural land (now less than 0.28 hectares per person) and energy resources, better pollution control, and the need for recycling and sustainability. They point out that the world already faces significant problems of poverty, malnutrition, conflict, and environmental deterioration, and they raise the question, "Does human society want 10 to 15 billion people living in poverty and malnourishment or 1 to 2 billion living with abundant resources and a quality environment?"

The key question in the above consideration, of course, is how do we define a reasonable standard of living or relative prosperity. Although there can be much debate on such terms, most of us could probably agree on some common features of a reasonable standard of living: a satisfactory job, adequate housing, a varied diet without fear of food and water shortages, a sense of safety and security, at least one automobile for each family, other transportation and travel opportunities, and satisfactory access to education, health care, recreation, and personal development. If we consider these features to constitute a reasonable standard of living, say, for a typical middle-class family, it is apparent that most of the people in the world do not have this standard. If we cannot provide these amenities now for 6 billion people in the world, can we expect to provide them for 8 to 10 billion in the twenty-first century? A number of scientists believe we cannot, and some believe we cannot ever ensure adequate food supplies (Kendall and Pimentel, 1994).

Other scientists consider current human population growth an "ecopathological process" that is out of control and injuring the earth. In this sense, such population growth in some parts of the world, is carcinogenic, a "cancer-like growth" with the potential of destroying the global ecosystem (Eisley, 1961; Gregg, 1955; Hern, 1990, 1993). Dr. Warren Hern, a physician, anthropologist, and public health epidemiologist, has pointed out the striking similarities between human population growth in the world and the growth of malignant cancer in an individual. Both are characterized by (1) rapid, uncontrolled growth, (2) invasion and destruction of adjacent tissues or environments, (3) metastasis, or spread by colonization, and (4) dedifferentiation, or loss of distinctiveness in individual components. Furthermore, the malignant process in an individual or an ecosystem often involves the production of toxic metabolites. In the human body, any two of these symptoms are suggestive of cancer; in the human population, all five are now occurring in the global ecosystem.

This idea that population growth is a malignant process is very unpopular, of course, because no individual wants to be compared to a cancer cell, and many objections to this view can be raised. There are obvious differences between cancer in an individual and population growth in the world. Whereas cancer, if untreated, usually kills the patient (unless there is a spontaneous remission), world population growth will not necessarily "kill" the earth. The earth will survive a human onslaught, but be so drastically changed that modern civilization as we know it could collapse. Human history contains examples of societies that have self-destructed, and the potential for self-destruction on a larger, even global, scale cannot be dismissed out of hand. The world of the twentieth century has certainly had frequent warnings of both human and environmental tragedies if we continue to conduct "business as usual" on a course of continuing growth.

These facts and theories on human population growth focus our attention on a number of related questions: Why is this growth occurring? How is it occurring, and where is it occurring?

WHY IS HUMAN POPULATION GROWTH OCCURRING?

The simplest explanation of the rapid growth of the human population is that three births are occurring in the world for every death. Human reproductive potential evolved at a time when mortality was high and longevity was low. Even in the time of Aristotle and Plato, relatively recently in terms of the complete span of human existence on earth, average life expectancy at birth was only about 30 years. In most of the world, life expectancy at birth is now over 60 years. In many of the world's poorest countries life expectancies remain very low, sometimes less than 50 years of age, but in most industrialized nations life expectancies are over 70 years. In the United States, life expectancies increased from 42 years in 1900 to 76 years in 1995. Female life expectancy in this country is now 78 years, and that of males, 72 years.

The primary fact of human populations in both developed and developing nations over the past 120 years is that death rates have fallen more than birth rates (Fig. 15.2). In the seventeenth and eighteenth centuries, the crude birth rates (the number of births per year per 1,000 people) in the total population were around 38 to 42, and death rates were almost as high, at 35 to 39. Hence, the rates of natural increase were relatively small. With the beginning of modern medicine and economic growth spurred by the industrial revolution in the late nineteenth century, however, death rates began to fall, life expectancies increased, and population growth accelerated. In the world population of 1991, the crude death rate had fallen to 9 per 1,000 while crude birth rate declined to only 27 per 1,000, so the rate of increase was 18/1,000, or 1.8 percent (Miller, 1992).

Crude death rates fell in both more-developed and less-developed nations, but the decline of birth rates started sooner in industrialized nations, more or less paralleling the decline in death rates (Fig. 15.2). In less-developed nations, birth rates remained high into the late twentieth century.

Within these general patterns of birth and death rates which have occurred worldwide, there is a tremendous disparity between nations and regions. A wide spectrum of demographic or population patterns exist, depending in large extent on the state of economic development. Some of the most industrialized countries such as Japan, United States, and those in western Europe have reduced birth

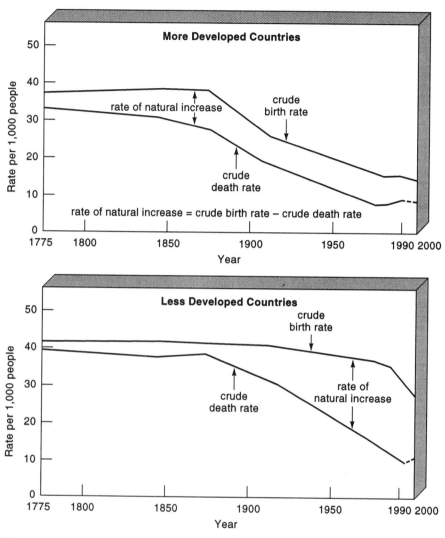

Figure 15.2 Changes in birth and death rates since the eighteenth century in nations of different states of economic development (from Miller, 1992).

rates substantially and now show population growth rates of 0.6 percent or less. In Europe, 12 or 13 nations have reached zero population growth or are very near ZPG. One of their major population concerns is the aging of the population, leaving reduced numbers of young people to enter the work force. At the other extreme, some of the world's poorest and most agrarian nations still have very high birth rates. Even with relatively high death rates, their rates of population growth are high, often over 2 percent and sometimes as high as 3 or 4 percent per year. Table 15.1 shows the remarkable difference between an annual growth rate of 1 and 3 percent in terms of growth per century. This table shows that even 1 percent annual growth almost triples the population in 100 years if this growth rate is

TABLE 15.1
Relationship of population growth per
year and per century

Rate per Year	Rate per Century
1%	270%
2%	724%
3%	1,922%

sustained, and a 3 percent annual growth rate leads to an incredible 19-fold increase over 100 years.

Crude birth rates and death rates have many specific components, including age-related birth and death rates and survivorship patterns, or life expectancies. These are subjects of demography, which we cannot fully explore here, but it will be helpful to look briefly at a few demographic factors in human population growth.

The major demographic factors in all human population growth are the birth rate and infant death rate. Crude birth rates typically vary from 10 to 50 births per year per 1,000 total population. These rates, of course, are influenced by many factors including the number of females of reproductive age in the population, the culture, the educational and economic status of the population, the health facilities available, and so on. Among the least-developed nations of the world, rates of 40 to 50 occur; the most industrialized nations usually have crude birth rates around 10 to 15.

Another way of expressing birth rate, or fertility, is the number of children per woman with children. In most industrial nations, this usually averages around 2 to 2.5. In many less-developed nations, this often averages from 4 to 10. In most nations of the Middle East and south Asia, for example, the average number of children per married woman is five to seven. In China, where a strict governmental policy of only one child per family was enforced for many years, but some exceptions were allowed, the average number of children per married woman is just slightly over one.

Another aspect of reproductive rate is infant mortality. Infant mortality rates vary greatly from lows around 5 to 15 of every 1,000 live births to highs of 150 to 200. Notice, however, that the denominator is different than it is for crude birth rates. Infant mortality rates are usually expressed per 1,000 live births; crude birth rates are usually expressed per 1,000 total population, so the resulting numbers are not directly comparable.

Infant mortality rates have shown remarkable declines with the progress of economic development. Before the advent of modern medicine in the nineteenth century, often 50 percent or more of all newborn infants died in their first year. The risks of childbirth and infectious disease were the big killers. Now in most of the world's countries, infant mortality is less than 2 percent. In the United States, for example, infant mortality is less than 1.0 percent; only 9 to 10 infants out of 1,000 die within their first year. This is not the best standard in the world—in some European countries, only 5 to 6 infants per 1,000 births die within their first

TABLE 15.2
Examples of basic demographic data in poor, moderate, and rich nations in terms of economic development (The *World Almanac, 1995*).

	Crude Birth Rate[a]	Infant Mortality[b]	Average Life Expectancy	Annual Rate of Population Growth
Afghanistan	43	156	45 yrs.	2.5%
Bangladesh	35	107	55	2.3
Cambodia	45	111	50	2.9
Chad	42	132	41	2.2
Ethiopia	45	106	52	3.1
Kenya	42	74	53	3.1
Mozambique	45	129	48.5	2.9
Rwanda	49	119	40	2.8
Somalia	46	126	54.5	3.2
Brazil	21	60	62	1.3
India	28	78	58.5	1.8
Indonesia	24	67	61	1.6
Mexico	27	27	74	2.2
Peru	27	51	65.5	2.0
Thailand	19	37	68.5	1.3
France	13	7	78	0.4
Germany	11	7	76.5	−0.01
Japan	10	4	79	0.3
Sweden	14	6	78	0.3
Switzerland	12	7	78.5	0.3
United States	15	10	75.5	0.7

[a]Number of live births per 1,000 total population.
[b]Number of infant deaths in first year per 1,000 births.

year. This is a remarkable achievement of modern medicine and prenatal care. At the other extreme, some of the world's poorest countries have 10 to 20 percent infant mortality; that is, 100 to 200 out of 1,000 children die in their first year. Even this rate may be exceeded in times of famine or epidemic disease. Still, short of such catastrophes, the high rates of infant mortality shown in some poor nations (Table 15.2) are better than those prevailing throughout much of human history and prehistory.

The numbers in Table 15.2 indicate that the world population is a very complex mosaic of national and regional patterns. Some nations, such as Ethiopia and Kenya, have high birth rates and, despite high mortality rates and short life expectancies, are growing very rapidly. Other nations, such as Germany, have very low birth rates, low mortality rates, long life expectancies, and are at zero or even negative population growth. If we were to collect these facts for the more than 190 nations in the world[1] and couple them with the diverse economic, ecologic, and cultural backgrounds of each population, we can appreciate how complex this topic of world population growth is. Nonetheless, some definite trends and patterns are emerging on the global scene.

1. There were 192 nations in the world as of January 1995, with 184 of these members of the United Nations.

WHERE IS WORLD POPULATION GROWTH OCCURRING?

The combination of high rates and large numbers means that the overwhelming preponderance of world population growth is occurring in the less-developed countries (LDCs). Figure 15.3 diagrams the world's population by large regional divisions as of 1975 and projected to the year 2000.

More than 90 percent of the projected increase in the world's population over the next 25 years will occur in the developing nations of Asia, Africa, and Latin America (Corson, 1990). India and China alone, due to their large base populations of 930 million and 1.2 billion, respectively, are contributing over 30 million people per year to the world's population. Among the industrialized nations of the world, the United States is one of the fastest growing. At our current population of 260 million and rate of 0.7 percent annually, we are adding nearly 2 million people annually in the United States, even more considering immigration..

Regrettably, the fastest population growth and the greatest increases in total numbers are occurring in the countries least able to support this growth. The relationship between poverty and high population growth rates has been interpreted in two different ways.

Some scientists view poverty as the result of high population growth. They see crowded, rapidly growing populations as exceeding their resource bases, so there are inevitable shortages of both material and economic resources, that is, shortages of space, housing, food, jobs, and social services. This is the traditional Malthusian view; there are too many people for the available goods, so poverty results. Many nations must maintain an economic growth rate of 2 or 3 percent just to stay at the same level of economic condition if their population is growing at this rate. This situation allows no real improvement in living standards, and frequently the resource base of the country does not allow any economic growth, so essentially the people lose ground.

Other scientists see poverty as the cause of high population growth. They attribute high population growth to lack of education, lack of health care, and lack of a reasonable standard of living. In poor agrarian societies, children are put to work at an early age herding livestock, gathering firewood, chasing birds and monkeys away from grain fields, and so on. In such societies, young children, even 4 and 5 years of age, are seen as an asset in the menial chores of village life.

In addition, parents view children as their only source of old-age security. Villagers in India insist they must have six to eight children, so they can be assured that two or three of them will survive and stay home to take care of them in their old age. In fact, the World Health Organization projects that population growth can only be slowed when infant mortality rates are reduced. One study projected a world population of 10.5 billion by the year 2100 at the current rate of child mortality (89/1,000), but a world population of only 8.5 billion by the year 2100 if child death rates could be reduced 50 percent to 45/1,000 (WHO, 1985). This conclusion seems to be a paradox—a smaller population if more children survive? The projection is based on the basic human response to poverty and insecurity, and it does fit the data from many developing countries.

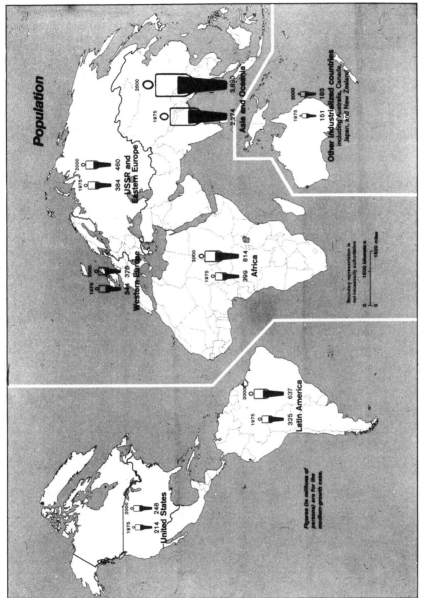

Figure 15.3 Diagrammatic representation of global population distribution in 1975 and projected for 2000 (from Barney, et al., 1982).

URBAN GROWTH

An important aspect of world population growth is the disproportionate share of growth occurring in the world's cities. The human species is aggregating in urban centers at rates greatly exceeding population growth alone. In the United States in 1800, only 5 percent of the population lived in towns of more than 2,500 people. By 1960, 65 percent of our people lived in towns of this size; by 1975, 73 percent lived in cities of more than 100,000.

In the twentieth century, Los Angeles has grown from a city of 102,000 people in 1900 to 1.9 million in 1950 to 8.8 million in Los Angeles County in 1990. Phoenix, Arizona, has grown from a town of 5,544 in 1900 to a metropolitan area of 2.1 million in 1990. Dallas, Texas, had only 42,638 people in 1900; by 1990 its metropolitan area included 3.8 million.

This is the story around the world. Cities in developing nations have shown spectacular growth in numbers, and many are now at the limit of habitability. Since 1900, Calcutta, India, has grown from a city of 850,000 to one of 12.5 million, and it is projected to reach 19.7 million by the year 2000. In just 50 years, from 1940 to 1990, Mexico City has grown from 1.4 million to 18.2 million. It is projected to become the largest city in the world by 2000 with 31.6 million people. Tokyo, Hong Kong, Bangkok, Jakarta, Bombay, Delhi, Karachi, Tehran, Cairo, Lagos, Rio de Janeiro, Sao Paulo—all are becoming giant cities in nations straining to meet basic human needs for housing, employment, education, and health services (Ehrlich and Ehrlich, 1990). Table 15.3 shows simplified data on urban populations in selected cities over the past 30 years. In 1850, there were only 4 cities in the world with more than 1 million people; by 1950 there were 100 such cities, and by 2000 there will be 1,000 cities with more than 1 million inhabitants.

TABLE 15.3
Urban growth estimates (in millions of people) in developing countries, from *The Global 2000 Report* (Barney, Pickering, and Speth, 1982).

City	1960	1970	1975	2000
Calcutta	5.5	6.9	8.1	19.7
Mexico City	4.9	8.6	10.9	31.6
Bombay	4.1	5.8	7.1	19.1
Cairo	3.7	5.7	6.9	16.4
Jakarta	2.7	4.3	6.9	16.9
Seoul	2.4	5.4	7.3	18.7
Delhi	2.3	3.5	4.5	13.2
Manila	2.2	3.5	4.4	12.7
Tehran	1.9	3.4	4.4	13.8
Karachi	1.8	3.3	4.5	15.9
Bogota	1.7	2.6	3.4	9.5
Lagos	0.8	1.4	2.1	9.4

Source: Barney, Pickering, and Speth, *The Global 2000 Report* (1982).

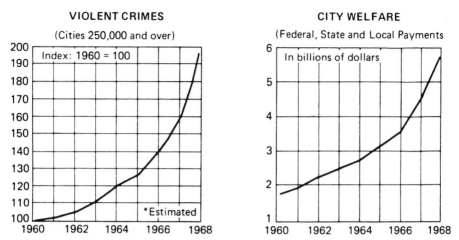

Figure 15.4 Violent crimes and welfare costs in American cities in the decade of the 1960s (from Weiss Associates, New York).

Worldwide urbanization presents a number of environmental problems both regionally and globally. Ecologically, cities are parasitic on the surrounding landscape. For life support, cities must import food, water, air, construction materials, and most natural resources. They must export waste products in gaseous, liquid, and solid form. If cities were covered by giant plastic domes, they would either suffocate in a matter of days or die more slowly from their own toxic products.

The ecological impact of cities extends beyond their immediate regions. Many of the world's cities import forestry products, mineral resources, and food from around the world. The world's oceans are fishing grounds for all nations. Pollution plumes from great urban centers are visible from space and certainly contribute to global atmospheric conditions. Cities modify local and regional weather, and may even have an impact on global climate patterns.

Cities are known as centers of homelessness, crime, unemployment, social welfare, and poverty. Figure 15.4 shows the trends in urban crime and welfare in the United States in just one decade, the 1960s. By the 1990s these trends have led Americans to rank crime as the number one problem in the United States. Many of the same problems occur in developing nations.

Cities also have some of the world's worst air and water pollution. Mexico City, Los Angeles, Tehran, Calcutta, Jakarta, and many of the others we have listed with growth problems are famous as well for terrible air quality or grossly fouled waters.

Why then do people flock to cities, if they find crowding, pollution, crime, and social disorder? The answers are complex, but the positive side of cities is that they offer cultural excitement and entertainment, education, jobs, communication, sports, sophistication, and the eternal hope for prosperity and even fame. Cities are where the action is; they are the centers of business and finance as well as many other human endeavors. In developing nations, urban migration is also a reflection of an exhausted and hopeless hinterland—a worn out landscape that can no longer offer a livelihood. In India in the 1950s and 1960s, cities offered food handouts from international aid. People flocked to cities to get free food; they became centers of survival.

Cities should certainly not be portrayed as all bad. They offer many efficiencies and ecological advantages. They can accommodate large populations in relatively small areas. They promote efficiencies in transportation, housing, utilities, distribution of essential goods, manufacturing, and the provision of essential services such as education and health care, that rural areas cannot supply. Imagine the problems that would occur, environmental and otherwise, if all the urban-dwellers of the world were scattered over the entire landscape. In fact, a major problem in many metropolitan areas is that many people who work in cities and want all the benefits of cities nonetheless want to live in the country. Commuting distances of 50 miles are not unusual. This situation, however, puts a great strain on resources as well as both the rural and urban landscape.

We will explore some of the relationships between economic status, population growth, urban and rural populations, and ecological conditions later. Here we note that these relationships are so interdependent that it is often impossible to determine cause and effect. Without question, poverty, population growth, and urban crowding are closely interrelated in many ways.

FUTURE PROSPECTS

There is little doubt that the world could make considerable progress in the quality of life for most people, and the quality of the global environment, if we could slow down and ultimately stabilize human population growth. If we can go to the moon, orbit the earth in space for weeks at a time, send television images around the world in a matter of seconds, and transplant hearts, why can't we achieve a better balance between people, resources, and the environment? Why can't we bring human populations into a more reasonable state of ecological equilibrium?

The complete answers to these questions lie deeply within the complex realms of science, philosophy, religion, economics, and politics. Simple answers do not exist. Nonetheless, we can consider some aspects of these problems within the field of global ecology.

First of all, it is apparent that world populations cannot come into stable balance in the short term because they have great momentum for further growth. The age structure of human populations in many nations indicates that a large proportion of the population is either in or just entering reproductive age or close to it. In Nigeria, for example, 48 percent of the population is under the age of 15; over 70 percent of the population is within the reproductive age. In Mexico, over 80 percent of the population is under the age of 45.

An indication of the momentum of population growth is shown in the age pyramids for three different nations (Fig. 15.5). Mexico, with an annual population growth of 2.5 percent, has a large percentage of its population in or entering reproductive age groups. Programs of family planning directed at adults may have some immediate effects on birth rates, but for long-term results these programs

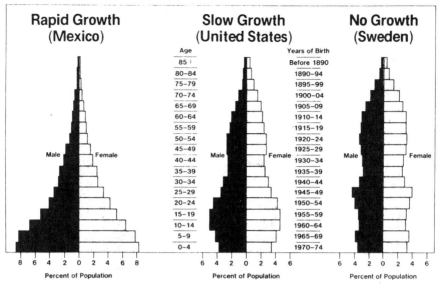

Figure 15.5 Recent age structures of three nations with different rates of population growth (Population Reference Bureau, Washington, DC).

must also influence adolescents and children entering reproductive age. Education and health programs, if highly successful, could modify future growth somewhat, but the age structure of the population virtually guarantees considerable momentum in reproductive rates for another generation.

Age pyramids show demographic reasons why human population growth cannot be slowed quickly, but they are merely statistical and graphic representations of population structure. The real reasons for population growth lie within human nature. Human nature is an intricate mixture of biology and evolution, philosophy and religion, education and culture. First of all, we, like most species, are intensely sexual beings for the most part, as evidenced by our social customs, business activities, entertainment, politics, and virtually every other aspect of our lives. Second, this sexuality, coupled with the drive for survival and reproduction, is influenced by religious beliefs, philosophical attitudes, and cultural mores. Religious injunctions against contraception, for example, seem a long way from ecological science, but they are part of the human population picture. Ultimately, they will come into conflict with ecological reality. A number of recent publications have dealt with the delicate matters relating to the human dimensions of global change (Lowe, 1991; McDowell, Wilhelm, and Wolf, 1992). The subject is so sensitive, however, that even the Earth Summit in Rio de Janeiro in 1992, sponsored by the United Nations, generally avoided the cultural aspects of population growth (Glantz, 1992).

A more hopeful development occurred at the UN population conference in Cairo in 1994. Despite many cultural and religious conflicts, the conference achieved agreements on the need for improvements in women's rights, education, health care, and population planning (Sachs, 1995). Most of the developing nations emphasized the importance of family planning; many of them, such as India, have had governmental programs promoting family planning since their independence (in India's case, for more than 40 years). There is now wide recognition that hundreds of millions of women in developing nations want access to modern contraceptives to limit their families. The entire world, especially the wealthier nations, must respond to this need with increased assistance. The Cairo conference was a step in the right direction.

POPULATION MODELING

Worldwide data on economic and ecological trends coupled with modern computer modeling have enabled scientists to project current trends in population. Among the most noteworthy attempts have been those initiated by the Club of Rome, a group of international businessmen and -women concerned about world population trends. They contracted with a group of scientists under the leadership of Donella Meadows, Dennis Meadows, Jurgen Randers, and William Behrens, a team of scientists at MIT. Their first book, *The Limits to Growth* (Meadows et al., 1972), provided a series of computer models for projecting human population in the future based on current trends and conditions.

It is important to distinguish between projections and predictions. The latter are statements of belief about what will happen; the former are statements about what will happen *if* current trends, or certain assumed trends, continue. Trends can change, however, and projections do not dogmatically assert what will happen. For example, social, political, and economic conditions can change, and such changes have a major impact on trends.

The *Limits to Growth* study used very complex models of global trends, linking economic and environmental trends in cybernetic, or feedback, models. The simplest feedback models showed the relationships between population size, births and deaths, and industrial capital, investment and depreciation (Fig. 15.6).

In these models, positive feedback loops, such as births per year (a reflection of fertility rates), increase the population, whereas negative feedback loops, such as deaths per year, decrease the population. Similarly, in economic terms, industrial output provides a positive feedback on investment and industrial capital, whereas depreciation is a negative feedback on industrial capital.

These simple models were tied into more complicated ones showing relationships between population, agriculture, industry, and pollution (Fig. 15.7). Finally, the world model was assembled relating many aspects of human activities and environmental trends (Fig. 15.8).

The basic procedures used in the Meadow's models were to enter the best data available on world economic, population, and environmental trends and let the computer programs project these trends into the future given certain assumptions. The basic assumptions were that there would be (1) no major change in the trends in birth rates, death rates, and rates of agricultural and industrial production, (2) no startling new technological developments such as fusion power, (3) no World War III involving the major world powers. After the standard models were run,

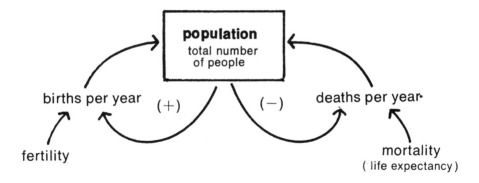

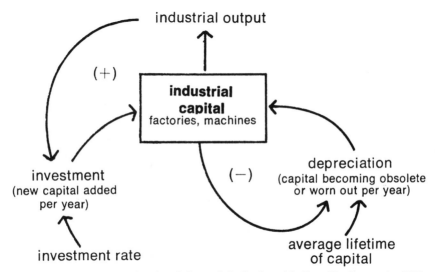

Figure 15.6 Population growth and capital growth feedback models (from Meadows et al., 1972).

various contingencies were entered into the computer programs, such as what would happen if natural resources were doubled, if food production kept ahead of population growth, if perfect pollution control were achieved, and so on.

In 1972, the standard model and virtually all the contingency models showed domed population growth curves for the human population, in which growth continued into the twenty-first century to various levels of 8 to 12 billion people, and then declined to more sustainable levels as human populations were overcome by resource shortages, pollution, or a variety of other factors (Figs. 15.9, 15.10, and 15.11).

The Meadow's models and *The Limits to Growth* were heavily criticized almost immediately as models of gloom and doom (Cole et al., 1973). The standard world run model of Meadows et al. (Fig. 15.9) assumed no major change in the physical, economic, or social trends that have governed the world in the twentieth century. Food, industrial output, and population grow exponentially until the di-

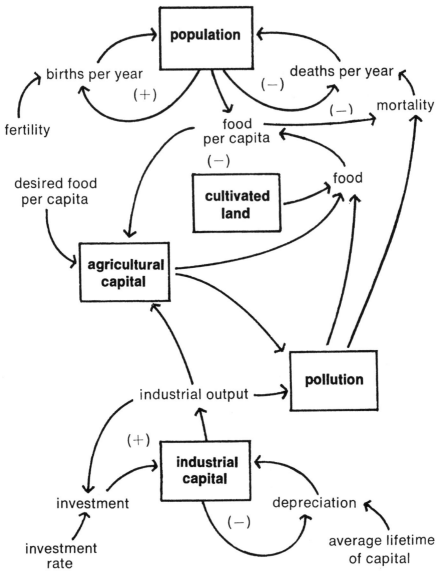

Figure 15.7 Feedback models of population, capital, agriculture and pollution (from Meadows et al., 1972).

minishing resource base forces a slowdown in industrial growth. Population growth is halted by a rise in the death rate due to decreased food and medical services. Critics of *The Limits to Growth* study claimed that the world is much too complex to be described by simple unitary models that treat global affairs as homogeneous. They pointed out that trends change throughout history, that we cannot model scientific achievements, political events, or social trends.

Nonetheless, many of the techniques used in *The Limits to Growth* have objec-

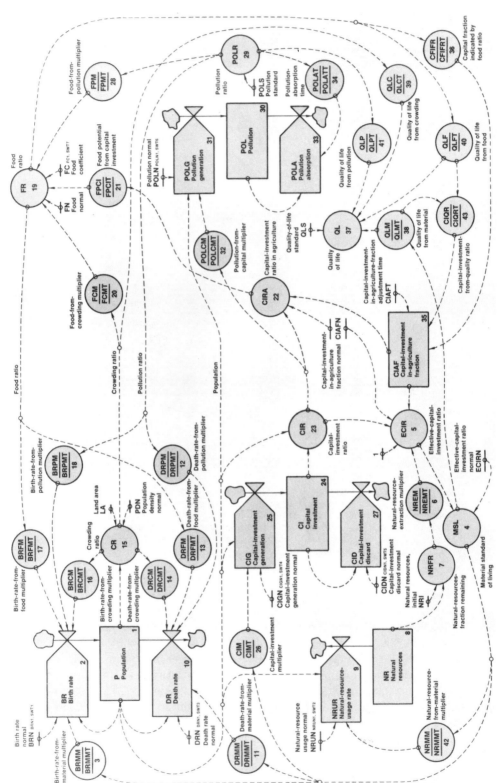

Figure 15.8 The World 2 Model initially used by Meadows et al., 1972 (from J. W. Forrester, 1971).

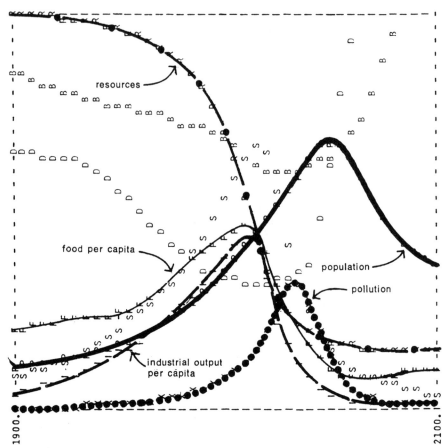

Figure 15.9 Projections based on the standard world model of Meadows et al. (1972), which assumed no major change in the physical, economic, or social trends that have governed the world in the twentieth century. Food, industrial output, and population grow exponentially until the diminishing resource base forces a slowdown in industrial growth. Population growth is halted by a rise in the death rate in the twenty-first century due to decreased food and medical services. B = crude birth rate, D = crude death rate, S = services per capita (from Meadows et al., 1972).

tive value, and many continued efforts have been made to refine these techniques, obtain better data, and strive for the most accurate projections possible.

The latest modeling efforts of the Meadow's team have produced a sequel, *Beyond the Limits: Confronting Global Collapse and Envisioning a Sustainable Future* (1992). In this research study they have sharpened the models and added more and better data from the last 20 years. The standard run comes out with a similar domed population curve (Fig. 15.12), and doubling resources shows the same projected pattern as in 1972. Some other contingency models are more encouraging, however. If population planning, pollution control, and erosion control are factored into the models, a logistic pattern is achieved (Fig. 15.13).

A point to emphasize again is that these models are not predictions of what will happen; they are projections of what will occur if certain trends continue and certain policies are put in place. Their value is in showing potential outcomes, in emphasizing the need for certain types of action, and in showing the importance of data on key topics relating to population, economics, and environment. The models are also of vital importance in showing the consequences of not changing historical trends; they highlight the inevitable dangers of unlimited exponential growth in population and industrial pollution and of unchecked exploitation of life-support systems.

GOVERNMENTAL RESPONSES

There have been varying degrees of political response to human population growth. Government responses can generally be grouped into six or seven different categories, although these responses are not necessarily mutually exclusive.

Figure 15.10 Projections based on the world model of Meadows et al. (1972) with natural resource reserves doubled. Pollution rises very rapidly causing an increase in the death rate and a decline in food production. Population decline is greater. B = crude birth rate, D = crude death rate, S = services per capita (from Meadows et al., 1972).

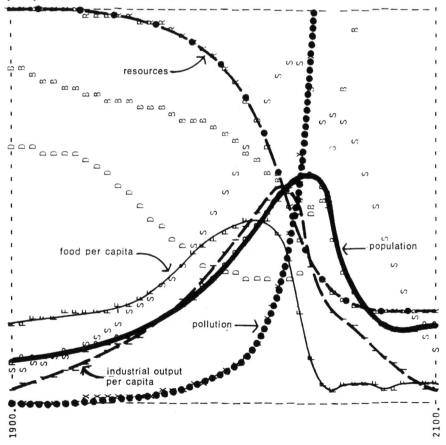

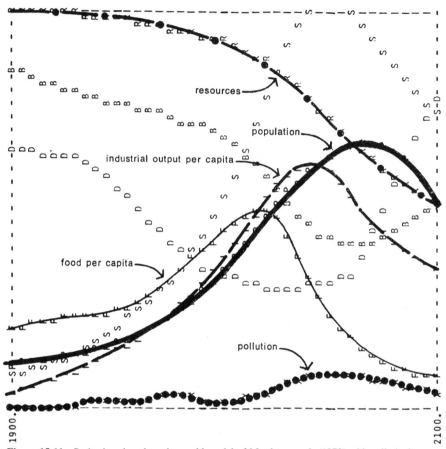

Figure 15.11 Projections based on the world model of Meadows et al. (1972) with unlimited resources and pollution control. These changes allow population and industry to grow until the limit of arable land is reached. Then food per capita declines, industrial growth is slowed, and gradual population decline occurs as birth rates fall and death rates increase. B = crude birth rate, D = crude death rate, S = services per capita (from Meadows et al., 1972).

First, some governments have ignored the problem, either by arguing that a human population problem does not exist or by refusing to do anything about it because action would involve moral choices in sensitive areas. This was the official U.S. position in the 1980s. It changed in the early 1990s to a more realistic and enlightened policy, which hopefully will continue. There is a danger, however, that our own government may return to a more backward policy. A bill was introduced in the U.S. Senate in 1995 to block all U.S. funding for international population planning. If politicians from one of the most powerful and presumably advanced nations on earth can show this kind of short-sighted thinking, we can see why the problem of global population growth is so intractable.

Second, some governments have seen population growth as the root cause of many of their national development problems, and they have therefore established official programs of family planning. As far back as 1954, the government of

India has had a cabinet-level post to oversee population planning, and it has taken all reasonable actions to foster birth control. Mexico, Indonesia, Thailand, and Taiwan have had major programs of family planning for more than 10 years. These programs have emphasized contraceptive education, distribution, and use. Many of these programs had some success, but obviously they have not been entirely effective. Without them, however, the present conditions of crowding and poverty in these countries would be much worse.

Third, some governments have focused on strong economic incentives to limit family size, such as better housing for small families and lower taxes for couples having no more than two children. Singapore has followed this route, in addition to providing birth control education, family clinics, and pre- and postnatal care.

Fourth, some governments have used coercive action. The People's Republic of China has strictly enforced a policy of only one child per couple, requiring abortion if additional pregnancies occur. This has been a highly repressive and totalitarian program in which only few exceptions were allowed. In effect for many years, the program has drastically reduced China's population growth, but more and more exceptions and violations are appearing in the system since it represents a drastic assault on human rights.

Fifth, some nations have relied primarily on advanced economic and educational development to bring about a "natural" decline in birth rates and population growth. Most of the nations of western Europe have achieved or nearly achieved

Figure 15.12 Standard world model, "Scenario 1." The world proceeds along its historical path as long as possible without major policy change. Population and industrial output grow until a combination of environmental and natural resource constraints eliminate the capacity of the capital sector to sustain investment. Industrial capital depreciates faster than new investment can rebuild it. As it falls, food and health services also fall, decreasing life expectancy and raising the death rate (from Meadows et al., 1992).

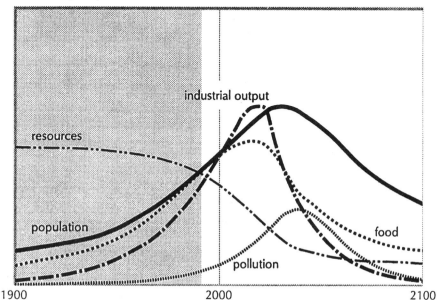

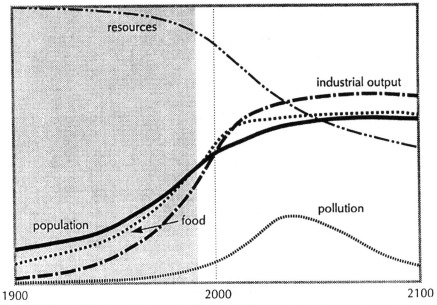

Figure 15.13 Stabilized population model, "Scenario 10," with technologies to reduce emissions, erosion, and resource use and with population planning adopted in 1995. The resulting society sustains 7.7 billion people at a comfortable standard of living with high life expectancy and declining pollution until at least the year 2100 (from Meadows et al., 1992).

zero population growth largely through improved economic and educational development. When the standard of living rises, educational programs reach certain levels of quality and sophistication, and medical services are universally available, birth rates have fallen. Germany and the Scandinavian nations are prime examples.

A sixth type of response to population growth has been increased research on contraceptive techniques to make birth control easier, cheaper, safer, and more widely available. In most nations of the world, biomedical research in the area of basic reproductive biology has been a rather low priority, certainly compared to research related to cancer, cardiovascular disease, and infectious disease. Some nations, however, particularly France, have persisted in attempting to develop new and better methods of birth control. A great deal needs to be done in this field, however, especially in regard to male contraceptives and disease prevention.

A seventh response, shown by some nations with declining birth rates, has been to provide incentives to couples to have more children. In recent years, Japan has been concerned about its aging population as fewer young people enter the work force. Government programs to encourage birth have been brought forth. The problems of achieving population stability are very difficult, and either rapidly increasing populations or the threat of declining populations represent serious economic situations.

The main message here is that the world must put more emphasis on programs of education, research, social services, and medical care to deal with issues of human reproduction in a humane, responsible, and balanced way. Our most difficult human problem may be trying to achieve population planning and fertility control in reasonable, ethical and culturally acceptable ways.

SUMMARY

The human population has exhibited a J-shaped, or geometric, growth curve over the past two thousand years, and the final course of this growth curve for the world as a whole cannot be predicted accurately. The global pattern is a complex mosaic of many different regional and national patterns. The highest rates of population growth are occurring in developing nations with the poorest economic conditions. The lowest rates are in more highly industrialized nations with high per capita incomes. Most of the nations of western Europe have achieved or will soon achieve zero population growth.

Population growth since the seventeenth century has occurred primarily because mortality rates have fallen more quickly than birth rates. In the least-developed countries, birth rates remain especially high, apparently due to the lack of education, medical care, women's rights, and contraceptive materials, all associations with poverty. The strong statistical relationship between high population growth and poverty has led some scientists to proclaim that population growth is the cause of poverty, whereas the relationship between development and reduced birth rates has led others to propose that poverty is the cause of high population growth.

The world has substantial momentum for continued population growth due to the age structure of many nations in which the great majority of people are of reproductive age and large percentages of the populations are young.

Some scientists feel that the world has already exceeded its optimal human population, which they calculate to be between 1 and 2 billion people. This estimate is based on the assumption of relative prosperity, or a reasonable middle-class standard of living, for all people in terms of the basic necessities and amenities of life.

A harsher view considers world population growth an ecopathological process comparable to cancer in an individual. This comparison is based on four or five basic system similarities between carcinogenesis in a living organism and population growth in the global ecosystem. Although this idea may be repugnant to our human psyche, it remains a powerful challenge for us to prove it fallacious.

Statistical models of human populations tend to project domed population growth, peaking in the twenty-first century and then declining due to resource depletion, pollution, or failing life-support systems. Such models do not predict what will happen, but they do project the logical outcome of present trends and policies.

There is no doubt that the issue of human population growth is one of the most important aspects of global change, but because it is a complex product of many philosophical, cultural, and emotional factors, it is not being dealt with adequately on the world scene. The subject must be approached more objectively and scientifically. There is an urgent need to bring human populations into better ecological balance with the life-support capabilities of the earth, and to enhance the quality of life for all people without ravaging our environment.

REFERENCES

Barney, G., T. R. Pickering, and G. Speth. 1982. *The Global 2000 Report to the President*. Washington, D.C.: U.S. Government Printing Office.

Cole, H. S. D., et al. 1973. *Models of Doom: A Critique of "The Limits to Growth."* New York: Universe Books.

Corson, W. 1990. *The Global Ecology Handbook*. Boston: Beacon Press.

Ehrlich, P., and A. Ehrlich. 1990. *The Population Explosion*. New York: Doubleday.

Eisley, L. 1961. "The House We Live In," WCAU-TV, Feb. 5, 1961. Quoted in I. McHarg, *Design with Nature*. Garden City, New York: Doubleday.

Forrester, J. W. 1971. *World Dynamics*. Cambridge, MA: Wright-Allen Press.

Glantz, M. 1992. What was unsaid at UNCED. *Boulder Camera*, July 2, 1992.

Global 2000: Technical Report. 1982. Washington, DC: Council on Environmental Quality.

Gregg, A. 1955. A medical aspect of the population problem. *Science* 121: 681–682.

Hern, W. 1990. Why are there so many of us? Description and diagnosis of a planetary ecopathological process. *Population and Environment* 12: 9–39.

Hern, W. 1993. Is human culture carcinogenic for uncontrolled population growth and ecological destruction? *BioScience* 43: 768–773.

Kendall, H., and D. Pimentel. 1994. Constraints on the expansion of the global food supply. *Ambio* 23: 198–205.

Lowe, M. 1991. *Shaping Cities: The Environmental and Human Dimensions*. Worldwatch Paper 105. Washington, DC: Worldwatch Institute.

McDowell, N., L. Wilhelm, and K. Wolf. 1992. The human dimension of global change. In *Pathways of Understanding: The Interactions of Humanity and Global Environmental Change*. Ann Arbor, MI: Consortium for International Earth Science Information Network.

Meadows, D. H., et al. 1972. *The Limits to Growth*. New York: Universe Books.

Meadows, D. H., D. L. Meadows, and J. Randers. 1992. *Beyond the Limits: Confronting Global Collapse and Envisioning a Sustainable Future*. Post Mills, VT: Chelsea Green.

Miller, G. T. 1991. *Environmental Science: Sustaining the Earth*. 3rd ed. Belmont, CA: Wadsworth.

Miller, G. T. 1992. *Living in the Environment: An Introduction to Environmental Science*. 7th ed. Belmont, CA: Wadsworth.

Pimentel, D., et al. 1994. Natural resources and an optimum human population. *Population and Environment* 15: 347–369.

ReVelle, P., and C. ReVelle. 1992. *The Global Environment*. Boston: Jones and Bartlett.

Sachs, A. 1995. Population growth steady. pp 94–95 in *Vital Signs*, eds. L. R. Brown, N. Lenssen, and H. Kane. New York: W. W. Norton.

Simon, J. 1992. There is no environmental, population or resource crisis (guest editorial). In Miller, G. T. 1992. *Living in the Environment: An Introduction to Environmental Science*, pp. 29–31. 7th ed. Belmont, CA: Wadsworth.

Weeks, J. R. 1992. *Population: An Introduction to Concepts and Issues*. 5th ed. Belmont, CA: Wadsworth.

WHO Bulletin. 1985. Geneva, Switzerland: World Health Organization.

World Almanac and Book of Facts. 1995. Mahwah, NJ: Funk and Wagnalls.

World Population Data Sheet. 1995. Washington, DC: Population Reference Bureau.

Chapter 16

WORLD FOOD
SUPPLIES

One of the first considerations in thinking about the pros and cons of human population growth is whether global agriculture can provide food for increasing numbers of people, over 85 million more people per year. Starvation conditions in Somalia, Sudan, Mozambique, and other nations of Africa have captured world attention in recent years, raising the question of ecological limits. Do these conditions reflect a degradation of local ecosystems to the point of collapse, or are they symptoms of strictly human economic and political failures? These are complicated questions, easily debated, not readily solved.

Some regions of the world produce surpluses of food, to the point where governmental curtailments of production are put in place, whereas other parts of the world suffer from chronic malnutrition. Most experts agree that 20 to 25 percent of the world's population, over a billion people, do not receive an adequate diet in either quantitative or qualitative terms (Kendall and Pimentel, 1994). One hundred and eighty million children in the world are underweight and malnourished (IFPRI, 1994).

Even in the United States, which produces a food surplus and where farmers are paid by the government not to grow crops in order to reduce the surplus, hunger is widespread. According to Dr. Larry Brown of Tufts University, 12 million children and 30 million adults in the United States are hungry (Grossfeld, 1993). One in 10 Americans is fed by some kind of government assistance. Those children and adults in the United States who do have access to food often eat too much of the wrong foods; such fat-rich, high-sugar diets contribute to the national problem of obesity in all age groups.

Despite the existence of famine and malnutrition, most of the world has shown remarkable success in producing food. On a global scale world agriculture has confounded dire predictions of widespread famine. Most of the nations of Asia, including China, Vietnam and India, have become self-sufficient in food production, and even export basic food grains, whereas 25 years ago they depended heavily on imported grains.

On the other hand, total world production of food grains per person has been unstable for the past 15 years, and has actually declined since 1984. For example, production in 1984 was 346 kg per person, whereas it was only 316 kg per person in 1992 and 305 kg in 1993 (Brown, 1995). This per capita decline was a result of population growth despite a slight increase in total grain production from 1.649 billion tons in 1984 to 1.697 billion tons in 1993. Thus the global picture of food supply is far from simple. Preliminary data for 1994 showed a slight rebound in world grain production to 311 kg per person, and both soybean production and meat production took an encouraging jump in 1994 (Brown, 1995).

This chapter will consider some of the problems, failures, successes, and future prospects of world agriculture.

MAGNITUDE OF THE PROBLEM

The dimensions of world food requirements are awesome. The agricultural biosphere is expected to produce daily food for 5.8 billion people plus 1.7 million new individuals per week. The human food requirement to provide basic nutrition is over 3.5 million tons of food per day, increasing at the rate of 83,000 tons daily (Brown et al., 1994). All of this must be done on the 11 percent of the earth's land surface that is suitable for agriculture. Since only 30 percent of the earth's surface is land, we are dependent on approximately 3 percent of the earth's total surface to meet human food requirements (Miller, 1992).

Currently, 1.5 billion hectares are cultivated for agriculture, less than 0.3 hectare per person (Meadows, Meadows, and Randers, 1992). This is less than one acre per person, barely enough to support a sparse Asian vegetarian diet, and certainly not enough to support an American or European diet or one with adequate protein, minerals, and vitamins. Many experts agree that about 0.5 hectare per person is needed for an adequate diet (Giampietro and Pimentel, 1993).

Theoretically, 2 to 4 billion hectares are cultivatable, but we have no guarantees that these additional lands have adequate soil and water resources for sustainable crop production. In the meantime, many agricultural lands under cultivation are declining in productivity due to soil erosion, water shortages, salination, desertification, and pollution.

In fact, some estimates indicate that approximately one-third of the world's arable land and forests have been lost in the past 40 years due to mismanagement and degradation (Giampietro and Pimentel, 1993). Based on the estimate that 0.5 hectare of farmland is necessary per person for a suitable human diet, the carrying capacity of the earth in terms of food that will supply a varied diet including meat is approximately 3 billion people (Pimentel et al., 1994). According to this estimate, the world is already seriously overpopulated. Many other factors enter into this equation, however. Global carrying capacity for a reasonable standard of living requires materials for housing and clothing, fresh water, energy, and many other resources. Hence, some scientists estimate that the true carrying capacity of planet earth if people are to enjoy a relatively prosperous lifestyle like that of middle-class America is only 1 to 2 billion people (Pimentel et al., 1994). Other options are available, of course. We could all convert to vegetarian diets, and thereby eliminate beef cattle, pigs, goats, sheep, and chickens, which consume grain and primary production. We could also choose to survive at a standard of living characteristic of India or China, where the annual per capita income is less than $500 and where the majority of people do not have automobiles, refrigerators, televisions, or telephones. Some of these relationships between ecology, economics, and the quality of life will be discussed in Chapter 21.

THE CONSEQUENCES OF FOOD SHORTAGES

The most dramatic result of food shortages is famine, the tragic starvation of a human population, as we have seen recently in several African countries. Two types of starvation conditions that are obvious are marasmus and kwashiorkor. *Marasmus* results from total caloric restriction and is manifested in its most dramatic form in a terrible wasting away of muscle and fat; both children and adults become virtual walking skeletons of skin and bone. *Kwashiorkor,* which is most acute in children, results from protein deficiency even when caloric intake per se is sufficient. Persons with kwashiorkor often have bloated stomachs or a puffy or bloated body resulting from edema. If children with marasmus or kwashiorkor are saved from starvation death by food supplementation, they may have long-term deficits in physical and mental development. The brain of a child requires adequate caloric and protein nutrition during critical stages of development; if food is not available at these times, the resulting deficits may be permanent.

Proper nutrition involves many ingredients other than simply calories and proteins. Lipids, minerals, and vitamins are all essential for normal growth and physiology. Table 16.1 shows in simplified form the functions, sources, and deficiency symptoms of the basic vitamins. For example, a tragic consequence of severe vitamin A shortage in children is permanent blindness, a condition sometimes seen in eastern India where the diet, dominated by rice, lacks sufficient fresh fruits and vegetables to provide requisite amounts of vitamin A. A few dollars' worth of vitamin A supplement per year can save the sight of children suffering from this deficiency. Similarly, adequate amounts of vitamin D in fresh milk can save children from the disabling bone deformities of rickets.

A BRIEF HISTORY OF FAMINE

Famines are severe food shortages that lead to starvation deaths. The earliest written reference to famine was in Egypt in 3,500 B.C., though certainly famines occurred before this time. The causes of famine have varied greatly. Often caused

TABLE 16.1

A synopsis of vitamins in human nutrition (Raven, Berg and Johnson, 1995).

Vitamin/Name	Function	Dietary Source	Deficiency Symptom
A (retinol)	Development of epithelium and visual pigments	Green vegetables, milk, liver	Night blindness, full blindness, skin problems
B-complex			
B-1	Coenzyme in CO_2 production in cellular respiration	Grains, legumes, vegetables, meat	Beri-beri, heart problems, edema
B-2 (riboflavin)	Coenzymes in metabolism	Many different food sources	Inflammation of skin and eyes
B-3 (niacin)	Coenzymes in energy transfer	Liver, lean meats, grains	Pellagra, inflammation of nerves, mental problems
B-5 (pantothenic acid)	Coenzymes in carbohydrate and fat metabolism	Many different food sources	Fatigue, loss of coordination
B-6	Coenzymes in amino acid metabolism	Vegetables, cereals, meats	Anemia, convulsions, irritability
B-12 Cyanocobalamin	Coenzyme in DNA/RNA production	Dairy products, red meat	Pernicious anemia
Biotin	Coenzyme in metabolism	Vegetables, meat	Depression, nausea
Folic acid	Coenzyme in metabolism	Green vegetables	Anemia, diarrhea
C	Development of collagen, connective tissue, immune system	Fresh fruits, especially citrus, green leafy vegetables	Scurvy; breakdown of skin, blood vessels, and connective tissue
D (calciferol)	Calcium absorbtion, bone formation, skin synthesis	Dairy products, cod liver oil	Rickets, bone deformities
E (tocopherol)	Antioxidant	Green leafy vegetables, seeds, margarine	Rare (aging?)
K	Blood clotting	Green leafy vegetables	Severe bleeding

by drought, they can also result from severe storms such as hurricanes or typhoons, flooding, volcanic eruptions, earthquakes, insect outbreaks, plant diseases, warfare, excessive population numbers in relation to landscape resources, or any combination of human and natural disasters.

Most of the famines we know about have occurred in the Old World, but this may be due primarily to historical recording. In 310 A.D. a famine that resulted in 40,000 deaths was described in Britain. In 1235 A.D. London alone recorded 20,000 deaths from a famine in which people ate grass and the bark from trees. Over 200 famines were recorded in Great Britain from 10 A.D. to 1846 A.D. The famous Irish potato famine of the mid-nineteenth century resulted in the death of more than 12 percent of the adult population of Ireland between 1846 and 1851.

Some of the worst famines in history occurred in China. In 1333 A.D. one region tallied 4 million dead, and in 1876, 9 to 13 million famine deaths occurred in China. Certainly many other famines occurred in China in the intervening years. India has also experienced numerous famines. In 1702, 2 million famine deaths occurred in India; in 1769–70, 10 million deaths; and in 1876–78, at least 5 million people died in a famine that affected an estimated 36 million people.

Throughout the Middle Ages in Europe famine was also recurrent. In Italy in 1347, famine followed epidemics of the Black Death because few workers were available to plant, till, and harvest crops. During the Thirty Years War in Europe, between 1618 and 1648, 30 percent of the peoples of Germany and Bohemia died as a result of war, plague, and famine.

Throughout the twentieth century, famines have recurred with tragic frequency. In 1920 China suffered severe food shortages that affected an estimated 20 million people, of which 500,000 died from starvation. In 1921–22 in the USSR, which was emerging from the Russian Revolution, starvation affected 20 to 24 million people, killing somewhere between 1.5 and 5 million. In 1928–29 China recorded 3 million starvation deaths, and during World War II, in 1943–44, India recorded 1.5 million. Even in the latter half of the century, famines have continued, primarily in Africa and Asia (see Table 16.2).

These recurrences of Malthusian predictions seem out of place in a modern world where agricultural technology produces food surpluses in some countries. Our response now is to meet famine crises with military action and airborne food supplements, as in Somalia in 1993. This is clearly a humane and sympathetic type of action, but it does not solve the long-term problem. We tend to attribute modern famines to political and economic mismanagement, but these are often only precipitating factors, sometimes masking the true causes.

Underlying human folly, however, there are frequently long-term patterns of environmental degradation, brought to the surface by short-term events such as droughts, floods, or wars. No permanent solutions to the problems of malnutrition and starvation will be found until we can deal with the environmental problems as well as the political and economic ones. In fact, the interconnections between environment, economics, and politics are so close and so complex that all three must be considered to solve the problems of famine.

TABLE 16.2
Some countries experiencing famines since 1950
(Brown et al., 1984).

Year	Country or Region	Estimated Deaths
1960–61	China	8,980,000
1968–69	Nigeria (Biafra)	1,000,000
1971–72	Bangladesh	430,000
1972	India	830,000
1973	Sahelian Countries	100,000
1972–74	Ethiopia	200,000
1974	Bangladesh	330,000
1979	Kampuchea (Cambodia)	450,000
1983	Ethiopia	30,000
1990s	Somalia, Sudan, Mozambique	unknown

WORLD FOOD PRODUCTION AND HUMAN POPULATION GROWTH SINCE 1950

The remarkable fact about global agriculture from 1950 to the late 1970s is that world food production increased more rapidly than human population growth. Local and regional food shortages occurred, as implied by the famine deaths in Table 16.2, but in the world as a whole food grains production per capita increased, actually achieving a peak in 1984 (Fig. 16.1). For example, food grain production from 1950 to 1984 increased at an annual rate of 2.9 percent, soybean production at an annual rate of 5.1 percent, meat production 3.4 percent annually, and world fish catch 4.0 percent annually (Brown et al., 1993). During this same period, world population growth averaged approximately 1.8 to 2.1 percent per annum, so that food grains production per person increased by almost 40 percent (Miller, 1992). There were and are exceptions to this favorable picture, but first we should consider how this remarkable circumstance was achieved. The accomplishment is known as the Green Revolution.

THE GREEN REVOLUTION

The Green Revolution refers to the application of science, technology, and management to world agriculture. The major features of the Green Revolution in the 1960s and 1970s were: (1) the development of new genetic strains of wheat, rice, and other crops to provide greater yield and faster growth, (2) increased use of fertilizers, insecticides and herbicides, (3) improved water management and irrigation, (4) increased use of mechanization, from human and animal tillage to the use of modern machinery in cultivation and harvesting, (5) larger fields and farms to increase efficiency in planting, tilling, and harvesting, and (6) better storage, distribution, and marketing systems for cash crops.

All of these developments involved the modernization and westernization of agriculture. They greatly increased efficiency but also carried a number of costs and risks. Among the costs were much higher capital expenses for machinery, energy, fertilizers, pesticides, and new stocks of hybrid seeds. In addition, mechanized agriculture was dependent on world petroleum prices. Inflation raised the cost of tractors, harvesters, and irrigation equipment, and all the paraphernalia of modern crop production became prohibitively expensive for many farmers. Energy expenditures skyrocketed, not just for gasoline, oil, and electricity but for transportation of fertilizers and pesticides, storage of temporary surpluses, and large-scale marketing of single crops rather than mixed crops for local consumption. Better able to meet these rising costs, as well as the costs of capital, agribusinesses moved in and took over from small landowners.

Nonetheless, the immediate benefits were enormous. Old, inefficient, and marginal operations became productive. Cash-crop surpluses were produced to provide great economic benefits. Electricity became more widely available, people were able to buy modern products, and housing and transportation improved. Farmers, especially middle-class or wealthy ones, experienced financial gains

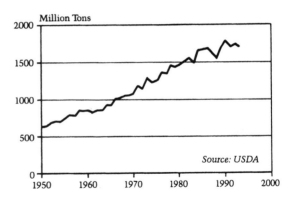

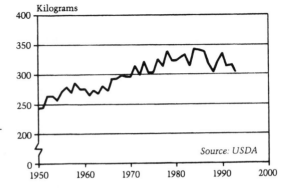

Figure 16.1 Total world grain production, 1950–93 (top) and world grain production per person, 1950–93 (bottom) (from Brown et al., 1994).

never before possible, and many aspects of village life improved. Improvements occurred provided everything worked according to plan, but often everything did not work according to plan.

RISKS OF THE GREEN REVOLUTION

Among the risks of the Green Revolution is the classic risk of putting too many eggs in one basket. Old-fashioned agriculture tended to be diversified; people raised many different crops on different planting and harvesting schedules. Even the same crop usually was grown from different strains of seeds. The result of this diverse small-scale agriculture was general inefficiency, but the system also had checks and balances that provided a certain type of insurance. In the event of natural disasters, for example, such as floods, insect plagues, or disease outbreaks, some crops or some strains might be affected, but others would usually survive. Also, the system did not depend on the price of petroleum on world markets or the availability of chemical fertilizers and pesticides. In other words, the system had some inherent safety in both biological and economic terms.

In contrast, the large-scale crops produced by the Green Revolution, usually consisting of one genetic strain planted on the same schedule over a broad region,

were often more vulnerable to short-term disasters such as hurricanes and typhoons out of season, monsoon seasons out of sync, disastrous floods, insect and
disease outbreaks, and wars. These events could wipe out an entire monoculture.
Old systems could adapt more readily to stochastic events. This vulnerability of a
modern system is not unique; many of our most advanced systems are technologi-
cally vulnerable; cities, for example, are subject to power outages. Such weak-
nesses do not prompt us to give up electricity. We need to be aware, however,
that when we increase efficiency and production, we may also increase risk.

Another problem with the Green Revolution is that it has not always proven to
be sustainable in some areas. In some localities it has produced surface and ground
water pollution, led to soil salination, or created environmental health problems
from excessive pesticide use (Kendall and Pimentel, 1994).

REGIONAL SUCCESSES AND FAILURES
OF THE GREEN REVOLUTION

The Green Revolution has had some of its greatest successes in Asia and Latin
America, and some of its worst failures in Africa. India, China, and Vietnam
have all seen remarkable increases in grain production, especially rice, and are
now major rice exporters. India relied on massive food grain imports from the
1950s through the early 1970s; for more than 20 years, the United States sent
India 20,000 tons of wheat daily through what was known as the Wheat Loan
Program under Public Law 480. India is now one of the top ten rice exporters in
the world.

Although the population of India more than doubled between 1959 and 1990
(from 397 million to 844 million people), rice production increased by a factor of
4.3 (from 24.8 million tons to 107.5 million tons), and wheat production increased
by a factor of 7.05 (from 7.6 million tons to 54 million tons). By the mid-1980s
India became self-sufficient in food grains production, so the wheat loan program
was stopped, and by the late 1980s India was an exporter of rice.

Wheat and corn production have also increased worldwide, and global production increases in all three grains have been due to both increased acreages under
cultivation and increased yields per unit of cropland. Yields have increased in
some countries but not others. China has been notable for yield increases in all
three major food grains: rice, wheat, and corn (Fig. 16.2). China has adopted
some features of the Green Revolution, especially new genetic strains of rice and
wheat, but it has stayed with traditional hand tillage because of its huge labor
force, and it has developed biological pest control to a fine art rather than rely
primarily on chemical insecticides.

France has had the most spectacular increases in wheat yields, and the United
States has had increases in corn yields in most years except those of severe
drought. Virtually all aspects of the Green Revolution have been used in the
United States.

Iran and Tanzania are examples of countries that have shown practically no

Thousand kilograms per hectare per year

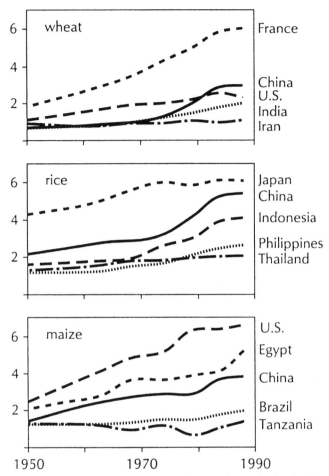

Figure 16.2　Grain yields for selected countries, 1950–90. Yields of wheat, rice, and maize (corn) tended to level off in the 1980s in the industrialized world, whereas they continued to rise in some developing nations such as India, Indonesia, and Egypt (from UN Food and Agriculture Organization).

improvement in wheat or corn yields, respectively, even though their populations have increased rapidly.

　In general, the greatest failures of the Green Revolution have occurred in Africa. Many nations in Africa have shown declining per capita crop production for the past 20 years (Fig. 16.3). Per capita grain production in Africa is down 22 percent since 1967. Both ecological disasters and human politics have contributed to this tragic record. In fact, many nations in modern Africa, afflicted by famine, disease, poverty, and warfare, are classic examples of the Malthusian scenario. Some scholars blame these problems on political mismanagement, racial injustices, tribal warfare, or a history of colonial subjugation; others point to droughts and meteorological catastrophes; and still others cite population pressures and land abuse. Probably all of these factors have contributed to Africa's problems; various combinations of human and natural events are leading Africa into a deteriorating situation that even the Green Revolution cannot relieve.

THE FUTURE OF WORLD AGRICULTURE

Can the Green Revolution continue? Can it build upon its successes to alleviate hunger and malnutrition? Can it provide food for over a quarter of a million more people in the world every day?

Technological optimists will say yes. They believe that we can meet world food demands if we now begin to capitalize on agricultural science and genetic engineering. Most of the advancements in the Green Revolution to date have relied on traditional techniques of plant breeding to produce superior varieties of crops, and on standard methods of modern agriculture, such as irrigation, fertilizer, pesticides, and mechanization. We have barely begun to use molecular biology and genetic engineering. The potential of genetic engineering in agriculture is tremendous. We can now engineer crops with greater yields, faster growth, early maturation, tolerance to water and temperature stresses, and resistance to diseases and insects. We can now design plants and animals for particular climates, for greater efficiency, for specific physiological traits such as nitrogen fixation, high protein yield, high fiber content, high vitamin content, low fat content, salt tolerance, low cholesterol content, and so on. The future is unlimited, and many exciting advances are possible. The otimists feel we have a great agricultural future ahead of us.

Not so, say the pessimists. No matter how many marvels emerge from the genetic laboratory, the realities of the world are much harsher, they say. Agricul-

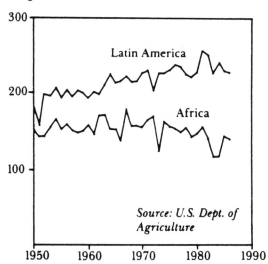

Figure 16.3 Per capita grain production in Africa and Latin America, 1950–86 (from USDA).

ture still requires water and soil, but water is becoming more fouled in much of the world, and soil is getting thinner and more depleted every day. Huge areas of cropland are being abandoned ever year.

The Food and Agricultural Organization of the United Nations reported in June 1993 that high-quality soil will soon be lost from 345 million acres of the world's cropland (an area twice the size of Texas) unless better farming practices are put in place (Stone, 1993). Even the great breadbaskets of the world, such as mid-western United States, which leads the world in wheat, corn, and beef production, is losing soil at a tremendous rate, withdrawing its ground water faster than it is replaced, and polluting its surface water unacceptably. The dual threats of flood and drought can devastate the best of modern agriculture!

In the meantime, say the pessimists, we are barely keeping up in the fight against plant pathogens and destructive insects. Both groups of organisms are constantly evolving resistance to pesticides and fungicides. Our present strategies of chemical control create increasing pollution and potential health problems.

Even most agricultural scientists, including Dr. Norman Borlaug, the Nobel laureate who developed the high-yield wheat strains that launched the Green Revolution, now grant that the Green Revolution cannot fulfill its promise without equal advances in family planning and population stability, and better management of global resources.

RECENT TRENDS

Another concern for the pessimists, or perhaps realists, is that the growth in agricultural productivity seems to be leveling off. Wheat and rice production in China showed a classic logistic curve in the 1980s, as did rice production in Indonesia and wheat production in France (see Fig. 16.2). Worldwide, the growth of food grains production in the late 1980s and early 90s slowed considerably. The annual rate of grain production from 1984 to 1992 was only 0.7 percent, compared to 2.9 percent from 1950 to 1984. Similar declines in annual growth rate have occurred in soybean and meat production (Brown et al., 1993).

Many questions emerge. Are these declines caused by a basic deterioration in agroecosystems through the loss of soil and water? Are they caused by pollution or toxic chemical accumulations? Are they a response to weather changes, or are they the result of human management problems that we can correct? No one can be sure. Locally, all of these situations may exist, but the global trends in food production should certainly alert us to the possibility of serious ecosystem problems.

Paul Vitousek (1992) and his colleagues at Stanford University estimated that agriculture and human activities are already utilizing or wasting 40 percent of all terrestrial net primary productivity in the world. Considering the extent of forests and natural grasslands, and the fact that the human population could double in less than 50 years, this is an astonishing figure, and it smothers comforting thoughts about the future ability of agriculture to supply food for 10 to 15 billion people on earth. This estimate raises again the issue of global carrying capacity for people and livestock. Perhaps this carrying capacity is not as great as we may think. Kendall and Pimentel (1994) estimate that global food production is adequate to support 7 billion people if we all adopted a vegetarian diet and fed no grain to livestock.

FOOD FROM THE OCEANS

There has always been a certain scientific and economic faith that the oceans will provide great sources of food if we can properly develop and harvest these sources. Undoubtedly the oceans can be a boon to us, and limited examples of aquaculture with salmon, oysters, and shrimp have demonstrated some of their potential. Both fresh and saltwater aquaculture showed increases in 1994, most of which occurred in China (Platt, 1995). However, most nations are still fishing the oceans as roving predators rather than cultivators. This predatory mode of taking all we can get is showing troublesome signs of decline or even collapse in some areas.

First of all, the total world fish catch per person has been declining since 1988. From 1950 to 1988 the annual growth in world fish catch averaged 4 percent; from 1988 to 1992 it averaged *minus* 0.8 percent (Brown et al., 1993). This occurred despite intensified fishing efforts. The Food and Agricultural organization of the UN estimates that "all 17 of the world's major fishing areas have either reached or exceeded their natural limits" (Weber, 1993).

A number of important fisheries have suffered disastrous declines (ReVelle and ReVelle, 1992). The California sardine industry collapsed in the 1950s, the North Sea herring industry in the late 1960s, the west African pilchard industry in the early 1970s, and the Peruvian anchovy in the mid-1970s when its catch declined from over 13 million tons per year to less than 1 million tons (Weber, 1994). Many coastal and estuarine food sources, including oysters and shad in coastal United States, have declined throughout the twentieth century. Many of the fisheries that have been depleted by a combination of overfishing, pollution, and habitat loss will never recover, or if they can be restored, such recovery may take many years. The great promise of unlimited food production from the sea has not been realized. Some of the ecological factors responsible for this will be discussed in Chapter 19.

SUMMARY

Global food production has shown some remarkable advances in the last 50 years, but there are many exceptions to this general trend. Some areas of the world produce food surpluses and arbitrarily reduce production to maintain market prices. Nonetheless, hunger and famine still occur in many places at different times. Approximately 20 to 25 percent of the world's people have inadequate diets. There are now ominous signs that global food production can no longer be counted on to keep up with population growth. World food grains production per person has actually declined since the mid-1980s.

The successes of world agriculture thus far have been largely due to the Green Revolution, the application of modern science, technology, and management to agroecosystems. The Green Revolution has involved new genetic strains of high-yield crops; greater use of fertilizers, irrigation, and pesticides; mechanization; and the development of agribusiness. The main benefit has been a remarkable increase in crop yields; the main costs have been high capital expenditures, increasing dependency on world energy supplies, increases in risk, increasing pollution, and losses of small farms. The Green Revolution faces great challenges to maintain its record of remarkable growth; its future will most certainly involve the application of molecular biology and genetic engineering. World agricultural growth has been slowing, or even receding, in recent years, and we do not know if this is due to the deterioration of agroecosystems, changing weather patterns, or human management problems. In any case, current trends are worrisome.

Hopes for unlimited food supplies from the oceans are not being realized, and actual declines of seafood production have occurred worldwide. It is apparent that world food supplies will be a limiting factor on the quality of life in many countries, and perhaps on population numbers as well.

REFERENCES

Brown, L. R., et al. (eds). 1984. *State of the World, 1984*. New York: W. W. Norton.
Brown, L. R., et al. (eds). 1988. *State of the World, 1988*. New York: W. W. Norton.
Brown, L. R., et al. (eds). 1993. *State of the World, 1993*. New York: W. W. Norton.
Brown, L. R., H. Kane, and D. M. Roodman. (eds.) 1994. *Vital Signs*. New York: W. W. Norton.
Brown, L. R. 1995. Grain production rebounds. pp 26–27 in *Vital Signs*. eds. L. R. Brown, N. Lenssen, and H. Kane. New York: W. W. Norton.
Ehrlich, P., and A. Ehrlich. 1990. *The Population Explosion*. New York: Doubleday.
Giampietro, M., and D. Pimentel. 1993. The tightening conflict: Population, energy use, and the ecology of agriculture. Paper presented at NPG Forum, Oct. 1993, Teaneck, NJ.
Grossfeld, S. 1993. Wasting away: America's losing battle against hunger. *Boston Globe*, July 25, 1993, pp. 36–39.
IFPRI. 1994. Washington, DC: International Food Policy Research Institute.
Kendall, H., and D. Pimentel. 1994. Constraints on the expansion of the global food supply. *Ambio* 23: 198–205.
Meadows, D., D. Meadows, and J. Randers. 1992. *Beyond the Limits: Confronting Global Collapse and Envisioning a Sustainable Future*. Post Mills, VT: Chelsea Green.
Miller, G. T., Jr. 1992. *Living in the Environment*. 7th ed. Belmont, CA: Wadsworth.
Pimentel, D., et al. 1994. Natural resources and an optimum human population. *Population and Environment* 15: 347–369.
Platt, A. 1995. Aquaculture boosts fish catch. pp. 32–33 in *Vital Signs*. eds. L. R. Brown, N. Lenssen, and H. Kane. New York: W. W. Norton.
Raven, P., L. Berg, and G. Johnson. 1995. *Environment*. Philadelphia: Saunders College Publishing.
ReVelle, P., and C. ReVelle. 1992. *The Global Environment: Securing a Sustainable Future*. Boston: Jones and Bartlett.
Stone, R. 1993. FAO sounds soil loss siren. *Science* 261: 423.
Vitousek, P. M. 1992. Global environmental change: An introduction. *Ann. Rev. Ecol. and Syst.* 23: 1–14.
Weber, P. 1993. *Abandoned Seas: Reversing the Decline of the Oceans*. Worldwatch Paper 116. Washington, DC: Worldwatch Institute.
Weber, P. 1994. *Net Loss: Fish, Jobs, and the Marine Environment*. Worldwatch Paper 120. Washington, DC: Worldwatch Institute.

Chapter 17

AIR POLLUTION

In the previous chapters we have considered several ways in which human populations are impacting planet earth—land degradation, desertification, deforestation—and the historical aspects of environmental destruction. In the next three chapters on pollution, we look at human impacts on the atmosphere, hydrosphere, and climate. Air, water, and soil pollution affect food production, human health, and virtually every aspect of our lives. Pollutants also play key roles in global warming (or global cooling) and climate change.

Everyone is familiar with pollution in daily life and in newspaper and television coverage. But despite its familiarity, pollution is not that easy to define. Some ecology books don't try to define it (ReVelle and ReVelle, 1992; Smith, 1992; Tudge, 1991), perhaps assuming that everyone knows what pollution is. Others don't even list it in the index (Christensen, 1984; Rambler, Margulis, and Fester, 1989). Still others define pollution primarily in human terms: "Environmental pollution is the unfavorable alteration of our surroundings, wholly or largely as a by-product of man's actions" (Kormondy, 1984).

Is there, however, pollution from nonhuman sources? What about ash from a volcanic eruption, smoke from a forest fire caused by lightning, oil from a natural seep, sulfurous odors from a thermal vent or hot springs? Most scientists would consider these pollution, although they have nothing to do with human activities.

A generally accepted definition of pollution is "an unwanted change in the atmosphere, water, or soil that can harm humans or other organisms" (Raven, Berg, and Johnson, 1995). Most definitions of pollution use the terms "unwanted," "undesirable," or "unfavorable" in referring to environmental alterations. All of these words involve a value judgment, but in most cases pollution is obviously undesirable. It smells or looks bad; it assaults our senses in various ways. Pollution can also be less obvious. We may not become aware of it until some undesirable symptom occurs, such as cancer caused by years of exposure to side-stream smoke or some type of airborne radiation.

SOURCES OF AIR POLLUTION

Most of the sources of air pollution are well known: automobiles, trucks, and buses; factories and power plants; oil refineries, gas stations, and petrochemical facilities; dumps and landfills; airports, docks, marinas, and bus stations; mines and construction sites; farms, feedlots, and fallow fields; forest, grassland, and brush fires; overgrazed fields and desertified landscapes; and an endless array buildings, houses, yards, lawn mowers, and household activities. The city of Los Angeles has even proposed limiting the use of backyard barbecues because they are sources of urban air pollution. In much of the world, considerable amounts of air pollution come from village cooking and household fires.

Sources of air pollution can roughly be grouped into two types: "point-source" and "non-point-source." *Point-source pollution* comes from an easily identifiable, usually fixed source, such as a factory's or power plant's smokestacks. Point-source pollution is usually the easiest to recognize and do something about. *Non-point-source pollution* comes from innumerable small sources, often mobile, such as cars, trucks and buses, or from broad extents of landscape, such as eroded fields that produce clouds of dust in windstorms, or thousands of miles of city streets that produce polluted runoff in rainstorms. Admittedly, the distinction between point-source and non-point-source pollution is not always clear. A jet airplane, for example, is a point source of air pollution, but because it moves across the country and because there are hundreds of aircraft criss-crossing the skies at any one time, air travel, like automobile travel, can be thought of as a non-point-source of air pollution. In general, transportation is the greatest source of air pollution in United States, followed by industrial sources (Raven, Berg, and Johnson, 1995). In personal exposures, however, smoking exceeds virtually all environmental sources.

TYPES OF AIR POLLUTION

There are basically four types of air pollution, with innumerable varieties of each and combinations of the four: (1) particulate, (2) gaseous, (3) photochemical, and (4) radioactive.

Particulate air pollution is the most obvious. It includes smoke, dust, and certain forms of haze or mist. Some particles, such as chimney soot and agricultural dust, are heavy and settle out if not kept aloft by air currents. Others, such as fine pollen grains and mold spores, are so fine that they remain in colloidal suspension in the air.

Gaseous air pollution is infinitely complex and usually invisible unless it has a particulate component. It may be odorless, or it may have a foul smell. We cannot see or smell carbon monoxide or most of the oxides of nitrogen, but they are serious air pollutants. Hydrogen sulfide is an example of an air pollutant that we cannot see, but it has unpleasant odor, so we are at least strongly aware of its presence. In general, many of the gaseous air pollutants are the most difficult to control and produce the most serious ecological effects. Table 17.1 lists some common air pollutants, their sources, and effects.

Both photochemical smog and radioactive air pollution can be a combination of gaseous and particulate air pollution. Photochemical smog represents an interaction of nitrogen oxide and hydrocarbons such as methane or ethane, stimulated by

TABLE 17.1
Examples of some common air pollutants.

Pollutant	Sources	Some General Effects
Particulates	Smoke from combustion of coal, petroleum, wood, volcanoes, etc. Dusts from agriculture, mining, ranching, milling, etc. Metallic dusts (e.g., lead, beryllium) Aerosols from mists, fogs, sulfates, etc.	Irritant to living organisms Reduce sunlight Form "droplet nuclei" for precipitation
Sulfur dioxide	Combustion of coal and petroleum	Toxic to plants Forms acid rain Respiratory irritant
Nitrogen oxides	Automobile exhaust	Respiratory irritants
Carbon monoxide	Automobile exhaust	Asphyxiant (reduces oxygen in blood)
Ozone	Power generation Industrial sources	Powerful oxidant Toxic to plants Respiratory irritant
Hydrocarbons	Automobile exhaust Petrochemical plants	Contribute to smog Potentially carcinogenic
Hydrogen fluorides	Petroleum Glass Aluminum and fertilizer industries	Respiratory irritants
Hydrogen sulfides	Petroleum refineries Chemical industries	Respiratory and eye irritant
Ammonias and phosgenes	Chemical industries	Respiratory irritants
Arsines	Metal industries	Respiratory irritants
Chlorines	Textile industries	Respiratory irritants
Aldehydes	Thermal decomposition of fats or oil	Respiratory irritants
Photochemical smog	Nitrogen oxides, hydrocarbons, and sunlight	Eye, throat, and respiratory irritant
Benzene	Gasoline Petroleum	Leukemia
Benzo-a-pyrene	Coal Petroleum Soot	Carcinogenic

the energy of sunlight to form peroxyacetyl nitrate, or PAN, under the following general equation (an oversimplification, and not a balanced equation):

nitrogen oxide + methane + ethane + sunlight → peroyxacetyl nitrate

$$NO_4 + CH + C_2H_6 + sunlight \rightarrow C_2H_3O_5N$$

PAN is an irritating gas that combines with urban dust and mists to form the characteristic smog of Los Angeles and the brown cloud of Denver. Urban smog can also result from the combustion of fuel oil to form the typical gray smog of cities in the eastern United States and Europe.

Radioactive air pollution also comes in a gaseous form, as with radon, a radioactive gas resulting from the decay of uranium minerals in rocks and soil, or a particulate form, as the machine milling of plutonium from former nuclear weapons factories such as Rocky Flats, Colorado.

OTHER WAYS OF CLASSIFYING AIR POLLUTION

There are several other ways of classifying air pollution. One is by its geographic extent: local, regional, intercontinental, or global. An example of local air pollution is that surrounding an industrial plant or downwind of factories emitting pollution. Staten Island is downwind of petrochemical plants in New Jersey, and its air carries a lot of gaseous pollutants. Urban air pollution can be local or regional depending on its extent. (See Box 17.1). In and around Los Angeles, where pollution extends to the San Gabriel and San Bernardino mountains east of the Los Angeles basin, it becomes a regional problem.

International air pollution occurs, for example, when industrial pollution from Chicago and Detroit impacts Ontario and eastern Canada, or when air pollution from the industrial Midlands of England causes problems in Scandinavia.

Examples of global air pollution are found in the effects of carbon dioxide and methane on global warming or of chlorofluorocarbons on the ozone shield. We will consider these effects in Chapter 18.

Air pollution can also be classified as primary or secondary. Primary air pollution is caused by the original pollutant entering the air. An example is nitrogen oxide or hydrocarbons from automobile exhaust. Secondary pollution, in this case, is caused by PAN, which results from the chemical reaction of the two original pollutants in the air. In another example, sulfur dioxide from coal burning is a primary pollutant, which in the presence of fog droplets or rain becomes weak sulfuric acid under the following simplified reaction:

$$SO_2 + H_2O \rightarrow H_2SO_4$$

Thus, acid rain and sulfate formation is a secondary air pollutant (Fig. 17.1). This example also illustrates the close linkage between air pollution and water pollution.

BOX 17.1
Daily Life and Urban Air Pollution in Developing Nations, July 10, 1996

Dateline: Beijing, China

At 12:00 o'clock noon on a hot summer day with no clouds, the sun is barely visible. A grey haze envelops the city, dispersing direct sunlight, creating a drab appearance to the entire cityscape. High-rise apartments penetrate the haze, and construction cranes stick up like weeds. The streets are crowded with buses, trucks, and cars, not as bad as in Calcutta or Cairo or Jakarta, but a hundred times worse than just 15 years ago when bicycles dominated city streets as well as country roads. Yu Lin winds her way through the stalled traffic just to cross the boulevard, her eyes smarting and the gauze over her mouth already grey from dust and smoke. Her throat feels sharp and sore despite the gauze mask, but this is a regular part of life in Beijing.

Dateline: Mexico City

The evening is cool, but Pedro feels hot as he labors just trying to breathe. He is 14, the fifth child in a family of 11, but he looks more like a 9-year-old. He sits on the edge of a cot, hands on his forehead, each breath a stifled attempt to get enough air. His ribs ache, and he wishes only for the chance to breathe freely and lie down. Why do these attacks come on in the evening? Why does the hint of rain bring them on? Why can't he get some fresh air? Why doesn't he get better as he grows older? Why can't someone help him? Why is life so tough in the largest city in the world?

Dateline: Calcutta, India

Rain and wind lash the city for the tenth hour, flooding streets to the curb. Most underpasses are blocked with deeper water. Pedestrians and rickshaw wallahs try to navigate between gridlocked cars on Chittaranjan Avenue and Lower Circular Road, amid a terrible din of horns, shouts, and driving rain. Mukherjee knows it will be many hours before he can get home in the Alipore civil lines, long after dark. He can usually make the commute from his office on Park Street in just under two hours by a combination of subway, bus, and rickshaw. His clothes are soaked, he has a bothersome cough, he feels alternately feverish and chilled, but these are all part of life in Calcutta. The monsoon clears the air temporarily, but the soot and grime will be back on Mukherjee's handkerchief tomorrow when the sun will return to steam-broil Calcutta's 14 million inhabitants.

CHS

ENVIRONMENTAL INFLUENCES ON AIR POLLUTION

The distribution and severity of air pollution is greatly influenced by geography and weather. Cities and industrial areas provide concentrated sources of air pollutants. Wind can spread pollutants to other regions or concentrate it in vortices. Hills and valleys can restrict airflow, and air-temperature patterns can determine whether pollutants disperse upward or accumulate at ground level.

The most obvious accumulations of air pollution occur in cities during stagnant-air episodes or inversions. Normally air is stratified in terms of temperature with warm air at the earth's surface and cooler temperatures occurring at higher elevations. This layering permits warmer, polluted air to rise and diffuse into upper levels of the atmosphere. Temperature inversions occur when layers of warmer air overlay cooler layers at the ground level, trapping polluted air near the ground. If

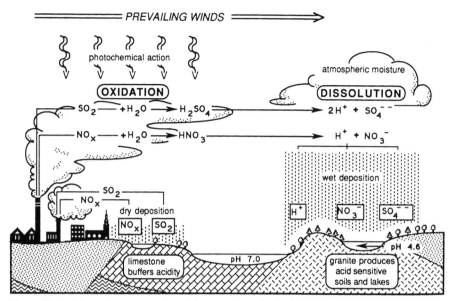

Figure 17.1 Schematic representation of the formation, distribution and impact of acid rain (from Kemp, 1994).

cities lie in basins, or have mountains on the leeward or downwind sides, as Los Angeles does, air inversions and stagnant air can trap pollutants for several days, causing dangerous buildups of foul air.

A common form of air pollution often overlooked is the problem of indoor air. Since most people spend the majority of their time in buildings, indoor air quality is an important consideration. Indoor air quality is affected by many things: ventilation, building materials, paints, carpets, upholstery, drapes, synthetic furniture, and human activities such as smoking. One of the greatest concerns in indoor air quality is the radioactive gas, or radon. Radon is a naturally occurring decay product of uranium in the earth's crust. It is odorless and tasteless but readily inhaled in air or ingested in water. It emits ionizing radiation in the form of alpha particles that do not penetrate the human skin to an appreciable extent but enter the body through the lungs or the gastrointestinal tract. Over a long period of time radon emissions are carcinogenic. Outdoors radon is so diluted and dispersed by air currents that it is of no particular concern. In houses, however, it may accumulate to potentially dangerous levels. Indoor levels are usually measured in picocuries (a picocurie is one-trillionth of a curie). The EPA has set 4 picocuries as a common and safe level. If indoor air is measured at 20 picocuries, the lung cancer risk is estimated to be equivalent to smoking one pack of cigarettes a day; at 100 picocuries equivalent to two packs per day, and at 200 picocuries equivalent to 4 packs of cigarettes per day (Miller, 1993). Of approximately 140,000 human deaths per year from lung cancer in the United States, approximately 5,000 to 20,000 of these can be attributed to radon exposure in homes and workplaces (Raven, Berg, and Johnson, 1995).

The geographic and topographic distribution of radon in rocks and soil varies nationwide. Radon testing has now become a factor in real estate sales and building construction.

Building materials are also of importance in indoor air pollution. Asbestos is probably the most famous example, since it was widely used as insulation in older buildings before its fibers were known to be carcinogenic. Asbestos represents an expensive problem, whether it is removed or sealed in place so that fine asbestos fibers do not become airborne. Sometimes removal creates greater problems than sealing it in place, and sealing is often recommended if the material is already dusty or shredded.

ECOLOGICAL EFFECTS OF AIR POLLUTION

The effects of air pollution can be grouped into four broad categories: (1) effects on natural ecosystems, (2) effects on agroecosystems, (3) effects on human health, and (4) effects on man-made materials. Each of these may be subdivided as we will see in the following material.

Effects on Natural Ecosystems

Ecosystems can be affected directly by air pollution or indirectly by acid rain. It is difficult and often impossible to separate the effects of these agents since they act together in nature. In laboratory studies, however, a clear distinction can be made. Dramatic direct effects have been seen in the pine forests of the San Bernardino and San Gabriel mountains east of Los Angeles. Trees are very active in gaseous air exchange, absorbing ambient air and modifying it as they utilize carbon dioxide and release oxygen. Hence, forests have frequently been called "the lungs of the world" (Caldicott, 1992; Maini, 1990). All components of gaseous air enter the leaf interior and may affect the plant. Oxides of sulfur and nitrogen, fluorides, aldehydes, ozone, hydrocarbons, and many other pollutants have actual or potential toxic effects on plant physiology. Ponderosa pines *(Pinus ponderosa)* in the San Bernardino mountains have been weakened by air pollution and acid rain, making them more susceptible to insect attacks and plant pathogens. The same process has occurred in some forests of eastern United States, Central Europe, and Scandinavia. One of the most drastic examples of air pollution totally destroying vegetation occurred in Copperhill, Tennessee, where fumes from a copper smelter killed all trees and other plants over a sizable area downwind of the smelter (Odum, 1993).

In the Sudbury District of Ontario, Canada, two million tons of air pollution annually from mining operations have killed many trees in an area of 1900 square kilometers. White pine *(Pinus strobus),* one of the native timber trees, now exists on only 8 percent of the productive land (Goudie, 1990). In Swansea, Wales, air pollution from coal burning has caused extensive loss of vegetation.

In Norway, air pollution from aluminum smelters has caused the deaths of pines *(Pinus sylvestris)* along a valley for 26 kilometers. In the San Bernardino mountains, ponderosa pine deaths due to air pollution from Los Angeles have occurred as far as 129 kilometers to the east of the city (Goudie, 1990).

In Germany, forest death, known as "waldsterben" or "waldschaden," has occurred extensively as a result of air pollution and acid rain. In 1985, 55 percent of German forests reported air pollution damage. Different tree species are affected differentially. Fir trees in western Europe are highly susceptible, and Wellburn (1988) reported that 87 percent of white fir *(Abies alba)* showed air pollution damage in western Germany, whereas 51 percent of Norway spruce *(Picea abies)*, 59 percent of Scotch pine *(Pinus sylvestris)*, 50 percent of beech (*Fagus* spp.), and 43 percent of oak (*Quercus* spp.) were also damaged. Smaller percentages of damage were seen in France, Switzerland, Austria, and Italy. In Switzerland, 25 percent of the fir trees and 10 percent of the spruce have died in recent years as a result of air pollution or acid rain. Eastern Europe has areas even more severely damaged by air pollution, but data are hard to find and compare.

Areas of forest destruction due to air pollution can be found throughout the world. Even a remote portion of western Tasmania, south of the mainland of Australia, has extensive forest death from mining and smelting activities. Once-forested hillsides near Queenstown present a picture of total devastation.

The alteration or destruction of forests is just one aspect of air pollution's effects on natural ecosystems. Ecological ramifications occur throughout the biotic community, affecting soil, water, microbial populations, and animal life. If forests are completely destroyed as they were in Copperhill, Tennessee, or Queenstown, Tasmania, then, of course, the entire ecosystem is lost. Usually, the changes are less apocalyptic but still drastic.

Sulfur dioxide emissions in some localities have produced acid rain with pH values as low as 2.3, near Kane, Pennsylvania, in 1978, and 2.4, in Scotland in 1974 (Gates, 1985). These levels are more acidic than vinegar. In New England and Scandinavia, acid precipitation has often been in the range of 4.0 to 5.0, still acidic enough to increase the leaching of toxic aluminum, lead, mercury, and cadmium from rocks and soils, so that all living organisms are affected. The toxic effects of acid rain are closely related to natural concentrations of these metallic ions in soil and rock substrates.

Acid rain effects have been particularly evident in the Adirondack and Green mountains of New York and Vermont and in the lakes of Norway and Sweden. In 1975, 51 percent of Adirondack lakes above 610 meters elevation had pH values less than 5.0, and 90 percent of these lakes had no fish populations (Re-Velle and ReVelle, 1992). In 1930 only 4 percent of these lakes were without fish. In Norway and Sweden, 16,000 lakes have lost their fish populations in the last 50 years, and in Canada 14,000 are fishless or virtually without fish, apparently due to acidification. In Minnesota, Wisconsin, and the upper Great Lakes, 80 percent of the lakes are threatened by excessive acidity (Miller, 1993).

As lake and stream acidity increases, virtually all other aquatic organisms decline: rotifers, bryozoans, sponges, turbellarians, annelids, molluscs, crustaceans, insects, salamanders, frogs, and turtles. The loss of these animals also affects waterfowl populations, and many of these acidified regions have experienced 60 percent declines in waterfowl populations (ReVelle and ReVelle, 1992). There are other reasons for declining waterfowl numbers, and declining bird populations in general, but acid precipitation is one among several. In 1995, waterfowl populations have partially rebounded in the U. S., but this may be only a temporary increase due to favorable rainfall patterns.

Effects on Agroecosystems

Many of the same toxic effects of air pollution and acid precipitation on forests and natural vegetation occur on agricultural crops. High concentrations of ozone, sulfur dioxide, fluorides, and the oxides of nitrogen can cause direct leaf injury to corn, wheat, soybeans, and many vegetable and fruit crops. In addition, acid precipitation dissolves vital nutrients from the soil, especially calcium, potassium, and magnesium, to further reduce productivity. Recent estimates of air pollution damage indicate that U.S. crop production is lowered 5 to 10 percent, representing agricultural losses in the United States due to air pollution in the range of $1.9 to $5.4 billion per year (Miller, 1993).

Ozone, a common urban air pollutant that is windborne from cities into surrounding countryside, is particularly damaging to agricultural crops. Even in sparsely populated rural areas, ozone concentrations have been measured as high as 0.038 to 0.065 ppm. Experimentally it has been found that increasing ozone levels from 0.04 to 0.09 increases damage to soybean plants from 7 to 31 percent (Brown and Young, 1990). The 1987 grain harvest in North America, Europe, and China was estimated to be reduced from 913 to 865 million tons due to ozone pollution alone (Brown and Young, 1990).

The mention of ozone in the preceding sentence leads us into the air pollution effect with the greatest potential to alter both natural and agricultural ecosystems. Ozone is a confusing compound. At the earth's surface, this three atom-molecule of oxygen is an active oxidant, a serious pollutant damaging to humans, plants, animals, microorganisms, and materials. In the upper atmosphere, however, ozone provides a protective shield against excessive ultraviolet radiation. The destruction of ozone in the stratosphere by chlorofluorocarbons (CFCs) is one factor in global warming. Coupled with increased carbon dioxide, methane, and nitrogen oxides, the destruction of the ozone layer will contribute to significant climate change. We will discuss the possibilities and probabilities of this in the next chapter.

Effects on Human Health

The effects of air pollution on human health can be considered in two broad categories: acute, or short-term, and chronic, or long-term. The two categories overlap, and acute episodes may play a role in manifesting or exacerbating some of the long-term effects of air pollution.

History provides many examples of acute air pollution episodes that caused illness or death. In 1948 a persistent blanket of air pollution over Donora, Pennsylvania, resulted in respiratory illness in 14,000 people; coughs, shortness of breath, chest pain, eye and nose irritations were common, and 20 deaths were attributed to the pollution. In 1952 a famous smog hovered over London, causing distress in hundreds of thousands of people and approximately 4,000 deaths (Godish, 1990). London air quality at that time was much worse than it is now, since there was widespread burning of coal to heat individual homes and virtually no controls on industrial and automobile emissions. Many deaths were attributed to pneumonia, asthma, bronchitis, and heart problems, all of which were aggravated by bad air. Hence, many deaths and illnesses were a combination of an acute stress acting on more long-term effects of air pollution.

An acute episode of a less serious nature showing the interaction of pollution and infectious disease occurred in New York in 1962, when stagnant air produced a sudden increase in sulfur dioxide concentrations in lower Manhattan. A sharp increase in colds, coughs, and pharingitis occurred (Fig. 17.2). An epidemiological investigation linked the onset of upper respiratory infections to an increase in sulfur dioxide air pollution (McCarroll et al., 1966).

The link between urban air pollution and chronic lung disease has been studied and debated at great length, and it is not as clear as that between smoking and lung disease. However, there is increasing evidence that urban air pollution can contribute to emphysema, asthma, chronic bronchitis, and even lung cancer. Some of these effects are particularly evident in smokers. Chronic bronchitis and asthma are characterized by restricted airways, as irritants in the air cause inflammatory swelling in the bronchi and bronchioles. Emphysema results from damaged alveoli which impair oxygen transfer. In all three types of lung disease, breathing is a frightening struggle. Both particulate and gaseous air pollution can stimulate the development and onset of these diseases. Urban air pollution causes a fibrosis (thickening) of the alveolar wall and an impairment of ciliary function in the upper respiratory tract.

Air pollution may also be a contributory factor in lung cancer, and heart disease, although the evidence for this is not entirely clear. Cardiovascular diseases and cancer are the two leading causes of death in the United States, the former accounting for approximately 890,000 deaths per year, and cancers approximately 530,000 deaths per year. Together, these two categories of illness represent over 60 percent of mortality in the United States. Of all cancer deaths, lung cancer alone accounts for more than 100,000. Without question, it is one of the leading types of cancer.

Evidence for the role of ambient air pollution in heart disease and cancer is mixed. Some studies based on comparisons of urban and rural populations show higher rates of heart disease and cancer in cities. But many other factors are involved, including crowding and stress, so it is difficult or impossible to assess the effects of air pollution alone. Other studies are based on statistical analyses of disease rates in urban areas with various indices of air pollution. For example, one study in Los Angeles showed a correlation between heart disease rates and levels of carbon monoxide air pollution (Bodkin, 1974). Urban areas in Great Britain show higher lung cancer rates than urban areas in New Zealand with similar population densities but less pollution (Godish, 1990).

On the basis of a study of human health and air pollution in Philadelphia, Joel Schwartz of the EPA and Douglas Dockery of the Harvard School of Public

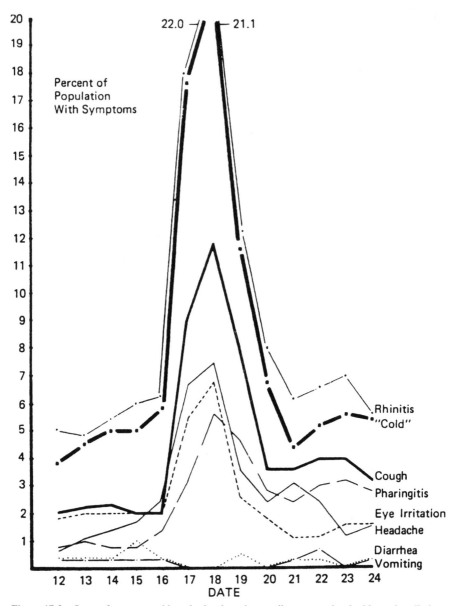

Figure 17.2   Onset of common colds and related respiratory disease associated with an air pollution episode in New York City in December 1962 (from McCarroll et al., 1966).

Health estimated that air pollution is responsible for 60,000 deaths in the United States each year, primarily among people with concurrent lung disease (Schwartz and Dockery, 1992).

Mortality rates in New York city increase 2 percent on high pollution days and decrease 1.5 percent on clear air days. Air pollution episodes exacerbate asthma attacks, emphysema, pneumonia, and tuberculosis, and cause additional heart strain. The National Academy of Science estimates that 4,000 premature heart disease deaths occur annually from air pollution.

The connection between cancer, heart disease, and smoking has been evident for more than 30 years, and now evidence implicates passive smoke, or sidestream smoke, for individuals who have frequent exposure to tobacco smoke even though they do not smoke. For example, sidestream smoke affects children whose parents smoke in the home and car. The tobacco and cigarette industry challenges this finding, but extensive reviews by the National Academy of Science, Environmental Protection Agency, and World Health Organization all indicate that passive smoke is a health risk. The tobacco industry markets cigarettes worldwide, directing quite a bit of its advertising effort at young people. Whereas cigarette smoking by adults in the U.S. has been decreasing, throughout the developing world it has been increasing markedly.

Although the overall data on urban air pollution and chronic disease are difficult to evaluate and sometimes confusing, there is no doubt that certain components of air pollution are carcinogenic. These components include asbestos in insulating materials, benzene and benzo(a)pyrene from petroleum fumes, radon, aromatic hydrocarbons, B-napthylamine, arsenic dust, plutonium dust, and many other synthetic and natural materials that can become airborne.

Although air pollution control is expensive, studies in environmental and health economics have estimated that each $1 billion spent in the United States on the control of air pollution could potentially save $1.6 billion in medical costs.

Effects on Materials

In the 1970s it became tragically evident that air pollution and acid rain were damaging buildings and outdoor art treasures in Europe at an alarming rate. The president of the Greek Academy of Science reported in 1971 that the Parthenon had deteriorated more in the past 50 years than in the previous 2,000. The statuary on many of Europe's finest cathedrals and castles show the corrosive effects of air pollution. It is common now to see prominent buildings, such as Notre Dame in Paris, Parliament in London, and the Parthenon in Athens, shrouded in scaffolding as workers repair the damage suffered by stonework, brickwork, and sculpture from air pollution (Fig. 17.3).

The physical damage caused by air pollution is diffuse and extensive. It includes the deterioration of paint, metals, stonework, plastics, rubber, fabrics, and virtually all other physical materials. This damage, coupled with the effects of air pollution on natural and agroecosystems, human health, and even global climate, makes air pollution expensive to deal with, but even more expensive to ignore.

CONTROL OF AIR POLLUTION

The control of air pollution is a multidisciplinary subject combining many aspects of the environmental sciences, including engineering, atmospheric chemistry, sociology, and economics. The engineer and atmospheric scientist must devise ways

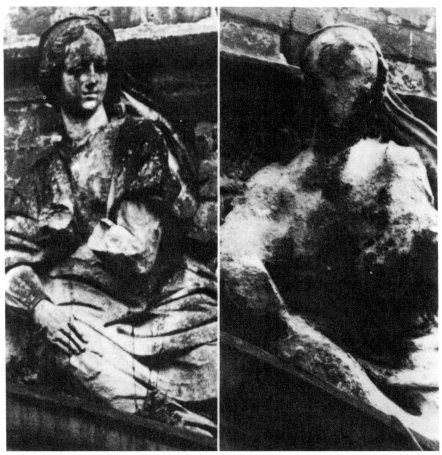

Figure 17.3 Corrosive effects of acid rain on stone statuary in the industrial region of the Rhein-Ruhr. This sandstone sculpture is at Herten Castle, built in 1702 near Recklinghausen, Germany. (left) Appearance in 1908, (right) appearance in 1969 (from AAAS, Washington, DC).

to inactivate, reduce, or remove air pollutants from industrial smokestacks, power plant emissions, automobiles, and agricultural, commercial, and domestic operations of all kinds. In many cases, they have done this with remarkable competence and efficiency (Chow and Connor, 1993). Control of pollution from many industrial and transportation sources has been achieved through the use of precipitators, filters, converters, and scrubbers. Hence, many of our power stations and industrial plants are much cleaner now than they were 30 or 40 years ago. Automobiles operate with greater efficiency and less pollution.

This is not to say that all of the engineering problems related to pollution control have been solved, but many have, and the main problems remaining are matters of public will and economics. Are we ready and willing to pay the bill for air pollution control? In some cases, legislation, such as the federal Clean Air Act of 1970, has forced the issue, but ways can be found to ease or defeat regulations if pollution control costs are perceived to be too high. A company can decide to close down a plant and go overseas. The general public can refuse to buy more expensive items, if pollution control costs are passed on to the consumer. The public can also initiate tax limitation measures to avoid paying for pollution control systems where public funding is involved. Public awareness of the costs and benefits of pollution control is thus a critical factor in its success. A major problem here, as in many environmental programs, is that the costs of pollution control are obvious and immediate, whereas the benefits are usually diffuse and long term.

Despite the delayed benefits of air pollution control, the economic advantages are substantial. A 1992 study by Dr. Jane Hall and her colleagues of California State University estimated that air pollution in the Los Angeles basin is responsible for 1,600 deaths per year and entails additional medical costs of $10 billion annually. The estimated costs of cleaning up this air pollution to the level of federal standards, which will presumably eliminate most of the additional medical health care costs, range from $5 billion to $21.5 billion and average $9.8 billion. Thus air pollution control could possibly pay for itself within a year or two based on health care costs alone, not to mention cost savings from reduced damage to natural ecosystems, agricultural crops, and physical materials (Hall et. al., 1992).

TRENDS IN AIR POLLUTION

Trends in air quality have both a bright and a dark side. From 1940 to 1980, the level of urban particulates in the United States improved substantially, from emissions of 21.9 million tons to 7.8 million tons. Most of our cities became noticeably cleaner in appearance. In some cases, such as Chicago, Pittsburgh, and Youngstown, part of this improvement was due to the collapse of the iron and steel industry, and related smokestack factories. In other cases, it was due to the use of cleaner fuels and better particulate control such as electric power stations. Without question, most of our cities, as well as those in western Europe, have better air now because of less industrial smoke.

The trend in gaseous pollution, especially from non-point sources such as automobiles, is generally worse, however. In the same 40-year period from 1940 to 1980, the annual emission of sulfur oxides increased from 17.4 million metric tons to 23.7, nitrogen oxides from 6.5 to 20.7 million metric tons, hydrocarbons from 13.9 to 21.8, and carbon monoxide from 74.7 to 85.4 million metric tons (Blumenthal, 1985). Figure 17.4 shows the intermittent rise of the oxides of sulfur and nitrogen from 1900 to 1980 in the United States.

Throughout the world, both particulate and gaseous air pollution has increased from many sources. The rapid industrialization of the developing world has been one major factor, along with tremendous increases in motor vehicles and agricultural machinery (United Nations Environment Programme, 1994). Other major factors include increased land scarring, overgrazing, desertification, and deforestation. Forest fires in both tropical and temperate nations, either accidental or deliberate, as a method of clearing forestland for pasture or cropland, have contributed substantially to both particulate and gaseous air pollution. Many areas, both urban and rural, are often enveloped in an atmospheric haze, which can be seen from

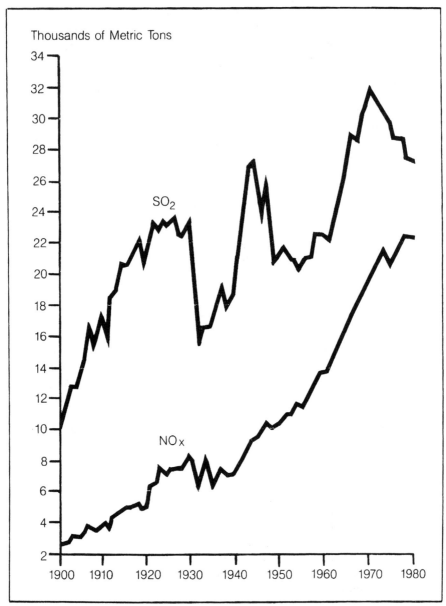

Figure 17.4 Annual Emissions of SO_2 and NO_x in the coterminous United States, 1900–80 (from U.S. Environmental Protection Agency).

aircraft and spacecraft. On the Indonesian portion of Borneo in the mid-1980s, smoke from forest fires was so pervasive that Garuda, the national airline of Indonesia, had to terminate scheduled air service to five towns because of poor landing visibility. Many airline pilots in the United States and Europe experienced increased smog problems around airports throughout the 1970s and 1980s. Despite general improvements in urban particulate pollution, photochemical smog became a problem. Astronaut Paul Weitz commented in a 1983 spaceflight: "Unfortunately, this world is becoming a grey planet. . . . Our environment apparently is going downhill. . . . We are fouling our own nest."

Air pollution becomes a global influence in its worldwide effects on atmospheric chemistry and physics. The rise in global carbon dioxide levels, nitrogen oxides, sulfur oxides, methane, and chloroflurocarbons all have global implications in climate change. These topics will be discussed in the next chapter.

SUMMARY

Air pollution occurs in four basic forms—particulate, gaseous, photochemical, and radioactive—as well as combinations of these. It comes from a variety of sources: industrial, agricultural, domestic, and natural. Air pollution can also be classified according to origin (point-source or non-point-source), geographic distribution (local, regional, global), and physical or chemical complexity (primary or secondary).

The impacts of air pollution can be measured on natural ecosystems, agroecosystems, human health, and physical materials. The ecologic and economic costs of air pollution damage are often insidious, diffuse, and long term. They include serious damage to forests, soils, lakes, streams, and all components of terrestrial and aquatic ecosystems; to human health, ranging from mild respiratory distress to fatal chronic disease; and to physical materials from fabrics to art treasures and buildings.

Air pollution control is possible in many cases, especially where the source of pollution is factories or automobiles. Costs are usually immediate and expensive whereas benefits are long term and may be slow to be realized.

Air pollution trends have shown great improvements in particulate pollution in the cites of industrialized nations in North America, western Europe, and Japan, but continued deterioration in many gaseous pollutants. In the developing nations of Africa, Asia, and Latin America, both particulate pollution and gaseous air pollution are generally getting worse. This is a result of rapid industrialization without adequate pollution controls, rapid increases in motor vehicles, expanding agriculture, and poor-quality fuels.

Air pollution has major influences on climate, and the next chapter will consider the controversial topics of the Greenhouse Effect and global warming.

REFERENCES

Bodkin, L. D. 1974. Carbon monoxide and smog. *Environment* 16(4): 35–41.
Blumenthal, D. L. (ed.). 1985. *Introduction to Environmental Health*. New York: Springer.
Brown, L. R., and J. E. Young. 1990. Feeding the World in the Nineties. In *State of the World, 1990*, ed. L. R. Brown. New York: W. W. Norton.
Caldicott, H. 1992. *If You Love This Planet: A Plan to Heal the Earth*. New York: W. W. Norton.
Chow, W., and K. Connor. 1993. *Managing Hazardous Air Pollutants*. Boca Raton, FL: Lewis, CRC.
Christensen, J. W. 1984. *Global Science: Energy, Resources, Environment*. 2nd ed. Dubuque, IA: Kendall-Hunt.

Gates, D. M. 1985. *Energy and Ecology*. Sunderland, MA: Sinauer Associates.

Godish, T. 1990. *Air Quality*. 2nd ed. Boca Raton, FL: Lewis, CRC.

Goudie, A. 1990. *The Human Impact on the Natural Environment*. 3rd ed. Cambridge, MA: MIT Press.

Hall, J. V. et al. 1992. Valuing the Health Benefits of Clean Air. *Science* 255: 812–817.

Kemp. D. D. 1994. *Global Environmental Issues: A climatological Approach*. 2nd ed. London: Routledge.

Kormondy, E. J. 1984. *Concepts of Ecology*. 3rd ed. Englewood Cliffs, NJ: Prentice-Hall.

McCarroll, J. R., et al. 1966. Health and urban environment: Health profiles versus environmental pollutants. *American Journal of Public Health* 56(2): 226–275.

Maini, J. S. 1990. Forests: Barometers of environment and economy. In *Planet Under Stress*, ed. C. Mungall and D. J. McLaren. New York: Oxford University Press.

Miller, G. T. Jr. 1993. *Environmental Science*. 4th ed. Belmont, CA: Wadsworth.

Odum, E. P. 1993. *Ecology and Our Endangered Life Support Systems*. 2nd ed. Sunderland, MA: Sinauer Associates.

Rambler, M. B., L. Margulis, and R. Fester. 1989. *Global Ecology: Towards a Science of the Biosphere*. New York: Academic Press.

Raven, P. H., L. R. Berg, and G. B. Johnson. 1995. *Environment*. New York: Saunders College Publishing.

ReVelle, P., and C. ReVelle. 1992. *The Global Environment*. Boston: Jones and Bartlett.

Schwartz, J., and D. Dockery. 1992. Increased mortality in Philadelphia associated with daily air pollution. *Amer. Review of Respiratory Disease* 145: 600–604.

Smith R. L. 1992. *Elements of Ecology*. 3rd ed. New York: HarperCollins.

Southwick, C. H. (ed). 1985. *Global Ecology*. Sunderland, MA: Sinauer Associates.

Tudge, C. 1991. *Global Ecology*. New York: Oxford University Press.

United Nations Environment Programme and World Health Organization. 1994. Air pollution in the world's megacities. *Environment* 36(2): 5–13, 25–37.

Wellburn, A. 1988. *Air Pollution and Acid Rain: The Biological Impact*. New York: John Wiley.

Chapter 18

WEATHER AND
CLIMATE CHANGE

Of all the controversial issues in global ecology, few have been as contentiously debated in recent years as the subject of global climate change. Some scientists have expressed great concern about the greenhouse effect and global warming. They believe that air pollution is leading to uncontrolled global warming that will modify climates throughout the world with serious ecologic and economic effects. They foresee the possibility of the western deserts of the United States expanding while other parts of the world will be devastated by rain and floods.

Other scientists deny any firm evidence for pollution-related global warming, and believe instead that world temperature changes are within natural fluctuations, or may actually be returning to a global cooling cycle. They foresee a possible return of the northern hemisphere to a little ice age. This was a common view 25 years ago, and still prevails in some scientific and popular circles.

Climate changes have been a part of the earth's history since long before the advent of human beings, and most scientists feel that it is difficult to distinguish at the present time between natural and human-induced climate change (Canby, 1994). There is no doubt that global temperatures have been increasing for the past 200 years (Thompson, et al., 1995), and this warming trend has accelerated in the last 2 decades. There is not agreement, however, on why this is occurring or how far it will go, but most scientific evidence now points to pollution-related global warming with serious ecological and economic impacts (UN, IPCC, 1995).

So where does this leave us? How should the ordinary citizen view these controversies? As in many areas where experts disagree, the best approach is to try to gain some background on the principles involved and look as objectively as possible at data and trends.

There seem to be at least two forces contributing to global warming and two contributing to global cooling. Those contributing to global warming are (1) the buildup of atmospheric carbon dioxide, and other so-called "greenhouse gases", and (2) the thinning or destruction of stratospheric ozone. Forces potentially contributing to global cooling are (1) the Milankovitch cycles of axial and orbital variation, and (2) particulate air pollution and atmospheric turbidity or aerosols.

FORCES FOR WARMING: THE GREENHOUSE EFFECT

Solar radiation, in the form of visible and shortwave ultraviolet radiation, enters the earth's atmosphere and strikes the surface of the earth. Much of this radiation is converted to heat in the form of longwave, or infrared, radiation which is reradiated into the atmosphere. Hence, the earth is warmed and life is possible.

The amount of infrared radiation and heat that is retained in the atmosphere surrounding the earth is a function of many factors, including the type and amount of cloud cover and the concentration of carbon dioxide and other gases in the air. Carbon dioxide, methane, and nitrogen oxides act as heat traps; the greater their concentration, the more heat retained. Cloud cover does not act quite so simply. Some types of clouds contribute to atmospheric heat retention, but other types reduce surface temperatures by blocking incoming solar radiation. We will consider this complication later in this chapter.

Although the greenhouse effect has gotten a bad name in recent years, a stable greenhouse effect is actually necessary for life on earth. Without the heat-retaining qualities of the earth's atmosphere, temperatures on earth would be well below freezing, as on Mars. Hence, carbon dioxide is an essential atmospheric component for both physical and biological reasons, that is, for heat retention and for photosynthesis.

The current problem in the greenhouse effect is increasing levels of atmospheric carbon dioxide and other "greenhouse gases." The classic graphs of carbon dioxide buildup recorded at monitoring stations in Hawaii and Antarctica are now included in almost every textbook of environmental science and global ecology. Ice core data from the Siple station in Antarctia show a rise in carbon dioxide from the eighteenth century to the present. The Hawaii records are more recent but show the same annual increase with seasonal variations: more carbon dioxide in the northern hemisphere in winter months when trees are not photosynthesizing and when more fuel is burned to heat buildings (Fig. 18.1). If this trend continues, a likely result will be greater atmospheric heat retention. The earth's atmosphere acts like a greenhouse; solar energy comes in, is converted to heat, and retained.

Over the past 200 years, from preindustrial times to the present, atmospheric carbon dioxide concentrations have risen from approximately 280 parts per million (ppm) to 350 ppm today, an increase of 25 percent, most of which has occurred in the last 50 years. This has resulted from increasing industrial activity in the world, more motor vehicles, increasing human and domestic animal populations, and increasing destruction of forests and grasslands. The chemical processes of oxidation, combustion, and decay all add carbon dioxide to the atmosphere, and the only major counteraction of carbon dioxide uptake and utilization is biological photosynthesis. We have been active in speeding up the formation of carbon dioxide while impairing its natural recycling.

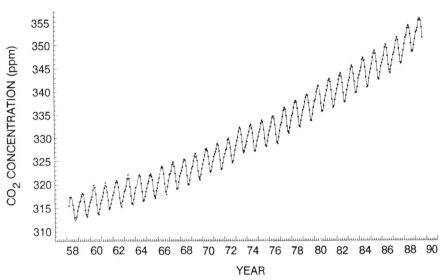

Figure 18.1 Mean monthly concentrations of atmospheric CO_2 at Mauna Loa, Hawaii, 1958–89 (from American Geophysical Union).

Carbon dioxide is not the only greenhouse gas. Other atmospheric gases that retain infrared heat are methane (CH_4), nitrogen oxides (N_2O), and ozone (O_3), all of which are increasing as a result of pollution and human activities. Water vapor is also a very significant atmospheric heat trap, a function of clouds and especially humidity.

Chlorofluorocarbons, or CFCs, from old refrigerators and air conditioners, industrial solvents, and styrofoam, and even some old aerosol cans, are considered greenhouse gases, but they act in a different way. The mode of action of the CFCs is to rise into the upper atmosphere where they are broken down by high-energy ultraviolet light to release chlorine. The free chlorine atom reacts with ozone in the following way to breakdown ozone:

$$CL + O_3 \rightarrow ClO + O_2$$

Chlorine monoxide then combines with a free oxygen atom to release chlorine and oxygen. The chlorine atom repeats this process in a chain reaction:

$$ClO + O \rightarrow Cl + O_2$$

Calculations in quantitative chemistry indicate that each atom of chlorine can destroy 100,000 molecules of ozone before it is finally neutralized. It takes about 10^{18} molecules of CFCs to make just one styrofoam cup. Thus the potential for ozone destruction from such products is tremendous.

The net result of these reactions is partial destruction of the ozone layer in the upper atmosphere. Although ozone is a pollutant and oxidant at the earth's ground level, it forms a protective layer in the stratosphere by filtering dangerous UV light. Thinning of the ozone layer, and the formation of definite ozone holes over the polar regions, allows more solar radiation to enter and thereby contributes to global warming. There is firm scientific evidence now that the ozone holes over the polar regions are caused by human-induced air pollution. Stratospheric ozone in the Antarctic has dropped by 50 percent since 1991, and in the arctic by 10 to 15 percent.

There are other consequences of destruction and increased radiation in addition to global warming. Increased UV radiation is detrimental to many organisms, both aquatic and terrestrial. The U.S. Environmental Protection Agency predicts decreased yields of corn, wheat, rice, and soybeans due to the destructive action of increased UV radiation. The EPA also reports danger to marine food chains from lethal effects of increased UV on ocean phytoplankton.

Human health effects, primarily in the form of skin cancer and cataracts, are also a concern. A significant increase in human skin cancer has already occurred in Australia; this is attributed to rises in UV radiation due to the proximity of the Antarctic ozone hole (Caldicott, 1992). In the United States, the EPA estimates that a 6 to 7 percent ozone depletion in the northern hemisphere would cause an additional 2 million new cases of basal-cell and squamous-cell carcinomas per year and an additional 30,000 cases of malignant melanoma per year (Miller, 1992).

The effects of global warming on human health involve other acute and long-term problems. Acute effects include increased deaths during heat waves. In July 1995, for example, a heat wave in midwestern United States with several days of temperatures around and above 100° F (39 to 40° C) caused more than 600 human deaths, primarily of elderly people within urban dwellings without air conditioning. Many of these people had predisposing heart and respiratory problems, but heat stress was a precipitating factor in their deaths.

Long-term consequences of global warming also include the potential expansion of tropical diseases into subtropical and temperate regions. This could occur with infectious diseases such as malaria, schistosomiasis, filariasis, trypanosomiasis, leishmaniasis, dengue fever, and yellow fever (Stone, 1995).

IS GLOBAL WARMING A REALITY?

Over the past 100 years, average global surface temperatures have risen erratically about one degree Fahrenheit, from approximately 58.5° F to 59.5° F (Fig. 18.2). Proponents of the greenhouse effect take this increase as evidence of global warming due to pollution and human effects. Computer modeling of CO_2 effects on surface air temperatures globally show an even more pronounced effect (Fig. 18.3, from Hansen et al., 1993). Subsurface earth temperatures also show a significant increase. Underground temperatures from thousands of boreholes all around the world at depths of 20 meters have shown average increases of 0.6° to 0.8° C from 1890 to 1990 (Pollack and Chapman, 1993). In North America, over 350 boreholes have shown temperature increases of 0.5°C to 2.5°C (Deming, 1995). Even the Arctic Ocean has warmed 1° C compared to a few years ago according to a joint U.S.–Canadian expedition (Travis, 1994). Most of the world's glaciers have

Temperature change (°C) compared to 1951–80 average

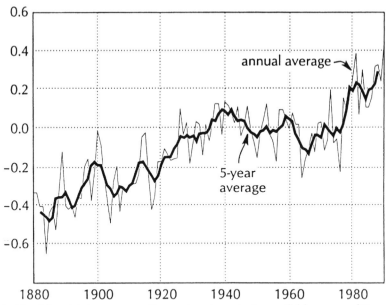

Figure 18.2   The rising global temperature. Temperature change in degrees Celsius compared to 1951–80 average (from Meadows et al., 1992).

receded in both the northern and southern hemisphere during this same period in places as widely separated as Alaska, Colorado, Bolivia, Nepal, New Zealand, and Peru (Hansen et al., 1993).

Critics point out that long-term historical records show many periods of increasing and decreasing global temperatures, and they maintain that the rise of the last 100 years is within the range of natural variation (Canby, 1994). CO_2 and temperature records from the Vostok Antarctic ice cores show remarkable correlation, with a cold period 150 to 160 thousand years ago, an abrupt rise in global temperatures about 140 to 130 thousand years ago, and an erratic decline leading to the northern hemisphere ice ages around 20 thousand years ago. Since the most recent ice ages, both CO_2 levels and global temperatures have risen substantially (Fig. 18.4). These historical changes have all occurred without the driving force of human impact, so far as we know. There is nothing new about global climate change; what is new is the speed of change and human impact (Stern et al, 1992).

Another point of criticism of anthropogenic climate forcing is that many temperature-monitoring stations are in or near cities, and it is well known that cities form local "heat islands" due to the higher temperatures generated by bricks, mortar, and pavement. These records do not necessarily represent the temperatures of surrounding countryside. This reliance on urban temperature readings has pushed global records falsely higher, according to critics (Balling, 1993).

Despite these criticisms, the best records we have show an increase in global temperatures over the last 200 years. The increase is not steady, but erratic, with many dips in any given year. The controversy centers around whether the overall increase in the past one hundred years is related to human-induced air pollution or whether it is entirely within the range of natural variation. At this point, we cannot say if an anthropogenic greenhouse effect is already here, or even if it will come. Most scientists feel it is likely if present trends continue (Schneider, 1993; Rosensweig and Hillel, 1993), and many feel it is inevitable (Hansen et al., 1993; Caldicott, 1992; McKay and Hengeveld, 1990), but others doubt its existence (Lindzen, 1993; Michaels, 1993). The doubters question the models on which human-induced global warmings are based.

If increased greenhouse warming is occurring, or is likely to occur, one global consequence often mentioned is rising ocean levels. A warmer earth would melt glaciers and polar ice caps, contributing to a rise in sea level. Ocean levels are very difficult to measure since they depend in part on shoreline stability. Around landmasses with sinking shorelines, the ocean level appears to be rising if measured from land; in land areas with rising shorelines, the ocean level appears to be falling. Hence, ocean level records in different parts of the world show different trends. In Baltimore over the past 75 years, the ocean level has apparently risen about 25 centimeters (approximately 10 inches), but in Stockholm on the Scandinavian peninsula, the ocean level appears to have fallen (Fig. 18.5).

Different statistical models of global warming and sea level rise by the year

Figure 18.3 Surface air temperature changes in a global climate model based on measured increases of greenhouse gases (CO_2, CH_4, CFC, N_2O, and O_3). The model has a sensitivity of 4° C for doubled CO_2 and includes the thermal inertia of the oceans (National Geographic Society, 1993).

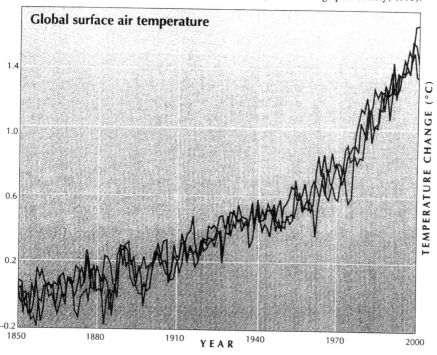

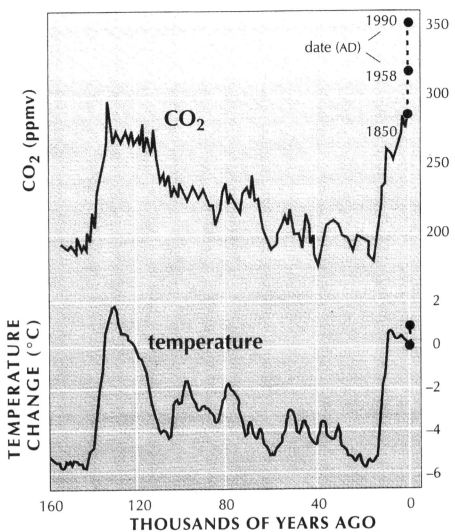

Figure 18.4 Carbon dioxide and temperature records from Antarctic ice cores over the past 160,000 years, based on the Vostok ice core. Recent atmospheric measurements have been added as dashed lines (from J. Hansen et al., 1993).

2050 project global temperature rises on the order of 1.5° to 4.5° C (3° to 9°F) (ReVelle and ReVelle, 1992). Rises of 3 to 7 feet could be disastrous for many shoreline areas, especially during hurricanes and typhoons, and many of the world's greatest coastal cities, including New York, Washington, D.C., London, Rotterdam, Bombay, Singapore, Tokyo, Rio de Janeiro, and numerous others would be in jeopardy. Some low-lying oceanic islands, such as the Maldives in the Indian Ocean, and many coral atolls in the Pacific Ocean would be inundated and virtually submerged if this were to occur.

The Caribbean island of Trinidad has been experiencing rising ocean levels. In some areas the sea has encroached 800 meters inland in the last 60 years, and significant amounts of land have been lost (Bloch, 1994). In Key West, Florida, however, nearly 150 years of sea level measurements have failed to show any significant change (Bloch, 1994). With half of humanity living within 40 miles of a seacoast, the economic stakes of ocean level change are incredibly high, and human concerns are understandable. The fact is that burgeoning human impacts on the earth are creating a first-time global experiment of unknown outcome.

The tropical Pacific Ocean began a warming phase in 1976–77 associated with

Figure 18.5 Ocean level changes on the North American east coast at Baltimore, and the Scandinavian peninsula at Stockholm (from Silver and DeFries, 1990).

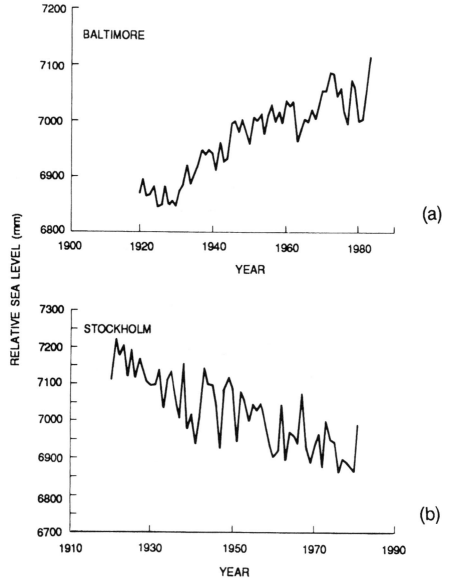

El Niño climatic oscillations, which have been especially pronounced in the southern hemisphere. These oscillations have been attributed to a vast warm-water pool in the southern Pacific that has recurred at periodic intervals to bring major climate changes to neighboring landmasses. The warming phases have usually resulted in extensive droughts in the western Pacific (Indonesia and Australia) and record-breaking rainfalls in the eastern Pacific (the west coasts of North and South America). Since 1976 there have been four, possibly five, El Niños, but inexplicably three of the last four years (from 1990 to 1994) have shown the warm oceanic waters characteristic of El Niño (Kerr, 1994). Scientists do not understand what is forcing or driving these episodes of Pacific warming, but some feel their increased frequency may be an indication of global warming. In any case, El Niños have global effects; the record temperatures of the 1980s, as well as the drastic California rainstorms and floods of 1995, were possibly triggered by the recurrent El Niños (Kerr, 1994). These warmer water temperatures have also been associated with serious declines in populations of marine plankton, fish, and sea birds (Hill, 1995).

FORCES FOR GLOBAL COOLING

At least two types of global forces can contribute to a global cooling trend: (1) Milankovitch cycles, broad cycles in the earth's movement within our solar system, and (2) atmospheric dust and turbidity, which shade solar input.

Milankovitch cycles are long-term astronomical cycles of axial and orbital variation which have been postulated as basic causes of glacial cycles in earth's geological history. An axial cycle of approximately 22,000 years is a cyclic tilting of the earth's axis that determines the direction of solar radiation. If the northern hemisphere axis tilts away from the sun slightly, it effectively reduces solar radiation to the northern hemisphere and results in a significant cooling effect. Since glacial and interglacial periods have lasted about 20,000 years, this axial cycle has traditionally been proposed as the basic cause of the northern ice ages (Crowley and North, 1991).

There are also geologic records of longer-term climate cycles, with deglaciations, or warm periods, occurring over spans of 100,000 years. Some attempts have been made to link these cycles to orbital variations, long-term cycles in the earth's elliptical orbit around the sun, which influence the amount of solar radiation striking the earth. Recent studies of geological deposits in Nevada, however, which were based on oxygen isotope ratios in calcite deposits over a 500,000-year period, fail to show an accurate correlation of climate changes with the presumed Milankovitch cycles (Winograd et al., 1992). These studies indicated that major climate shifts responsible for glaciation and intervening warm periods were aperiodic; that is, they did not occur on a regular cyclic basis, as the Milankovitch hypothesis predicts. Basically, the reasons for dramatic climate changes over the long span of geological time are unknown.

The lack of a close correlation between Milankovitch cycles and climatic periodicities does not negate the possibility that axial and orbital variations of the earth have strong climatic influences. In fact, it is likely that such variations are major driving forces in climatic change which would tend to lead us toward global cooling at the present time. But such astronomical variations within our solar system are not the only factors in historical climate changes. Winograd and his colleagues (1992) concluded that the earth's climatic cycles also originate from "internal nonlinear feedbacks" between the atmosphere, the ice sheets, and the oceans. In other words, what happens to the atmosphere is very important, but as we will see in Chapter 19, and as we have seen with the El Niño oscillations, the ocean is a tremendously powerful factor in many aspects of global change (Kagan, 1995). The interactions of the atmosphere and the ocean, along with changes in the biosphere and lithosphere, must be superimposed on axial and orbital variations. Collectively, they are the final determinants of global climates. Few, if any, ecological phenomena are simple, and we do not yet understand most aspects of global climate change.

ATMOSPHERIC TURBIDITY

Another potential cooling factor for the earth's climate is atmospheric dust, smoke, and haze. In Europe and the United States, industrial smoke has lessened in the past 50 years, but it has increased sharply in much of the developing world with the rapid growth of industry and motorized transportation. Smokestacks belch from hundreds of thousands of factories throughout Asia, Africa, and Latin America; millions of household fires of wood, coal, straw, and dung patties permeate the air; and most of the world's motor vehicles add dramatically to urban smog (Fig. 18.6).

Burning forests also add significantly to atmospheric turbidity. Forest fires in the Amazon basin have been visible from spacecraft for many years, fires in Borneo in the 1980s burned vast areas of tropical forest for more than five years, and extensive fires in Yellowstone National Park in 1988 had smoke plumes stretching over 600 miles eastward into Nebraska and southward to Denver. Mean

Figure 18.6 Diagrammatic representation of the sources and types of aerosols producing atmospheric turbidity. Volcanic ash is a major source following eruptions, such as that of Mount Pinatubo in the early 1990s, whereas agricultural dusts and smoke are becoming of greater importance on a continual, long-term basis (from Kemp, 1994).

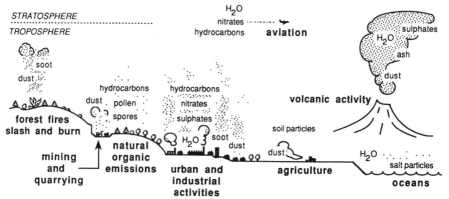

daytime temperatures in Denver showed a slight but measurable drop of one or two degrees attributed to the shading effect of smoke on solar input. The smoke plumes from burning oil wells in the Persian Gulf War of 1991 were detectable throughout India and even as far as Hawaii (Williams, Heckman, and Schneeberger, 1991), but they had no measurable effect on global climate.

Volcanic activity can also thrust incredible amounts of smoke and ash into the global atmosphere. One of the most famous historical cases was the eruption of Mount Tambora in Indonesia in 1815. The huge clouds of smoke and dust it produced blocked sunlight to the extent that temperatures in North America and northern Europe were reduced. In 1816 New England experienced the "year without a summer" when frosts occurred in July and August and virtually all agricultural crops failed. Global effects were also seen following the eruption of Krakatau between Java and Sumatra in 1883. More recently, Mount St. Helens in the state of Washington in 1980, El Chichon in Mexico in 1982, and Mount Pinatubo in the Philippines in 1991 have all had widespread regional impacts and possibly global effects as well. Mount Pinatubo has frequently been cited as a factor in northern hemisphere cooling and increased precipitation in 1992 and 1993. ReVelle and ReVelle concluded in 1992 that "particles from combustion and particles from volcanic eruptions may be affording us temporary protection from the initial stages of earth's warming."

Another major cause of atmospheric turbidity, perhaps the most important of all, is agricultural dust. Windblown dust from China is often detectable in the United States, and dust from the Sahel of Africa has frequently been detected in the Caribbean. With the increasing agriculturization of the earth's land surface as well as increasing desertification, wind erosion is a significant factor in atmospheric quality. The American Dust Bowl of the 1930s was a dramatic reminder of the environmental and economic consequences of drought and land abuse. Whether desertification is primarily due to natural climate cycles or anthropogenic factors makes little difference in terms of its atmospheric impact. The conversion of millions of acres of land annually from semiarid grasslands to desert contributes mightily to atmospheric turbidity.

Many modern agricultural practices also contribute to airborne dust. In the midwestern United States, the common practice of leaving grain fields fallow for much of the year increases soil loss and contributes excessive dust to the air. If you don't believe this, drive through eastern Colorado on a windy day in January.

The bottom line is that atmospheric turbidity from many sources—industrial smoke, urban smog, sulfate particles, forest fires, and agricultural dust—is a significant factor in reducing solar input. This turbidity has a cooling influence on the earth's surface, which may give us a temporary, but unstable, reprieve from global warming (Kerr, 1995).

FORCES IN CONFLICT

Global forces are obviously in conflict. Given the conflicting effects of greenhouse gases and the Milankovitch cycles, of ozone depletion and atmospheric turbidity, we don't know exactly where we are in terms of climate change, and no one can predict future climates with assurance. The prevailing evidence is in the direction of further global warming, but this conclusion has several uncertainties.

One of the main reasons for uncertainty is the confusing effects of clouds. The forces of global warming may increase hydrospheric evaporation and lead to greater cloud formation. High clouds, such as stratocirrus, trap earthly heat while still permitting most solar radiation to enter, and thus contribute to global warming. Other types of clouds, such as low, dense cumulonimbus, block incoming solar radiation and contribute to surface cooling. The interplay of surface temperatures, the hydrologic cycle, and cloud formation is very complex and unpredictable. Local and regional effects of global warming, or global cooling, have not been modeled with acceptable accuracy. The relationships between temperature changes and precipitation are very difficult to model.

In the clash of global forces, one of the most likely outcomes will be increasing variability. This will be both spatial and temporal, and the consequences can be serious. There will be places and times when the forces of regional warming prevail, for example, during natural drought cycles when no clouds form, and there will be other places and times, for example, during, prolonged monsoon periods, when regional cooling occurs. We can expect sharper weather fronts from such occurrences—greater extremes of temperatures and air pressures. This will mean hotter hot spells, colder cold spells, wetter wet spells, and drier droughts. Long-term trends will be hard to detect because of high variability and record-breaking temperatures, precipitation, and winds. World weather in recent years has been fitting this pattern more and more, with record heat in the United States and record droughts in southeast Asia in the 1980s; record snowfall in the western mountains and record rainfall in the midwest in the United States in 1993. Records can always be broken, of course, and some may stand for a long time, but the frequent occurrence of new records in all directions in the last 10 years suggests many competing forces in action.

Another consequence of extreme weather fronts is more violent storms (Eisma, 1995). Sharply delimited temperature and pressure ridges are the driving force behind tornadoes, hurricanes, and typhoons. Temperature and pressure differentials generate the powerful movements of air that become raging windstorms (Figure 18.7). We can expect record blizzards as well as record heat waves.

Perhaps it is no coincidence that some of the most damaging hurricanes in history (Andrew, Agnes, Camille, and Hugo) have all occurred in recent years. Of course, there have been more people and buildings in their paths, and this is why they have been so economically destructive, but the fact remains that these storms had record-breaking wind speeds and record precipitation levels.

In Chapter 3, we discussed the biosphere as a great moderator of weather. One need only think of the temperature differential between a cool shaded forest and an asphalt parking lot on a sunny July day. On a larger scale, the same contrast exists between the deserts of the Sahara and the rain forests of Zaire. It is reasonable to hypothesize, therefore, that we will force greater extremes of weather and climate as we destroy the biosphere. As we cut forests, desertify grasslands, expand cities, foul the air, and pollute the seas, we are impairing the ability of planet earth to moderate extremes and maintain global homeostasis.

Figure 18.7   A tornado forms in southwestern Oklahoma as a swirling cloud mass descends earthward from a thunderhead. Atmospheric factors of global warming and global cooling, coupled with drastic modifications of natural ecosystems, may create sharper weather fronts of pressure and temperature, leading to more frequent and violent storms. (Photo by National Severe Storms Laboratory)

SUMMARY

Global climate change is a topic that evokes concern and controversy. Proponents of the greenhouse effect postulate global warming due to the accumulation of heat-retaining gases in the atmosphere and the partial destruction of stratospheric ozone which shields the earth from intense solar radiation. Others, however, point to two opposing forces that can theoretically contribute to global cooling: the Milankovitch cycles and atmospheric turbidity. According to the Milankovitch theory the earth should be returning to another ice age, at least in the northern hemisphere. Atmospheric turbidity, which is increasing from industrial, agricultural, and domestic sources, can have a cooling effect by blocking sunlight. Cloud formations represent the largest unknowns in global climate change. Certain types of clouds help to retain heat, while others have a cooling effect on earth's surface temperatures. Theoretically, global warming will increase hydrospheric evaporation which will lead to increased cloud formation. The outcomes of more clouds are not predictable.

El Niño oscillations in the southern Pacific have been characterized by vast warm-water pools. These have increased in frequency in recent years and have had major climatic influences on adjacent continents and major ecological impacts on marine organisms. Their fundamental cause is unknown.

Temperature records of the past 200 years show erratic global warming, but it is not known if this trend is within the range of normal variation or actually represents a pollution-caused enhancement of the greenhouse effect. If global warming is occurring, the environmental and economic consequences can be severe. They include disruptions to agriculture, rises in ocean levels, drastic changes in coastlines, increased skin cancer and cataracts, and major dislocations of global ecosystems.

One of the more likely outcomes of the clash between the forces of global warming and global cooling is increased variability and violence of weather. Sharp weather fronts with abrupt delineations of temperatures and pressures will develop, giving rise to extreme weather conditions.

REFERENCES

Balling, R. C. Jr. 1993. Global temperature data. *Research and Exploration* (National Geographic Society) 9(2): 201–207.

Bloch, N. 1994. The ocean comes ashore: Is Trinidad drowning? *Earthwatch* 13: 19–27.

Caldicott, H. 1992. *If You Love This Planet: A Plan to Heal the Earth.* New York: W. W. Norton.

Canby, T. 1994. *Our Changing Earth.* Washington, DC: National Geographic Society.

Crowley, T. J., and G. R. North. 1991. *Paleoclimatology.* New York: Oxford University Press.

Deming, D. 1995. Climatic warming in North America: Analysis of borehole temperatures. *Science* 268: 1576–1577. (June 16, 1995).

Eisma, D. 1995. *Climate change: Impact on Coastal Habitation.* Boca Raton, FL: Lewis.

Hansen, J. A., A. Lacis, R. Ruedy, M. Sato, and H. Wilson. 1993. How sensitive is the world's climate? *Research and Exploration* (National Geographic Society) 9(2): 142–158.

Hill, D. K. 1995. Pacific warming unsettles ecosystems. *Science* 267: 1911–1912 (March 31, 1995).

Kagan, B. A. 1995. *Ocean-Atmosphere Interaction and Climate Change.* New York: Cambridge University Press.

Kerr, R. A. 1994. Did the tropical Pacific drive the world's warming? *Science* 266: 544–545 (Oct. 28, 1994).

Kerr, R. A. 1995. Study unveils climate cooling caused by pollutant haze. *Science* 268: 802 (May 12, 1995).

Lindzen, R. 1993. Absence of scientific basis. *Research and Exploration* (National Geographic Society) 9(2): 191–200.

McKay, G., and H. Hengeveld. 1990. The changing atmosphere. In *Planet Under Stress,* ed. C. Mungall and D. J. McLaren. New York: Oxford University Press.

Michaels, P. J. 1993. Benign greenhouse. *Research and Exploration* (National Geographic Society) 9(2): 222–233.

Miller, G. T. Jr. 1992. *Living in the Environment.* 7th ed. Belmont, CA: Wadsworth.

Pollack, H. N., and D. S. Chapman. 1993. Underground records of changing climate. *Scientific American* 268: 44–50.

ReVelle, P., and C. ReVelle. 1992. *The Global Environment.* Boston: Jones and Bartlett.

Rosenzweig, C., and D. Hillel. 1993. Agriculture in a greenhouse world. *Research and Exploration* (National Geographic Society) 9(2): 208–221.

Schneider, S. H. 1992. Will sea levels rise or fall? *Nature* 356: 11–12.

Schneider, S. H. 1993. Degrees of certainty. *Research and Exploration* (National Geographic Society) 9(2): 173–190.

Silver, C. S., and R. S. deFries. 1990. *One Earth, One Future: Our Changing Global Environment.* Washington, DC: National Academy Press.

Stern, P. C., O. R. Young, and D. Druckman (eds.). 1992. *Global Environmental Change.* Washington, DC: National Academy Press.

Stone, R. 1995. Global warming: If the mercury soars, so may health hazards. *Science* 267: 957–958.

Thompson, L., et al. 1995. Late Glacial Stage and Holocene tropical ice core records from Huascaran, Peru. *Science,* 269: 46–50 (July 7, 1995).

Travis, J. 1994. Taking a bottom to sky "slice" of the Arctic Ocean. *Science* 266: 1947–1948 (Dec. 23, 1994).

UN Intergovernmental Panel on Climate Change. 1995. UN World Meteorological Organization.

Williams, R., J. Heckman, and J. Schneeberger. 1991. *Environmental Consequences of the Persian Gulf War.* Washington, DC: National Geographic Society.

Winograd, I. J., et al. 1992. Continuous 500,000-year climate record from vein calcite in Devils Hole, Nevada. *Science* 258: 255–260 (Oct. 9, 1992).

Chapter 19

WATER POLLUTION
AND OCEAN
ECOLOGY

Earth is appropriately known as the water planet. With 70 percent of the world's surface covered by water, and even a few more percent if we include frozen water, earth is indeed an aquatic planet. Rivers, lakes, and oceans constitute the earth's vascular system, transporting nutrients, disposing of wastes, generating weather, and moderating climates. Water is a vital component of all living organisms; water makes all life possible.

CLASSIFYING WATER POLLUTION

Water pollution, or the unfavorable alteration of water, may involve five basic types of contamination: (1) enrichment, or excessive nutrients, from sewage, fertilizers, plant and animal wastes, and so on, (2) toxic chemicals from industrial, domestic, military, and agricultural sources, (3) radioactive contamination from industrial, military, medical, and a variety of other sources, (4) particulate silt from soil erosion and construction, and (5) thermal pollution from power plants, industry, and domestic sources.

The ecological effects of these different types of pollution vary widely. For example, enrichment pollution, or eutrophication, as it is often called, may actually stimulate biological growth, but it is usually of an undesirable sort. Eutrophication, or eutrophy, usually involves excessive growth of algae, dinoflagellates, or other forms of plankton. These "plankton blooms" are often accompanied by high bacterial counts in the water. Plankton and bacterial blooms may discolor the water and make it unfit for swimming or other water-contact sports. Furthermore, plankton blooms of some species, especially the dinoflagellates, produce toxic metabolites in the water that may kill fish and benthic organisms. Most plankton blooms eventually lead to mass decomposition, which robs water of oxygen and suffocates fish if they are not first killed by toxic metabolites. Milder forms of eutrophication do not necessarily produce such drastic results, but they certainly make the water less attractive and more expensive to treat if it is destined to be used in a municipal water system. Thus, the consequences of eutrophy are far-reaching both ecologically and economically.

Toxic pollution does not stimulate biological growth, even of undesirable life forms, but it kills living organisms directly. Literally tens of thousands of toxic chemicals are used in our industrial, agricultural and domestic activities. Some of the most prominent are insecticides, herbicides, and fungicides such as chlordane, lindane, parathion, malathion, toxaphene, and kepone; industrial acids and alkalis such as sulfuric acid and sodium hydroxide; disinfectants such as chlorine, bromine, and iodine; solvents such as acetone, carbon tetrachloride, and trichloroethylene; petrochemicals such as crude oil, gasoline, benzene, toluene, and aliphatic hydrocarbons; heavy metals such as mercury, copper, chromium and lead; and innumerable synthetic chemicals ranging from dioxin to vinyl chloride. Toxic chemicals are found in both surface and ground water, including drinking water wells (Table 19.1).

Many of these chemicals are directly lethal to animals, plants, and microorganisms. Others may not cause death immediately, but they may be mutagenic, carcinogenic, or otherwise disrupt cellular processes. The ubiquitous use of highly dangerous and poisonous compounds in the factory, automobile, home, lawn, garden, and park is a fact of modern life. Water pollution—in some cases subtle and virtually unnoticed, in others disastrous—is the almost inevitable result.

The range of our industrial enterprises, from which we all benefit in terms of jobs and products, is suggested by Figure 19.1, illustrating some of the major industries along the lower Mississippi River near Baton Rouge and New Orleans, Louisiana, in the early 1970s. The concentration of chemical, petroleum, metal-working, rubber, and plastic industries produces a wide range of effluents entering the river. At least 140 major chemical plants discharged more than 296 million pounds of toxic chemicals into the entire Mississippi in 1991, and 620 municipal sewage plants discharge more than 1 billion gallons of wastewater yearly (Kanamine, 1994). Yet this river is the drinking water source for 18 million people.

Radioisotopes from natural and artificial sources also enter aquatic systems. The major artificial sources are off-site migration from nuclear weapons plants, fallout from nuclear bomb tests, and accidental releases from nuclear power plants and former nuclear weapons plants. Minor artificial sources are radioactive wastes from hospitals, laboratories, and industries using radioisotopes. Collectively, these sources provide more than 30 radioisotopes of biological significance (Table 19.2). Some of these are particularly important such as strontium 90, which enters living systems and is absorbed by bone; iodine 131, absorbed by the thyroid gland; cesium 137, absorbed by the gonads; and plutonium 239, absorbed by the lungs.

Biologically, ionizing radiation is damaging. Radiation impacts cellular nuclei and can damage the DNA of genes and chromosomes. This damage may be mutagenic (causing genetic mutations), carcinogenic (cancer causing), or teratogenic (birth defects) and is a major contributor to the aging process. Thus, the impacts of radiation are both somatic, as in causing cancer, and genetic, as in mutations.

TABLE 19.1

Examples of toxic organic chemicals found in drinking water wells, in parts per billion (Council of Environmental Quality, 1981).

Chemical	Ground-Water Concentration (ppb)	State[1]	Surface-Water Concentration (ppb)
Trichloroethylene (TCE)	27,300	PA	160
	3,800	NY	
Toluene	6,400	NJ	6.1
1,1,1-Trichloroethane	5,440	ME	
	5,100	NY	5.1
Acetone	3,000	NJ	ni
Methylene Chloride	3,000	NJ	13
Dioxane	2,100	MA	ni
Ethyl benzene	2,000	NJ	ni
Tetrachloroethylene	1,500	NJ	
	740	CT	21
Cyclohexane	540	NY	ni
Chloroform	490	NY	
	420	NJ	700
Di-n-butyl-phthalate	470	NY	ni
Carbon tetrachloride	400	NJ	30
Benzene	330	NJ	4.4
1,2-Dichloroethylene	323	MA	9.8
Ethylene dibromide (EDB)	300	HI	ni
Xylene	300	NJ	24
Isopropyl benzene	290	NY	ni
1,1-Dichloroethylene	280	NJ	0.5
	118	MA	
1,2-Dichloroethane	250	NJ	4.8
Bis (2-ethylhexyl) phthalate	170	NY	ni
Dibromochloropropane (DBCP)	137	AZ	
	95	CA	ni
Trifluorotrichloroethane	135	NY	ni
Dibromochloromethane	55	NY	317
Vinyl chloride	50	NY	9.8
Chloromethane	44	MA	12
Butyl benzyl-phthalate	38	NY	ni
gamma-BHC (lindane)	22	CA	ni
1,1,2-Trichloroethane	20	NY	ni
Bromoform	20	DE	280
1,1-Dichloroethane	7	ME	0.2
alpha-BHC	6	CA	ni
Parathion	4.6	CA	0.4
delta-BHC	3.8	CA	ni

[1]The states listed are examples where ground- and surface-water surveys provided accurate data in 1981; they are not the only states with toxic chemical contamination in drinking water. ni = not investigated.

Since most organisms are well adapted to their environments through natural selection, most mutations are deleterious.

Apparently, living organisms have adapted to natural radiation levels and have developed DNA-repair mechanisms to deal with natural background radiation. This does not mean that one should expose oneself knowingly to strong natural radiation sources, but it does mean that we need not worry excessively about

230

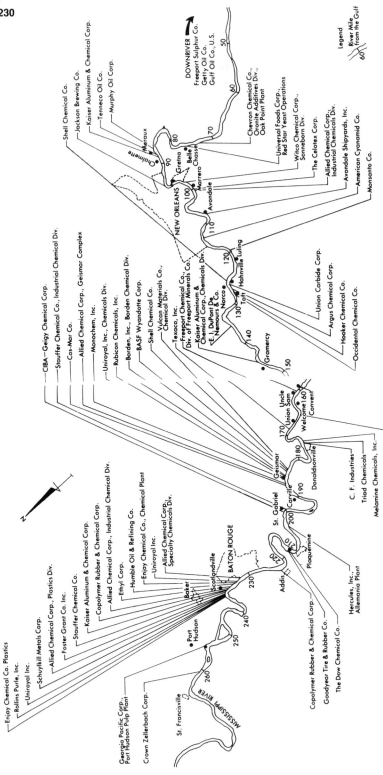

Figure 19.1 Industrial development along the Mississippi River near Baton Rouge and New Orleans, Louisiana, where the river affords abundant water for manufacturing and waste disposal. Of 36 chemicals in downriver drinking water after treatment, benzene, carbon tetrachloride, chloroform, dichloroethyl ether, ethyl benzene, and dimethyl sulfoxide have been found harmful to laboratory animals. Their effects on people are not fully understood (from McCaull and Crossland, 1974).

TABLE 19.2
Radionuclides of ecological importance (Odum, 1971).

Group A: Natural isotopes that contribute to background radiation

Radionuclide	Half-Life	Type of Radiation Emitted[2]
Uranium 235	7×10^8 yrs.	Alpha, Gamma
Uranium 238	4.5×10^9 yrs.	Alpha
Radium 226	1.6×10^3 yrs.	Alpha, Gamma
Thorium 232	1.4×10^{10} yrs.	Alpha
Potassium 40	1.3×10^9 yrs.	Beta, Gamma
Carbon 14 (see group B)		

Group B: Some nuclides of elements that are essential constituents of organisms and therefore important as tracers in biological metabolism studies as well as because of the radiation they produce[1]

Radionuclide	Half-Life	Type of Radiation Emitted[1]
Calcium 45	160 days	Beta
Carbon 14	5,568 yrs.	Beta
Cobalt 60	5.27 yrs.	Beta, Gamma
Copper 64	12.8 yrs.	Beta, Gamma
Iodine 131	8 days	Beta, Gamma
Iron 59	45 days	Beta, Gamma
Hydrogen 3 (Tritium)	12.4 yrs.	Beta
Manganese 54	300 days	Beta, Gamma
Phosphorus 32	14.5 days	Beta
Potassium 42	12.4 hrs.	Beta, Gamma
Sodium 22	2.6 yrs.	Beta, Gamma
Sodium 24	15.1 hrs.	Beta, Gamma
Sulfur 35	87.1 days	Beta
Zinc 65	250 days	Beta, Gamma

Group C: Nuclides important in fission products entering the environment through fallout or nuclear waste disposal

Radionuclide	Half-Life	Type of Radiation Emitted[2]
Strontium 90 and	28 yrs.	Beta
daughter yttrium 90	2.5 days	Beta
Strontium 89	53 days	Beta
Cesium 137 and	33 yrs.	Beta, Gamma
daughter barium 137	2.6 min.	Beta, Gamma
Cesium 134	2.3 yrs.	Beta, Gamma
Cerium 144 and	285 days	Beta, Gamma
daughter praseodymium 144	17 min.	Beta, Gamma
Cerium 141	33 days	Beta, Gamma
Ruthenium 106 and	1 yr.	Beta
daughter rhodium 106	30 sec.	Beta, Gamma
Ruthenium 103	40 days	Beta, Gamma
Zirconium 95 and	65 days	Beta, Gamma
daughter niobium 95	35 days	Beta, Gamma
Barium 140 and	12.8 days	Beta, Gamma
daughter lanthanum 140	40 hrs.	Beta, Gamma
Neodymium 147 and	11.3 days	Beta, Gamma
daughter promethium 147	2.6 yrs.	Beta, Gamma
Yttrium 91	61 days	Beta, Gamma
Plutonium 238	2.4×10^4 yrs.	Alpha, Gamma
Iodine 131 (see group B)		
Uranium (see group A)		

[1] Group B also includes barium 140 (^{140}Ba), bromine 82 (^{82}Br), molybdenum 99 (^{99}Mo), and other trace elements.
[2] Alpha particle radiation consists of a helium nucleus containing two protons and two neutrons; beta particle radiation consists of one electron; gamma rays are high-energy electromagnetic waves with very short wavelengths similar to x-rays.

normally low levels of background radiation. In any case, nothing can be done about such sources. We should avoid, of course, overexposure to non-ionizing radiation effects such as sunburn.

There is abundant evidence that radioisotopes from artificial sources are widely dispersed in air, water, and throughout the biosphere. Radioactive tritium from the Savannah River nuclear facility in South Carolina, largely waterborne, has been detected in cow's milk in cities 100 miles away. High levels of radioactive calcium 45, iodine 131, phosphorus 32, and other isotopes have been measured in the Columbia River and in its fish and aquatic birds downstream from the Hanford nuclear plant in Washington.

Some of the most dramatic dispersals of radioisotopes have been airborne from nuclear bomb testing. These isotopes then come to earth in fallout and rainstorms and thus enter the realm of water pollution. Strontium-90 is one of the most famous of such isotopes. It has entered aquatic systems and food chains across the northern hemisphere, showing up in the antlers of deer and elk and in the teeth of children and young adults growing up in the 1950s and 1960s when such testing was particularly active.

The impacts of these extraneous radiation sources on ecosystems are not well known. In terms of public health, the more dramatic radiation accidents and releases have been well studied, and in general the studies point to major medical problems for people receiving significant doses. In the aftermath of the Hiroshima and Nagasaki atomic bombings, the surviving population around the edges of the blast have had leukemia death rates 40 to 50 times above normal, breast cancer rates 6 to 8 times above normal, stomach cancer rates 4 to 5 times above normal, and lung cancer rates twice the normal (Blumenthal, 1985). Most studies of radiation effects on animals and people have shown cancer and mutation rates directly proportional to dose levels received above background. The fallout from the Chernobyl nuclear accident in the former Soviet Union in 1986 has produced major health problems in the neighboring populations. Both increased cancer deaths and increased birth defects have occurred in these populations. By the most conservative estimate, some 28,000 additional cancer deaths are expected to result from Chernobyl fallout in the 50-year period following the accident, approximately one-half of which will be in the former Soviet Union and the other half in portions of Europe receiving fallout (Hohenemser and Renn, 1988). Other scientists estimate as many as 475,000 additional cancer deaths worldwide from Chernobyl fallout (Gale, 1990).

The health effects of other nuclear power stations and nuclear weapons research and manufacturing plants are highly controversial and a source of legal contention. Several studies at nuclear facilities, such as Rocky Flats, Colorado; Oak Ridge, Tennessee; and Hanford, Washington, show increased levels of various types of cancer in the workers and in some cases in residents downwind from the plants (Lenssen, 1991). For example, a 1991 study of nuclear workers at Oak Ridge, Tennessee, found that leukemia rates were 63 percent higher for nuclear workers exposed to small doses of radiation over time than for non-nuclear workers. Similarly, a 1990 report on families of workers at the Sellafield nuclear reprocessing plant in England found that children of nuclear workers were seven to eight times more likely to develop leukemia if their fathers had low-level doses of radiation. Other research studies refute these findings, challenging the results on statistical and methodological grounds. The true picture is not yet clear. Perhaps the most serious aspect of the nuclear pollution of water, air, and even the oceans is the fact that its impact will extend far into the future. Radioactive plutonium, for example, from nuclear weapons manufacture, with a half-life of 24,000 years, remains strongly radioactive for thousands of years.

Another form of water pollution is turbidity. Turbidity simply refers to the amount of silt or sediment in water. Normal turbidity is greatly increased after rainstorms in erosion-prone watersheds. Bare earth fields, which may lose 30 to 40 tons of topsoil per year in the midwestern United States, contribute to excessive turbidity. So do overgrazed pastures, strip mines, and construction sites where high runoff washes soil and debris into streams, rivers, and lakes. The dramatic floods in the midwestern United States in the summer of 1993 gave tragic evidence of the destructive power of turbid water. Not only were crops leveled and houses destroyed but many areas of backwater were covered by inches of mud. The floodplain itself may eventually benefit from this mud, if it has nutrient qualities, but otherwise the mud smothers normal lake and river bottoms, totally altering their ecology. Beneficial aquatic vegetation is smothered, fish-spawning sites are destroyed, and aquatic productivity is reduced. The smothering effect of silt is most severe in coastal areas, where the high productivities of estuarine and coral reef communities can be completely destroyed.

Thermal pollution is a final major type of human alteration of aquatic ecosystems. Excess hot water from power plants and industrial operations, produced by cooling turbines, generators, and other equipment, is often released into lakes and rivers. Some nuclear power plants, for example, release water at 95° F into rivers or estuaries where the ambient temperatures may range from freezing in the winter to more than 75° or 80° F in the summer. The effects range from subtle to dramatic. Subtle changes include changes in the reproductive biology of aquatic organisms; more dramatic changes include plankton blooms and fish kills. Even worse, hot water can eliminate fish, molluscs, and aquatic insects entirely. The Mahoning River flowing through Youngstown, Ohio, was so heated by steel mill effluents when the town had an active iron and steel industry that it never froze in winter, nor did it have fish or other living organisms other than bacteria.

Thermal pollution often acts synergistically with other forms of pollution, for example, with toxic chemicals to increase the toxicity, with nitrates and phosphates to increase the damaging effects of eutrophication, with plankton blooms to accelerate oxygen depletion, and with microbial populations to increase epizootic diseases in aquatic animals or plants.

OTHER WAYS OF CLASSIFYING WATER POLLUTION

We have outlined five basic types of water pollution—nutrient (eutrophic), toxic, radioactive, turbid, and thermal—but there are several other ways of classifying water pollution. These concern primarily the site and extent of contamination.

Water pollution can be classified by site, origin, or extent as (1) surface water or ground water, (2) point-source or non-point-source, (3) local, regional, or international, or (4) well, spring, stream, river, pond, lake, coastal, or oceanic. Each of these might be subdivided further as special cases merit.

Surface water pollution is the most obvious, since we can sometimes see it or smell it, but ground water pollution is usually more harmful. First of all, polluted ground water is not subject to the natural remediation effects of sunlight and air exchange, so it is often more persistent. Furthermore, as with a leak in the ceiling in an old house, the origin of the problem may be some distance away from the place where the problem appears. Ground water pollution is particularly serious for individuals and municipalities obtaining drinking water from wells. Table 19.1 provides just a brief reminder of the extent of well water pollution in the United States.

Point-source pollution is that stemming from a specific point, such as the effluent pipe from a sewage treatment plant. Non-point-source pollution comes from a broad area; an example is the runoff from a city or from farmland. Urban runoff carries a variety of materials including urban dust, salt, petrochemicals, asbestos dust from automobile brake drums, particulate lead, and a host of other elements and compounds.

The distinctions between local, regional, and international water pollution are obvious. Pollution of the South Platte River running through Denver is a local problem, pollution of the Mississippi River is a regional problem, and pollution of Lake Erie, the North Sea, the Black Sea, or the Mediterranean is international. Similarly, distinctions may be made of spring, stream, river, lake, coastal or oceanic pollution. The key criteria are the sites and extent of the pollution.

The remainder of this chapter will focus on coastal, sea, and oceanic pollution, since the marine environment is the ultimate sink.

COASTAL AND OCEANIC POLLUTION

The hydrosphere, dominated by the world's oceans, is a dominant and dynamic component of planet earth. It is one of the distinguishing factors of earth, permitting the development and survival of the biosphere, and interacting continuously with the other major components of the earth: the atmosphere, lithosphere, and biosphere (Kagan, 1995).

The hydrosphere is also a complexly integrated component of the earth; every part from small streams to vast oceans are interconnected. Small streams depend on precipitation or ground water for their existence, precipitation depends upon atmospheric moisture that comes ultimately from lakes and oceans, and most streams finally deliver their contents to the oceans. Hence, the earth is called the water planet not just because 70 percent of its surface is covered by water but because water nourishes all life on earth, and provides the life blood of our existence.

Since oceans are the final destination for the world's lakes and rivers, and since much of all the material washing off the land ends up in the oceans, this section will focus on the health of the marine environment.

THE EXTENT OF THE OCEANS

While there are many places on the earth's land surface where one can get the feeling of overpopulation, especially in the world's crowded cities and traffic-packed highways, there are many places in the oceans where humans are rarely if ever found. A capsized sailboat recently drifted eastward from New Zealand with four survivors who desperately managed to cling to life without sighting a single vessel for nearly three months. They must have felt they were on an uninhabited planet.

In volumetric terms, the importance of the oceans is even more impressive. Whereas the water surface area of the earth is 2.4 times greater than the land surface area, if the world's land area were represented by 1 square meter, the ocean's volume would equal 8,300 cubic meters (Fig. 19.2). Ocean volume is even 7 times greater than atmospheric volume (8,300 m^3 compared to 1,260 m^3). Earth is indeed a water planet.

The vast extent of the oceans gives the impression that they are an unlimited frontier. We assume that we can dump everything into them we want, including our garbage, toxic chemicals, liquid and solid wastes, and even radioactive materials, with impunity. At the same time, we have great hope that the huge frontiers of the oceans will bail us out of our food shortages and will provide an untapped source of protein to feed the world. Given this great faith in the oceans, we need to look at them carefully to see what they currently provide and what their state of health is.

OCEAN PRODUCTIVITY

The harvest of fish from the world's oceans attained the level of about 80 million tons per year in 1989 and has declined slightly since then (Fig. 19.3). Total fish-

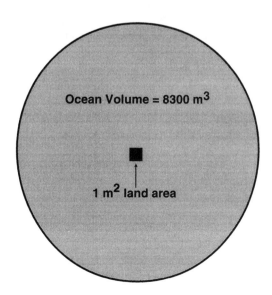

Figure 19.2 Schematic representation of the ratio of ocean volume to land surface. Ocean volume is estimated to be 8,300 cubic meters for each square meter of earth's land surface (drawing by D. Lorenz).

Million Tons

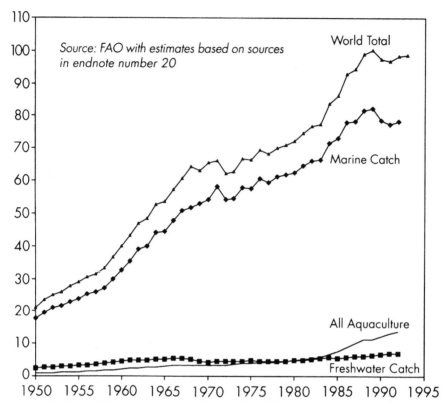

Figure 19.3 World fish harvest, 1950–1993 (from Weber, 1994).

eries production, including freshwater catch and aquaculture, achieved 100 million tons per year in 1989, dropped to 97 million tons in 1992 (Weber, 1994), but then increased to 101 million tons in 1994 (Platt, 1995a). Many experts feel that 100 million tons is a maximum sustainable fish yield (Corson, 1990). The problem of reduced fish numbers is manifold. First of all, the world's fishing fleet has increased greatly in size and technological efficiency. Fishing boats from many nations now use the latest in fish-finding sonar and the most efficient harvesting techniques. At the same time, coastal and oceanic pollution is affecting the spawning success of many species. As a result, populations of several commercial species are declining seriously (Holmes, 1994; Weber, 1994).

Examples of declining fish catches include species as diverse and Peruvian anchovies (−58 percent), Atlantic cod (−69 percent), Haddock (−80 percent), and Pacific herring (−71 percent) (Table 19.3). By 1992 the catches of 18 major species of marine fish had declined an average of 58 percent from their peak catch, providing clear evidence of ecological problems. Some of the larger fish are in jeopardy as well, including Atlantic and Pacific salmon, tuna, rockfish (striped bass), and swordfish (McGoodwin, 1990). Many benthic marine and estu-

TABLE 19.3
Fishery declines of more than 100,000 tons from their peak year to 1992 (Weber, 1994. Data from UN Food and Agricultural Organization).

Species	Peak Year	Peak Catch	1992 Catch	Decline	Change (percent)
		(millions of tons)			
Pacific herring	1964	0.7	0.2	0.5	−71
Atlantic herring	1966	4.1	1.5	2.6	−63
Atlantic cod	1968	3.9	1.2	2.7	−69
South African pilchard	1968	1.7	0.1	1.6	−94
Haddock	1969	1.0	0.2	0.8	−80
Peruvian anchovy[a]	1970	13.1	5.5	7.6	−58
Polar cod	1971	0.35	0.02	0.33	−94
Cape hake	1972	1.1	0.2	0.9	−82
Silver hake	1973	0.43	0.05	0.38	−88
Greater yellow croaker	1974	0.20	0.04	0.16	−80
Atlantic redfish	1976	0.7	0.3	0.4	−57
Cape horse mackerel	1977	0.7	0.4	0.3	−43
Chub markerel	1978	3.4	0.9	2.5	−74
Blue whiting	1980	1.1	0.5	0.6	−55
South American pilchard	1985	6.5	3.1	3.4	−52
Alarka pollock	1986	6.8	5.0	1.8	−26
North Pacific hake	1987	0.30	0.06	0.24	−80
Japanese pilchard	1988	5.4	2.5	2.9	−54
Total		51.48	21.77	29.71	−58

[a]The catch of the Peruvian anchovy hit a low of 94,000 tons in 1984, less than 1 percent of the 1970 level, before climbing up to the 1992 level.

arine organisms have also shown declining populations, including Alaska king crabs (Fig. 19.4) and Chesapeake Bay oysters (Fig. 19.5).

Although these tables and graphs show widely differing organisms in different geographic and ecologic locations, they all have one discouraging feature in common; they reveal serious declines in fish numbers in recent years or throughout the twentieth century. So significant are the declines that they constitute a "crisis in world fisheries," according to McGoodwin (1990).

In many species of fish and shellfish, populations and harvest yields have declined as fishing efforts have increased. This is true of the highly prized bluefin tuna (Fig. 19.6). Computer models of current data on total seafood harvests show continued declines in relation to human demand (Fig. 19.7).

Why are so many important species declining in populations and harvest? At least two major factors seem to be at work. In individual cases, natural causes such as El Niño may be operating as well. But the main reasons are certainly overfishing and pollution. Overfishing to meet the food demands of growing human populations is largely responsible for the decline of many commercial species in the North Atlantic (McGoodwin, 1990). Fish are pursued with fast boats, modern detection gear, and huge nets. International competition is keen. Many nations, especially Japan, Russia, and some of the European countries, exploit all the oceans of the world. Most nations have established exclusive fishing rights within 200 miles of their coastlines, and this limits international competition in

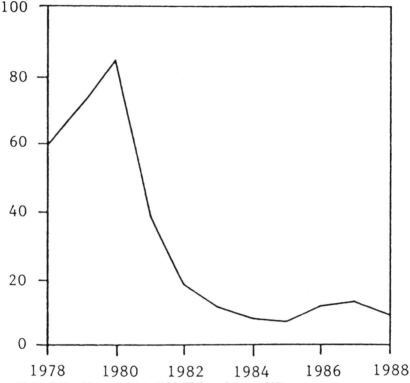

Figure 19.4   Alaskan king crab catch, 1978–87 (from Corson, 1990).

coastal zones, but depending on their catch limits, it still allows intensive fishing in continental shelf boundaries. The pressure is relentless and shows little signs of abating. Of the total marine harvest, over 90 percent is obtained by "hunting and gathering" and less than 10 percent by aquaculture (Corson, 1990). In the use of marine resources, we still follow frontier policies, but do so with the deadly efficiency of modern technology.

FURTHER EVIDENCE OF DETERIORATION IN MARINE ECOSYSTEMS

Apart from the statistics on commercial fish harvests, other indications of deteriorating ocean environments are available. In the Mediterranean Sea, the superb fish once available in the restaurants of Spain, Italy, Greece, and North African coun-

Thousand Tons

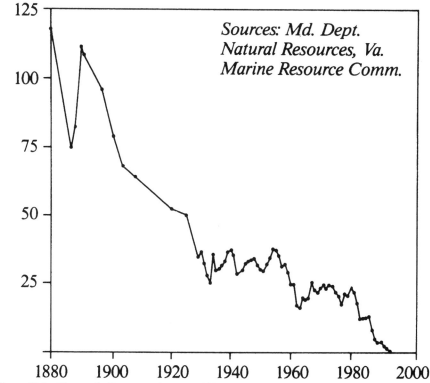

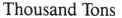

Figure 19.5 Chesapeake Bay oyster harvest, 1900–1993 (from Brown et al., 1995).

Figure 19.6 Bluefin tuna populations in relation to fishing effort in the Western Atlantic (from Meadows et al., 1992).

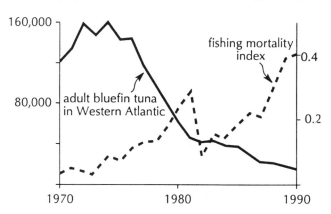

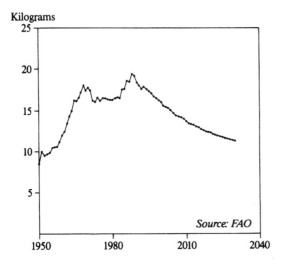

Kilograms

Source: FAO

Figure 19.7 World fish harvest per person, 1950–92, with projections to 2030.

tries are now seldom seen, or if found are exorbitantly expensive. The Bay of Naples, for example, used to produce elegant fish of large size. Now most of its catch consists of little anemic fish less than 10 inches in length. Off the island of Sulawesi in Indonesia, village fishermen used to bring in abundant catches of large reef fish, piling their outrigger canoes to the gunwales after a night of fishing. Now they wait throughout the night for a measly catch of 6-inch fish that barely fill a single bucket.

Most of the major seas of the world, including the Baltic, Bering, Black, Caspian, China, Java, Mediterranean, and Yellow, have serious ecological problems with declining fish populations. The Baltic Sea has high levels of toxic organochlorines, the Black Sea has excessive algal and bacterial blooms with anoxic conditions, the Bering Sea is depleted from overfishing, the South China and Yellow seas are plagued with oil and heavy metal contamination, and the Caspian Sea receives over 100,000 tons of wastewater and petrochemical wastes each year (Platt, 1995). All of these, and several other seas, have sharply declining seafood harvests.

The story is repeated throughout the world; fishing requires more work over longer distances for smaller catches. These problems of villagers and coastal fishers do not always show up in commercial statistics, but they are very real, affecting the livelihood of millions of people around the world.

The problems are especially severe in coastal regions, including both estuarine and coral reef habitats (Ray and Grassle, 1991). Coastal zones carry the highest pollution loads. Whatever comes from the land—toxic chemicals, sewage, animal wastes, fertilizers, silt, plastic debris, radioactivity, and thermal pollution—all of these have their greatest impact in coastal zones. Coastal zone waters are often shallow, and tidal action sloshes pollutants back and forth, repeatedly exposing the organisms living there to their harmful effects. The vulnerability of coastal zones to high pollution loads is shown by the distribution of oil slicks which outline major ship routes (Fig. 19.8). Although oil tankers also ply the open oceans, most oil spills take place near land, where ship accidents occur or where port and bunkering facilities are located. Toxic chemicals of all kinds, sediment runoff, fertilizer concentrations, thermal loading, radioactive contamination, and virtually every other kind of pollution is greatest in the coastal zone environment. This is not surprising, of course, since most human pollution originates on land.

Unfortunately, the coastal zones also contain the world's most productive marine environments: estuaries, coral reefs, and mangrove forests. These natural biological communities have high productivity and high diversity because they occur at ecotones, or ecological boundaries between the land and the sea. They usually have shallow water with high nutrient concentrations exposed to abundant sunlight, wave action, and oxygen. Estuaries and mangrove forests have complex mixtures of freshwater, brackish, and true marine organisms. It is precisely the conditions that make them beautiful and productive in their natural states that also make them highly vulnerable to pollution and destruction. Although we love the seashore, we overbuild the waterfront, often in ecologically valuable wetlands or on primary beachfronts. Such construction is highly vulnerable to coastal erosion and often contributes to it (Leatherman, et al., 1995). We also use the inshore waters as a dumping grounds for waste products from factories, farms, homes and cities. Even materials disposed of through so-called open-ocean dumping have a way of washing up on beaches and shorelines. Hence, everything from petrochemicals, plastics, and styrofoam to sewage sludge and medical wastes is washing up on our waterfronts.

Another indication of the deteriorating ocean environment is found in coral reefs. In the 1980s and extending to the present time, coral reefs from the Caribbean to the South Pacific and the Indian Ocean (Fig. 19.9) have been dying in a mysterious manner known as "coral bleaching." The coral looses living cells, becomes enshrouded in turbid clouds of dead algae and plankton, and turns a ghostly white (Bunkley-Williams and Williams, 1990). After repeated episodes of bleaching, the diverse and productive assemblage of marine plants and animals

Figure 19.8   Locations of visible oil slicks in the early 1980s (from Mysak and Lin, 1990).

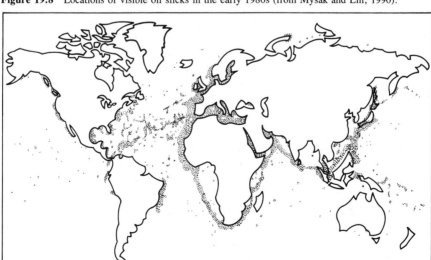

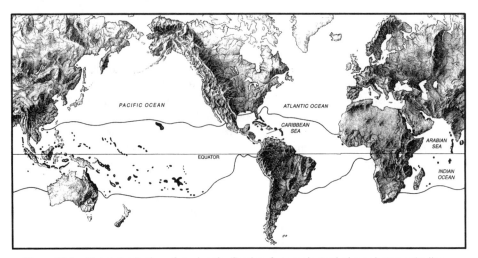

Figure 19.9   Global distribution of coral reefs. Coral reefs occur in tropical seas between the lines drawn on the map, primarily in the south Pacific, Caribbean, Indian Ocean, and Red and Arabian seas. They are among the most ecologically productive of all marine ecosystems (from Brown and Ogden, 1993).

associated with the reef begins to disintegrate; molluscs are lost, fish disappear, and the living community of the reef collapses. A reef ecosystem representing centuries and even millennia of growth and development can be destroyed in a matter of a few years (Weber, 1993).

The question is why? Why should this be occurring on a global scale, from parts of the Great Barrier Reef of Australia to the Red Sea to the reefs of Hawaii and Jamaica? Many ideas have been proposed, including temperature increases in ocean water due to El Niño oscillations or even more widespread global warming. Ocean surface temperatures have risen in many areas of coral bleaching, and we know that coral polyps and their symbiotic algae (known as zooxanthellae) are sensitive to even small temperature increases. Bleaching in many reefs was most prominently seen during the record heat years of the 1980s. But bleaching has also occurred in areas of normal water temperature, raising the possibility of turbidity, toxic pollution, or disease as precipitating factors, or perhaps some combination of the three (Brown and Ogden, 1993).

ARE THE OCEANS IN JEOPARDY?

Several scientists have expressed concern that pollution is affecting total ocean productivity, not just seas, coastal zones, coral reef communities, and commercial fish catches. More than 20 years ago, the prominent ocean explorer Jacques Yves Cousteau estimated that the deterioration of the marine environment was reducing oceanic life-support capabilities approximately 30 percent (Cousteau, 1971). More recently, a group of Russian oceanographers calculated that marine pollution has decreased the annual production of marine nekton (free-swimming marine organisms, including commercial fish species) at least 20 million tons, compared to that from healthy and intact oceanic ecosystems (Patin, 1982, 1985).

The problem, of course, is that most marine pollution, with the exception of oil spills and ship discharge, begins on land. Terrestrial pollution in most of the world shows little signs of abating, and a good many signs of increasing. Even if point-source industrial pollution is controlled in the most prosperous industrialized nations, the combined assaults of pollution from agriculture, unregulated industry, and domestic sources in burgeoning human populations means increasing pressures on coastal environments and the open oceans. It was demonstrated over 30 years ago that DDT residues could be found in Antarctic penguins, hundreds of miles from any known site of use, indicating global distribution of pesticides. Patterns of oceanic circulation, just like those of atmospheric circulation, ensure that pollutants entering the ocean at any point in the world may contaminate the entire global system. The answer to the question, "Are the oceans in jeopardy?" is unfortunately "Yes."

SUMMARY

The earth is indeed the water planet since 70 percent of the planet's surface is covered with water. The hydrosphere interacts continuously with the atmosphere, lithosphere, and biosphere in ecosystem circulation, weather generation, and climate modification.

Five basic types of pollution affect water quality and ecosystem function: enrichment (eutrophy), toxic contamination, turbidity and sedimentation, radioactive transport, and thermal pollution.

While streams, lakes, and rivers make up the vascular system of global ecosystems, oceans are the ultimate repository of water pollution. Coastal zones are especially vulnerable to high pollution loads since terrestrial runoff is deposited on the margins of continents and islands.

Most of the world's seas, partially enclosed marine areas, such as the Baltic, Black, Caspian, and Mediterranean seas, are in serious ecological trouble from both overfishing and pollution.

Coastal zone areas, especially estuaries, coral reefs, and mangrove forests are the most diverse and productive of all marine communities. Even pelagic, or open-ocean, productivity is related to coastal zone productivity, since many open-ocean organisms conduct essential parts of their life cycle in coastal zones. Coastal zone pollution, therefore, can have a major effect on ocean productivity. Estuaries, coral reefs, and mangrove forests are all facing serious environmental problems.

World fisheries production reached a peak of approximately 100 million metric tons in 1989, dropped to 97 million tons in 1992, and then rose again to 101 million tons in 1994 with advances in aquaculture, leading some experts to believe that 100 million tons is a maximum harvest under present conditions. Even this level may not be sustainable. Many commercial fish species have been declining in population and harvest as a combined result of overfishing and pollution effects. Some studies estimate that large-scale pollution of marine environments has reduced ocean biomass by 20 million tons.

REFERENCES

Blumenthal, D. S. 1985. *Introduction to Environmental Health.* New York: Springer.
Brown, B. E., and J. C. Ogden. 1993. Coral bleaching. *Scientific American* 268: 64–70 (Jan. 1993).
Brown, L. R. 1994. Facing food insecurity. In *State of the World, 1994,* ed. L. R. Brown et al. New York: W. W. Norton.
Brown, L. R. 1995. Nature's Limits. Chap.1 in *State of the World, 1995.* ed. Brown, L. R., et al. New York: W. W. Norton
Bunkley-Williams, L., and E. H. Williams. 1990. Global assault on coral reefs. *Natural History,* April 1990: 47–54.
Corson, W. 1990. *The Global Ecology Handbook.* Boston: Beacon Press.
Cousteau, J. Y. 1971. Statement on global marine degradation. *Biological Conservation* 4(1): 61–66.
Gale, R. P. 1990. Long-term impacts from Chernobyl in the USSR. *Forum for Applied Research and Public Policy* (Fall 1990).
Hohenemser, C., and O. Renn. 1988. Chernobyl's other legacy. *Environment* 30: 4–11, 40–45.
Holmes, B. 1994. Biologists sort the lesson of fisheries collapse. *Science* 264: 1252–1253.
Kagan, B. A. 1995. *Ocean-Atmosphere Interaction and Climate Modelling.* New York: Cambridge University Press.
Kanamine, L. 1994. The Mississippi's future: As muddy as the river. *USA Today,* Mar. 7, 1994, p. 3-A.
Leatherman, S., et al. 1995. *Vanishing Lands: Sea Level, Society and Chesapeake Bay.* College Park MD: University of Maryland.
Lenssen, N. 1991. *Nuclear Waste: The Problem That Won't Go Away.* Worldwatch Paper 106. Washington, DC: Worldwatch Institute.
McCaull, J. and J. Crossland. 1974. *Water Pollution.* New York: Harcourt and Brace.
McGoodwin, R. 1990. *Crisis in the World's Fisheries.* Stanford, CA: Stanford University Press.
Mysak, L. A., and C. A. Lin. 1990. The Tempering Seas. Chap. 6 in *Planet Under Stress,* eds. C. Mungall and D. J. McLaren. New York: Oxford.
Odum, E. P. 1971. *Fundamentals of Ecology.* 3rd ed. Philadelphia: Saunders.
Patin, S. A. 1982. *Pollution and the Biological Resources of the Oceans.* London: Butterworths.
Patin, S. A. 1985. Biological consequences of global pollution of the marine environment. *Global Ecology,* ed. C. H. Southwick. Sunderland, MA: Sinauer Associates.
Platt, A. E. 1995. Dying Seas. *World Watch* 8: 10–19.
Platt, A. 1995a. Aquaculture boosts fish catch. pp 32–33 in *Vital Signs,* eds L. R. Brown, N. Lenssen, H. Kane. New York: W. W. Norton.
Ray, C. G. and J. F. Grassle. 1991. Marine biological diversity. *BioScience* 41(7): 453–457.
Weber, P. 1993. Reviving coral reefs. In *State of the World, 1993,* ed. L. R. Brown. New York: W. W. Norton.
Weber, P. 1993. *Abandoned Seas: Reversing the Decline of the Oceans.* Worldwatch Paper 116. Washington, DC: Worldwatch Institute.
Weber, P. 1994. *Net Loss: Fish, Jobs, and the Marine Environment.* Worldwatch Paper 120. Washington, DC: Worldwatch Institute.

Chapter 20

THE CRISIS IN
BIODIVERSITY

The current loss of species is a major concern of modern biology and a rallying point for global conservation. Why all the fuss about biodiversity? Biodiversity refers to the number of species in a given area. In its largest context this means the total number of species in the world. Surprisingly, at least to nonbiologists, no one really knows how many species there are. About 1.4 million species of plants, animals, and microorganisms have been described and given scientific names. Almost 1 million of these consist of just two major groups of organisms: flowering plants (220,000) and insects (750,000) (Ehrlich and Wilson, 1991).

The known and named species in the world are thought to be only a small percentage of the true number. As early as 1952, Sabrosky estimated that there are as many as 10 million living species in the world. Thirty years later, Erwin (1982), working with insects in the Amazon basin, estimated the total number of living species of all organisms in the world to be 30 million. More recently, Stork (1988) has estimated that the number of forest arthropods alone may be as high as 80 million. If this last estimate is correct, the world may contain 100 million species. Thus, the fact is that we do not really know the true number of biological species on planet earth within an order of magnitude, that is, within a factor of ten. Many biologists place the probable number of living species in the world between 5 million and 50 million (Colinvaux, 1993). The consensus is more conservative, however, placing the approximate number of living species at 5 to 10 million (Raven and Johnson, 1992). Excluding bacteria and viruses, the most reasonable estimate for the number of species with eukaryotic cell types is in the range of 5 to 8 million.

WHAT IS A SPECIES?

Although biologists talk about species as though they are definite entities, there is no precise definition of the species concept. In general, a species can be defined as "a group of similar organisms capable of interbreeding with each other but not with other organisms" (Purves, Orians, and Heller, 1992). There are many exceptions to this general definition, however. Members of a species may have very different appearances. For example, dogs are all of the same species and thus potentially capable of interbreeding, although dogs of greatly different sizes could not do so. On the other hand, different species may be of very similar appearance. The hundreds of different species of the fruit fly *Drosophila* differ only in the color of their eyes, the veins on their wings, or the microscopic bristles on their legs.

Different species can interbreed in sympatric zones, or areas of distributional overlap. In most cases, hybrids of two species are not fertile. For example, a mule, which is a cross between a horse and a donkey, is sterile, but in other cases, fertile hybrids are known, as in the case of crosses between two different species of macaque monkeys. Species hybrids are more common in disturbed habitats or in areas of recent environmental change where two species are suddenly brought together.

All of these exceptions indicate clearly that there are no absolute criteria by which species can be defined (Raven and Johnson, 1992). In most cases, scientists are able to agree on a species designation, but disagreements often occur over the designation of different organisms as species or subspecies.

Although species are considered to be real biological entities, differences in classification are inevitable because the whole science of taxonomy or systematics (biological classification) is an attempt to pigeonhole organisms in specific categories when they often represent a continuous series. The fact that the Hawaiian Islands are known to have over 800 species of fruit flies is a remarkable example of biological diversity. This basically means that entomologists can recognize 800 consistent types of fruit flies in Hawaii. There may be even more than 800 if the criteria for distinguishing them are narrowed, or there may be fewer if criteria are broadened.

The criteria used to designate species may be based on morphology, physiology, biochemistry, behavior, or genetics. Originally, morphology, or structure, was the primary designator of species. Then physiological and biochemical differences, such as protein differences in body fluids, were found to be helpful and accurate indicators of different species. So too were behavioral differences such as mating behavior, activity patterns, communication, and display behaviors. Finally, chromosome patterns, genes, and DNA sequences have been found to be valuable designators of species differences. The bottom line is that species are distinct kinds of organisms that usually do not interbreed in nature with members of other species.

All of this shows us that taxonomy and systematics, the sciences of biological classification, are active and important aspects of modern science (*Systematics Agenda 2000,* 1994). They represent more than 18th and 19th century biology as sometimes alleged. The basic principles of biological classification were established by Carl Linnaeus, an eighteenth-century Swedish biologist who wrote the classic volume, *Systemae Naturae.* The scientific endeavors he started are still of vital importance today in modern ecology.

THE MEASUREMENT OF DIVERSITY

The study of diversity lies at the heart of modern biology. How and why organisms differ and how many different types there are in ecosystems are central themes of biological science. The most basic expression of biodiversity is the number of different species in a given location, but a more complete description of diversity should include some information on relative numbers of different species.

If, for example, two different forests each have 10 species of trees, and a given forest patch of each type has 1,000 individual trees, we might assume that both would have the same degree of biodiversity so far as trees are concerned. But this is not necessarily true; they could be quite different forests depending on the relative numbers of each species.

Suppose forest A, with 10 species and 1,000 individual trees, had 955 individuals of species 1 (say, an oak) and only 5 individuals each of species 2 through 10 (such as maple, elm, sycamore, pine, beech, and so on), the forest would certainly be considered an oak forest with only rare individuals of other species. On the other hand, if forest B, also with 10 species and 1,000 individual trees, had 100 individuals of each species, it would be a mixed forest, not dominated by oak.

Considering that the diversity and numbers of different species of trees would influence the types and numbers of animal species, such as insects, birds, and mammals, and also the numbers and types of fungi and microorganisms, such as bacteria and algae, the two forests would be very different. Forest B would most likely have a much greater biodiversity than forest A.

STATISTICAL REPRESENTATIONS OF BIODIVERSITY

Since we rarely have total counts of the numbers of species and individuals in any given biological community, various sampling methods must be employed to assess diversity. Many formulae have been proposed, and a complete discussion of these would fill an entire book. We can, however, describe three commonly used indices and show how they might apply.

The Shannon-Wiener diversity index is represented by the following:

$$H' = -\Sigma \, p_i \log p_i$$

where H' is diversity, p is the proportion of each species in the total number of individuals, and i is the number of each species (Colinvaux, 1993). Σ is the sum of the expression $p_i \log p_i$ for each species. The formula actually represents the probability that any given individual sampled at random from the population will be species i. Table 20.1 shows sample calculations of this index based on simple communities of 2 or 3 species. Notice that the community with three species in equal abundance has the highest diversity index. Some ecologists have pointed to statistical problems in the Shannon-Wiener index and have proposed their own

TABLE 20.1
Sample calculations of the Shannon-Wiener diversity index
(H') (Colinvaux, 1993).

No. of Species	Species			H'
	1	2	3	
2	90	10		0.33
2	50	50		0.69
3	80	10	10	0.70
3	33.3	33.3	33.3	1.10

Note: To produce strikingly different values the species in each "community" are
given extremely different relative abundances.

modifications. These problems generally deal with the nature of the community
samples and the randomness or potential bias of sampling methods.

The Margalef diversity index used a different formulation, based on information theory, to achieve a similar goal of qualifying diversity:

$$H = S - 1 \, / \, \log N$$

where H is diversity, S is the number of species, and N is the number of individuals. This formulation is based on the assumption of very large populations that
can be fully counted (Margalef, 1958).

An older and somewhat simpler index of diversity known as Simpson's index
is based on the probability that any two individuals sampled at random from a
community would belong to the same species. The representation of Simpson's
index is

$$D = 1 \, / \, \Sigma \, p_i^2$$

where D is diversity and p is the proportion of each species, i. This index gives
less weight to rare species and emphasizes common species in a community.

All of these indices attempt to describe biodiversity and community structure
in statistically definable terms. All have shortcomings and limitations, but they
bring mathematical rigor to the complex problems of sampling and community description.

PATTERNS OF BIODIVERSITY

Biodiversity varies geographically and temporally. Three of the most dramatic
variations in diversity are based on latitude, on area and isolation, and on successional stage.

Latitudinally, a fairly consistent generalization is that the greatest biodiversity
occurs in tropical forests near the equator, and as latitude increases toward the
poles, biodiversity decreases. A commonly cited example is that Ecuador on the
equator has 20,000 species of vascular plants, whereas Ohio, in an area about

the same size, has only 2,000 species (Colinvaux, 1993). Table 20.2 shows a similar comparison for birds. Panama, Columbia, and Venezuela, near the equator, have more than 1,000 species of birds, whereas Greenland and Antarctica, near the poles, have fewer than 60.

This same phenomenon has been documented for many groups of organisms including insects, fish, amphibians, reptiles, and mammals. Clearly, the greatest storehouse of biodiversity lies in tropical forests, especially tropical rain forests. Although tropical rain forests occupy less than 7 percent of the earth's land surface, they are thought to contain 50 percent of the world's species.

There are some specific exceptions to the generalization that the tropics have the greatest biodiversity. Many areas of the tropics, especially arid grasslands and deserts have lower biodiversity compared to areas in temperate zones with greater topographic diversity. In addition, many cold-adapted species, confined to high latitudes, show greater diversity toward the poles or upper temperate zones. Examples of these are seals, gulls, penguins (in Antarctica only) some shorebirds, whales, and bears. These groups do not negate the broad principle that biodiversity is usually greatest in the tropics.

Just as species diversity usually decreases with increasing latitude from the equator to the poles, species diversity also decreases with increasing altitude from near sea level to higher elevations. In ascending mountains into more rigorous climates, the biodiversity of most types of organisms declines. Thus, we find fewer species of plants, birds, and mammals in high alpine regions than in the surrounding foothills and lowlands. Because of varying topography, distinctions are sometimes made between alpha diversity (within habitats), beta diversity (between different habitats), and gamma diversity (over a broad region).

Another relatively consistent pattern of diversity in relation to geography is that relating to area and island size. In general, larger areas have greater diversity. All three forms of diversity—alpha, beta, and gamma—show increases in the numbers of species in land areas of larger size (Fig. 20.1). Even point diversity, in very small, restricted samples, shows some tendency to increase in relation to area size.

On islands, either those in oceans and seas or forest "islands" surrounded by

TABLE 20.2
Number of species of breeding birds in various parts of the western hemisphere (Southwick, 1976).

Region	Degrees Latitude	Breeding Birds: Number of Species
Antarctica	90–65 (S)	20
Greenland	80–60 (N)	56
Labrador	60–52	81
Newfoundland	47–52	118
New York	41–45	195
Florida	25–30	143
Guatemala	14–18	469
Panama	6–9	1,100
Venezuela	1–10	1,148
Colombia	4–12 (S-N)	1,395

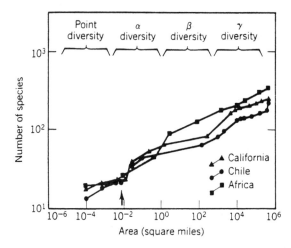

Figure 20.1 Diversity and species-area curves on three continents. Data are for numbers of bird species in chaparral habitat in California, Chile, and South Africa (from Colinvaux, 1993).

farmland or grassland, a similar relationship prevails, modified perhaps by such factors as the age of the island and its proximity to other areas of similar habitat. Bird species diversity in the West Indies of the Caribbean shows a definite increase in relation to island size, as do those of amphibians and reptiles (Fig. 20.2).

Data of this type have great relevance to conservation issues in determining how much area is required to support various populations and communities. Conservation biologists must decide what minimal areas should be protected and how to design nature reserves to ensure the survival of communities as well as endangered species.

A third interesting aspect of biodiversity is its temporal dimension. Biotic communities vary over time through the process of ecological succession. Thus, a cleared landscape originally occupied by a forest and allowed to revert to its basic community type will proceed over time through various pioneer stages (weeds and grasses) and intermediate stages (shrubs and young forest) to become a mature forest characteristic of the region's climate, soils, and topography, finally achieving what has been called the climax stage.

Diversity varies with these successional stages, also known as seres. Usually, biodiversity increases from the pioneer to intermediate stages; then it may stabilize or even decline toward successional maturity (Odum, 1993). Again, exceptions may occur in certain kinds of communities. In the mature tropical forest, species diversity may be at its greatest, especially if some edges occur along rivers or even small agricultural areas without excessive disturbance. In mature temperate forests, however, species diversity is usually lower than in areas with some intermediate growth. Natural disturbances such as natural fires, storms, and floods, if not devastating, may enhance diversity by resetting the successional clock.

WHY ARE THERE SO MANY SPECIES?

This simple question, of the type a child might ask, is actually profound and not fully answerable. It has occupied the thoughts of ecologists for many years with

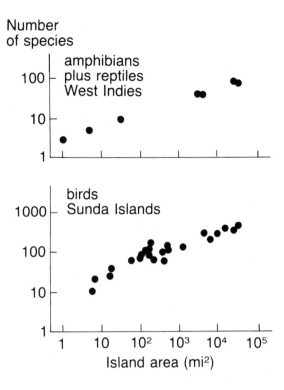

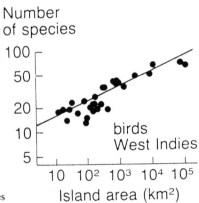

Figure 20.2 Species-area curves for birds, amphibians, and reptiles in the West Indies and for birds of the Sunda Islands, Malaysia (from Ricklefs, 1991).

no complete solution. The best answer we have at this time is that basic evolutionary processes, as we understand them, provide the driving forces for diversity. We know the earth is an infinite mosaic of environments. We further know that the elan vital of living organisms is based on the triad of survival, adaptation, and reproduction, collectively given the biological term *fitness*. Fitness has very subtle and precise parameters in which minor differences in structure, physiology, and behavior often make the difference between survival and death. Genetic variation provides the background for infinite changes in viability and fitness. Thus, variation and speciation is a continual phenomenon of life.

The classic paper in this field was written by the eminent ecologist G. E. Hutchinson in 1959. In "Homage to Santa Rosalia, or why are there so many kinds of animals?" Hutchinson speculated that speciation is driven by niche specialization due to competition for energy. The number of links in a food chain is limited by principles of energy flow and thermodynamics, so that most food chains have only three to six "links," or trophic levels, but the number of specialists in each trophic level is much less limited and capable of tremendous differentiation due to habitat complexity and environmental mosaics.

These considerations in biological philosophy leave open the question of why there are so many species in the humid tropics. Once again, a thorough answer is not available, but a general one may set us in the right direction, at least a direction for further research. We do know, however, that the physical and chemical processes of protoplasm function best at mild to warm temperatures with ample moisture conditions. Molecular and cellular processes have optimal temperature ranges for energy exchange, metabolism, growth, and reproduction. These optimal temperature ranges are typically from 10 to 40° Celsius, or about 50 to 105° Fahrenheit. These are precisely the year-round temperatures of tropical rain forests. Mild temperatures, coupled with abundant light and water, help to explain why evolution has flourished in the humid tropics. Of course, we have not answered the fundamental question of why molecular and cellular processes function best in these temperature and moisture conditions; that is perhaps a question for the biochemist. But we do have a basis for beginning to understand the exceptional wealth and diversity of life in the tropics.

HUMAN IMPACTS ON BIODIVERSITY

Virtually everything we do as human societies affects species diversity. Some of our activities enhance diversity; most reduce it. Among the beneficial activities are those aimed at conservation, such as the establishing of wildlife reserves and parks, species survival programs, and so forth. Many species, such as the gold lion tamarin, the California condor, the black-footed ferret, the whooping crane, the mountain gorilla, and quite a few others that we have driven to the brink of extinction would probably be extinct now without the dedicated efforts of conservationists to save them. In some cases these animals have been saved by protecting habitat, in other cases by captive breeding and reintroducing individuals or groups back into natural habitats. Botanical gardens, zoological parks, and both governmental and nongovernmental organizations are all playing important roles in the preservation of species and enhancement of biodiversity. Some species have made dramatic comebacks due to conservation programs; the bald eagle, peregrine falcon, brown pelican, rhesus monkey, and nilgai antelope are examples.

On the other hand, most human activities apart from those directed at conservation are detrimental to diversity. Deforestation, desertification, pollution, agriculturalization, urban sprawl, overgrazing, overfishing, excessive hunting and poaching, and simply high population pressure on the landscape—in most cases, all of these reduce natural biodiversity.

The most dramatic example is global deforestation. Tropical moist forests, which are thought to contain 50 percent of all the world's species, have been clearcut at the rate of 138,000 km^2 (53,000 square miles) per year, an area larger than the state of Pennsylvania (Houghton, 1994). By 1991, over half of the world's tropical forests had been destroyed. Currently, we are cutting tropical forests at the rate of almost 200 hectares per hour, equivalent to more than 400 acres per hour. This adds up to 42 million acres per year (Miller, 1993). Approximately 1.8 percent of tropical forests are being destroyed every year. Ehrlich and Wilson (1991) state:

> By the most conservative estimate from island biogeographic data, 0.2 to 0.3 percent of all species in the forests are extinguished or doomed each year. If two million species are confined to the forests, surely also a very conservative estimate, then extinction due to tropical deforestation alone must be responsible for the loss of at least 4,000 species annually.

The 4,000 species extinctions per year calculated by Ehrlich and Wilson might be higher, even by a factor of 10, if the number of forest species is higher than 2 million, as it may well be. Whatever this figure might be, if current rates of forest clearing are continued, one-quarter of all species on earth could be lost within the next 50 years.

The clear-cutting of tropical forests is not the only threat to earth's biodiversity. Some temperate forests have been destroyed at an even faster rate. In North America, less than 5 percent of our Pacific Coast redwoods are still standing, and old growth forests throughout the United States and Canada are being decimated because their timber is more valuable than that in secondary growth forests. We cannot take the moral high ground in preaching to Brazil, Indonesia, or African countries about their rates of deforestation when we are exceeding their rates of exploitation.

Other major forces in the loss of diversity are pollution, especially water pollution and the release of toxic chemicals, the destruction of grasslands and the spread of deserts, the uncontrolled expansion of cities, the agriculturization and commercialization of natural habitats, and unbridled overfishing and overhunting.

Regarding water pollution, two of the world's most diverse and most productive aquatic habitats, coral reefs and estuaries, are under assault from coastal pollution, including toxic chemicals, turbidity and siltation, eutrophication, sheer physical trauma from human activities, and possibly even global warming. The widespread phenomenon of coral bleaching (see Chapter 19) is an example of a global threat to marine diversity.

In freshwater pollution, limnological studies in Maryland's Patuxent River have shown that sewage outfalls reduce the diversity of a fish community 90 percent, from over 25 species to only 2 or 3 species capable of surviving in waters with high nitrate levels, high bacterial counts, and low oxygen.

Agricultural expansion, which is occurring so widely in developing nations to feed growing human populations, is also a large factor in the loss of species diversity. Agriculture, of course, is one of the driving forces in deforestation, but apart from this, the conversion of grassland to pasture and cropland inevitably causes a loss of diversity. Our own great midwestern plains are a clear example of this. In the western expansion of our frontier in the nineteenth century, we converted natural prairie grasslands which had over 100 species of grasses and forbs to monocultures of wheat and corn. In the process, we eliminated the passenger pigeon, the prairie wolf, and most of the great bison herds of the 19th century America. Flagrant overhunting certainly played a role in these exterminations along with habitat modification.

Much of the same process is in progress today throughout Africa, Asia, and Latin America, where a combination of overhunting and environmental destruction is reducing wildlife populations to small remnants of their former numbers. For example, both black and white rhinos now number less than 3,000 individuals, whereas just 30 years ago they numbered over 100,000. In 1970 the African elephant population was estimated to be 4.5 million; by 1990 only about 610,000 were left.

Still another way in which human activities reduce biodiversity is through the introduction of alien species. When Eurasian water milfoil was introduced into Chesapeake Bay, either intentionally or accidentally, probably around the early 1900's, it existed without much notice for 60 years. But in the 1960s, various ecological factors caused explosive population growth of this foreign species, so that it outcompeted many native plants including pondweeds, widgeongrass, water stargrass, wild celery, and a dozen or more other species. The diversity of aquatic vegetation in Chesapeake Bay was abruptly reduced from 20 species to only one or two. The native aquatic vegetation of Chesapeake Bay has been very slow to return, even with various replanting efforts and pollution control measures, and may never recover its former diversity.

The harmful effects of introduced species were first well documented in a book by the eminent British ecologist Charles Elton (1958). For example, the European starling was introduced into Central Park, New York, in 1891 and within 60 years, it had spread throughout the North American continent to the detriment and loss of many native birds, especially natural hole-nesting birds such as bluebirds. The aggressive and dominant starlings simply usurped the breeding niches of many native species. The European rabbit in Australia, the red deer in New Zealand, the muskrat in central Europe are additional examples of alien animals that have exploded in population numbers and distribution to the detriment of native species. Alien species often exploit new environments and new geographic ranges with extraordinary success because they may be free of the normal competitors, parasites, and predators that usually limit their numbers in native habitats.

Humans have also introduced pathogenic organisms to new environments. The fungal disease known as Chestnut blight, introduced to the United States accidentally on nursery stock from Asia in the early part of the twentieth century, was responsible for the extermination of native American chestnut in the United States.

The effects of alien organisms go well beyond the issue of biodiversity per se. Such organisms can create problems in agricultural economics, forestry, limnology, public health, and medicine, to mention just a few of the areas of human endeavor clearly affected.

IS THERE A CRISIS IN BIODIVERSITY?

Most ecologists believe that there is a crisis in global biodiversity in ecological terms. The time scale considered in reaching this answer is very important, however. In looking just one or two, or even four, years ahead, as many economic and political decision makers do, we might consider the loss of species a serious issue, but not a crisis. In looking 50 to 100 years ahead, however, the loss of biodiversity is clearly a crisis. Various authors have called this crisis an "extinction crisis" (Soule, 1986), an "extinction spasm" (Myers, 1987), a "megaextinction" (Myers, 1986), and "nothing less than a moral, scientific and economic tragedy" (Wilson, 1989, 1992).

Five mass extinctions are known to have occurred in earth's history, at the end of the Ordovician, Devonian, Permian, Triassic, and Cretaceous periods, spanning geologic time from 430 million years ago to 65 mya, when the dinosaurs met their final demise. Most of these occurred over a longer period than 50 to 100 years, however, and all were natural events. These events were very important in the earth's history, but they were quite different in cause and effect from the human-induced crisis in biodiversity that the world is now facing. The present extinction spasm is notable because of its extreme rapidity and the fact that it is the only one driven by a single species, human beings.

DOES DIVERSITY MAKE ANY DIFFERENCE?

Whether diversity matters depends on one's view of the human role on planet earth. Are we masters or members of the global community? Is the earth here for our exploitation, or are we guardians of its diversity and riches? Obviously, these questions are philosophical as well as scientific.

There are at least four kinds of reasons that we should be concerned about the conservation of biodiversity. These reasons are in the realms of ethics, esthetics, economics, and ecosystem function (Ehrlich and Wilson, 1991; Wilson, 1992).

First of all, most prevailing philosophies in many different religions consider human beings to be members of the global community, not its sovereign masters. All living organisms, which are products of millions of years of evolution, are considered by most of the world's religions to be under some divine guidance. As the dominant and most powerful living force on earth, we humans have an ethical responsibility to preserve the integrity and wholeness of living communities on earth. We certainly have the ability to destroy those communities, to devastate landscapes, and to turn whole ecosystems into wastelands, but most of us would agree that it is morally repugnant for us to do so. Our moral and ethical responsibility is to protect other species in the spirit of husbandry rather than destroy them in an attitude of conquest.

A second reason for concern over biodiversity is esthetics. Nature contributes immeasurably to the beauty of the world. Can we imagine a world without trees, grass, flowers, birds, and other animals? Perhaps yes, but such a world would be very artificial and impoverished. Professor Wilson of Harvard University, in his brilliant book *Biophilia* (1984), pointed out the almost universal need for contact with nature. We depend upon nature for beauty, renewal, and restoration. Even the most inveterate city dweller brings flowers into his apartment, displays pictures of wilderness scenes, and enjoys a day in the country. We are, for the most part, lovers of nature, and are troubled by a world without forests and fields, streams and lakes, beaches and coral reefs. We are concerned about a world without pandas or tigers or elephants or whales. We miss the sound of the nightingale or the call of the trumpeter swan. These concerns are not silly emotional worries; they are part of our basic nature. We need nature, just as we need love and recognition as a part of our humanity.

The world of economics also has some surprising ties to biodiversity that we tend to overlook. Most of our basic food supplies, many of our fibers, and a significant part of our medicinal products come from the natural world. To solve emerging world problems related to food supplies and health we will need to call increasingly on the world's storehouse of biological materials. We will need to find new food plants to meet changing environmental conditions, new medicines to meet the challenges of cancer, heart trouble, mental illness, and infectious disease. Some examples can emphasize the importance of biodiversity in meeting world problems.

Finally, biodiversity is central to ecosystem function, in all the complexities of biogeochemical cycles and the balance of nature. This is considered later under the topic of ecosystem services.

BIODIVERSITY AND HUMAN FOOD SUPPLIES

For human food supplies we rely extensively on the domestic forms of about 100 species of plants; these include grasses such as wheat, rice, millet and corn; legumes such as beans and peas; and vegetables such as potatoes and lettuce. Yet over the course of human history, 7,000 plant species have been utilized for food, and approximately 75,000 plants are known to contain edible parts (Wilson, 1989). Thus, the potential for increasing human food supplies from the world's storehouse of plants is great. In this storehouse there are undoubtedly plants with untapped potentials to meet the nutritional needs of expanding human populations.

One example is the winged bean, *Psophocarpus tetragonolobus,* which grows in New Guinea and is described by Wilson (1989) as a "one-species supermarket." Virtually the entire plant is edible—roots, stem, leaves, flowers, and seeds—and it has excellent nutritional value. Even the juice can make a coffee-like beverage. This plant can grow at the rate of 15 feet in a few weeks.

In many of the world's plants (and animals) there lies an incredible wealth of genetic material with the attributes of better growth, disease and insect resistance, drought and salinity tolerance, and the ability to grow on marginal soils, or withstand pollution. With modern genetic engineering, we can isolate some of these unique genetic capabilities and utilize them in the development of new food stocks. There are now vast losses of food crops to disease, insects, vertebrate pests, and inclement weather. By tapping the genetic resources of diverse species, we can make tremendous advances in developing plant and animal strains that can resist many of these onslaughts. The economic potential of the world's "genetic library" is virtually unlimited.

THE ROLE OF BIODIVERSITY IN FORESTRY AND ENVIRONMENTAL RESTORATION

What has been said for food plants applies equally well to trees for forestry, plants for fibers, and plants for environmental restoration. Again two approaches might be used to produce great economic benefits. One is to find new natural species that can be utilized for commercial purposes or used to develop domestic stocks through selective plant breeding. The other approach is to use the techniques of genetic engineering to insert special genetic qualities into existing stocks. Consider, for example, the economic benefits of new strains of Douglas fir or Norway spruce that would be genetically resistant to spruce budworm infestations, or strains of Ponderosa pine resistant to bark beetle and blue-strain fungus. Such strains would not only improve forest productivity, but they would result in significantly less pesticide and fungicide pollution of air, soil, and water. Imagine also the genetic capability to withstand urban air pollution and acid precipitation. If such genetic traits could be found and introduced to common forest trees in areas of high pollution damage, we would gain a little breathing room to solve pollution problems at their source. As it is now, many forests are dying before we can find any solution at all.

In a similar way, the world's natural biotic library undoubtedly contains species and genes with unique attributes to grow in damaged landscapes. Areas of salination, pollution, erosion, and soil loss are expanding throughout the world. Many of these areas are on a downward path of irreparable degradation. We need new species of plants that can vegetate degraded landscapes: species that can reverse the process of soil loss, species that will restore anthropogenic deserts to productive grasslands, species that will return forests to eroded mountains. Those in basic science and applied biology must work together in finding solutions to these problems. The plant and animal taxonomist analyzes and catalogues global biodiversity; the ecologist studies the distribution, abundance, and environmental tolerances of different species; the geneticist analyzes the genetic basis of taxonomy and adaptation; and the molecular biologist or genetic engineer considers procedures for developing the desired genetic qualities in the species under consideration.

BIODIVERSITY AND THE ENERGY INDUSTRY

We live in a society driven by fossil fuels. Although we use some energy from nuclear and alternative energy sources, including hydroelectric, solar, geothermal, wind-generated, and tidal, the daily energy needs of most industrialized countries are supplied primarily by hydrocarbons. In most agrarian countries, especially in Asia and Africa, the primary fuel source is wood and biomass. In the United States in 1850, 90 percent of the nation's energy came from wood and related biomass (Gates, 1985). Now about 85 percent comes from petroleum, coal, and natural gas. In many of the world's developing nations, the great majority of people's fuel still comes from wood. In India, China, and many nations of Asia and Africa, it is amazing how much human effort is expended to gather wood. Every stick and shrub is collected to provide the domestic energy source for cooking and heating. Such wood gathering is, in fact, a major cause of devegetation in much of the developing world.

Biodiversity can play a major role in meeting these relatively primitive fuel needs of agrarian nations, perhaps just in buying time until economic development can provide cleaner, better energy sources without denuding the landscape. For example, new species of microorganisms might be found to generate methane more efficiently from rice and wheat straw. New species and strains of plants can produce cooking oil to be burned instead of wood. For example, the babassu palm, *Orbignya phalerata,* from the Amazon basin, can produce 125 barrels of oil per year from 500 trees (Wilson, 1989). In the southwestern United States and Mexico, a desert shrub, *Euphorbia lathyris,* can yield up to 40 barrels of oil per acre on a sustainable basis (Chiras, 1985). Sunflowers, peanuts, soybeans, rapeseed, and various African palms can all produce oil that can be used in diesel engines, but at the present time existing strains are not productive enough in this regard to make them commercially valuable as energy sources. They illustrate the potential, however, for biodiversity to play a role in meeting the world's energy requirements.

BIODIVERSITY AND THE PHARMACEUTICAL INDUSTRY

Wild species of plants, animals, and microorganisms have been our greatest source of pharmaceutical products (Table 20.3). Medicines from natural species have yielded antibiotics, anticancer agents, antidepressants, heart stimulants, tranquilizers, muscle relaxants, pain-killing drugs, and anticoagulants, which are used in treating a wide variety of infectious and chronic diseases. We have barely tapped the potential of natural products. A wealth of new medicinals—perhaps

TABLE 20.3
Some important medicines derived from wild species of microorganisms, plants, and animals

Medicine or Pharmaceutical	Original Source	Medical Use
Penicillin	Penicillium mold	Antibiotic for treatment of bacterial infections
Chloromycin	Bacteria	Antibiotic for treatment of bacterial infections
Erythromycin	Bacteria	Antibiotic for treatment of bacterial infections
Streptomycin	Bacteria	Antibiotic for treatment of bacterial infections
Tetracycline	Bacteria	Antibiotic for treatment of bacterial infections
Quinine	Chinchona tree	Antimalarial
Resperpine	Rauwolfia plant	Treatment of high blood pressure and hypertension
Digitalis	Foxglove plant	Prevention of chronic heart failure
Vinblastine and vincristine	Rosy periwinkle	Treatment of cancer, Hodgkin's disease, and acute lymphocytic leukemia
Morphine and codeine	Poppy plants	Painkillers
Dicumerol	Alfalfa	Anticoagulant, used in treatment of heart attacks
Ancrod	Malayan pit viper	Anticoagulant, used in treatment of heart attacks
Cytabarine	Marine sponges	Antiviral
Bee venom	Honeybees	Arthritis treatment
Polio vaccine	Rhesus monkeys	Prevention of poliomyelitis and infantile paralysis

holding the cures for different types of cancer, neuromuscular diseases, degenerative diseases, and viral diseases—are still to be discovered in the earth's biota.

BIODIVERSITY AND ECOSYSTEM SERVICES

We take for granted the amazing fact that natural ecosystems "work." They produce diverse communities of plants, animals, and microorganisms that function in coordinated and productive ways. Forests, grasslands, marshes, lakes, estuaries, and coral reefs all work naturally to produce beauty and harmony. They perform all the functions and "services" outlined in Chapter 3 (see Table 3.1): they capture, convert, and store energy, they produce oxygen, they synthesize organic materials, they support species, they create and maintain order, they recycle wastes, they recycle essential minerals, they purify and store water, they create soil; they grow food, and they support life on earth.

A devil's advocate might say that we can accomplish all of these things with many fewer species and much less diversity. In fact, modern human life requires that we simplify natural diversity into agricultural ecosystems. Nonetheless, these agroecosystems depend upon soil, water, and air; they require subsidies in terms of energy and materials. They are not self-sustaining any more than a house, a garden, or a golf course is self-sustaining.

How many species does it take to make a self-sustaining system? That's a key question that we can't yet answer. The multimillion-dollar experiment known as Biosphere II outside Tucson, Arizona, attempted to answer that question by placing eight people and thousands of species in a sealed "biosphere" for two years to see if the system can be made to be self-supporting. The first experiment, which ran from September 1991 to September 1993, was generally successful, except the artificial Biosphere II ran low on oxygen, and some of the "Biospherians" had to be taken out of the system temporarily for medical treatment. The experiment did demonstrate how complex, difficult, and expensive it is to try to mimic the very things that nature does naturally. We take too many things for granted in nature, especially when our human egotism makes us feel we can do it all ourselves.

HOW MANY SPECIES DOES THE EARTH REALLY NEED?

The question of how many species we need is both a scientific and philosophical one. The answer, if one were possible, would involve consideration of an incredible number of related physical, chemical, and biological processes as well as complex ethical and aesthetic issues. Without doubt, the earth could probably get along quite well with fewer species, as long as we had the proper balance of producers, consumers, and decomposers, and as long as all basic biogeochemical cycles were in balance.

Unfortunately, we have no idea of what minimum number of species would be sufficient to maintain that balance. Indeed, we are engaged in a global, one-time experiment to see how far we can push the system.

Paul and Anne Ehrlich (1981) prefaced their book on extinction with the fable of the "Rivet Poppers." Suppose, they say, a financially strapped airline decides to remove and sell some rivets from its aircraft because it can make some extra money by doing so. After all, they reason, the airplane has more rivets than it really needs. Taking some rivets out also makes the plane lighter and more fuel-efficient. How many can be taken out before the wing falls off? We don't know since certain safety factors and redundancies were designed into the aircraft. Perhaps 10 percent could be removed and the plane would fly better, perhaps even 15 percent. At 20 percent, the plane might still fly better in good weather, but along comes some turbulence and the plane is in trouble. With a strong wind shear in a thunderstorm, the wings might disintegrate or the tail come off. Most of us are happy to have our airplanes as well as our automobiles engineered with provisions for safety.

The Ehrlichs point out that we can decide ourselves whether we want to fly in an airplane or ride in any given car. But we are all passengers without choice on Spaceship Earth. How far can we pop species out of the system before it might crash? We are playing with fire, and don't really know when the process of destruction will get out of hand. Both science and philosophical wisdom cautions us to be more prudent custodians of the earth's biosphere because it is our ultimate life-support system.

DIVERSITY AND STABILITY

The conventional wisdom in ecology is that within broad limits diversity contributes to stability. This is a logically attractive idea that makes common sense, but it has many exceptions and is based on relatively weak empirical data. This logic is plain when we compare the food chains in an Arctic and a temperate community. In a simplified Arctic terrestrial community, if *Cladonia* lichens die due to a disease or environmental problem, the entire community is jeopardized. Caribou, lemmings, and snowshoe hares suddenly lose their food resources, and all higher levels including human beings are at risk (see Fig. 6.1). In a temperate or tropical community with complicated food webs, the failure of any one producer does not necessarily jeopardize the entire community, since alternative producers can take over. There are obvious analogies in economics, where one-industry towns may be wiped out if that industry fails, whereas a town with multiple industries may survive.

Another source of evidence supporting the diversity-stability hypothesis is the fact that Arctic and desert regions with low biodiversity are the sites of some of the most dramatic population instabilities in nature. Lemming and snowshoe hare cycles and plagues of desert locusts are just two examples.

Nonetheless, there are many exceptions to the diversity-stability hypothesis, and the entire concept of ecological stability is even under question. Temperate and tropical ecosystems are not necessarily stable—they can change over time and space—so it is not entirely correct to say that diversity consistently leads to stability. Perhaps it is more correct to say that diversity leads to ecological resilience.

SUMMARY

Global biodiversity refers to the total number of species of living organisms on earth, a number that is unknown but currently judged to be in the general range of 5 million to 10 million.

Attempts to quantify species diversity in different communities involve various sampling procedures which combine the numbers of known species and their relative numerical balance. These statistical indices, of which the Shannon-Wiener, Margalef, and Simpson's indices are examples, provide numerical descriptions of community diversity. Different communities may have the same number of species but differ in diversity due to the proportional distribution of species.

Two major natural factors influencing biodiversity are latitude and habitat size. In general the highest species diversity occurs in the tropics, with decreasing diversity in higher latitudes toward the poles. Altitude influences diversity in the same way as latitude.

Habitat size (or island size) usually affects species diversity directly; that is, the larger the habitat or island, the greater the diversity, and the smaller the island, the less the diversity. This assumes that other factors such as habitat complexity are similar. Alpha diversity refers to diversity in small homogeneous habitats; beta diversity, across a variety of habitats; and gamma diversity, across a broad region.

Most human impacts including agriculture, urbanization, deforestation, desertification, pollution, and the introduction of alien species tend to reduce diversity. Human activities that increase or maintain diversity include preservation, conservation management, and reintroduction.

Many biologists believe the world is facing a human-induced extinction crisis, potentially greater and more rapid than any that has occurred in the past. A conservative estimate projects the loss of at least 4,000 species per year, with an accelerating rate of loss. The greatest loss is in tropical rain forests, the areas of highest diversity. Five previous mass extinctions have occurred in geologic time, but the current loss of diversity differs in being human-induced and very rapid.

The major areas of concern about biodiversity are ethical (do we have the moral right to eliminate species?), esthetic (the world will be impoverished without many plants and animals), economic (we will lose valuable genetic resources for extending food supplies, restoring habitat, and so on), and functional (we cannot stress ecosystems to the point where basic life-support functions fail).

REFERENCES

Chiras, D. 1985. *Environmental Science*. Menlo Park, CA: Benjamin Cummings.
Cody, M. L. 1975. Towards a theory of continental species diversity: Bird distributions over Mediterranean habitat gradients. In *Ecology and Evolution of Communities*, ed. M. L. Cody and J. M. Diamond, pp. 214–257. Cambridge, MA: Harvard University Press.
Colinvaux, P. 1993. *Ecology 2*. New York: John Wiley and Sons.
Ehrlich, P., and A. Ehrlich. 1981. *Extinction: The Causes and Consequences of the Disappearance of Species*. New York: Random House.
Ehrlich, P., and E. O. Wilson. 1991. Biodiversity studies: Science and policy. *Science* 253: 758–761.
Elton, C. 1958. *The Ecology of Invasions by Animals and Plants*. London: Methuen.
Erwin, T. L. 1982. Tropical forests: Their richness in Coleoptera and other arthropod species. *Coleoptera Bulletin* 36: 74–75.
Gates, D. M. 1985. *Energy and Ecology*. Sunderland, MA: Sinauer Associates.
Houghton, R. 1994. The worldwide extent of land use change. *BioScience* 44: 305–313.
Hutchinson, G. E. 1959. Homage to Santa Rosalia, or Why are there so many different kinds of animals? *Amer. Naturalist* 93: 145–159.
MacArthur, R. H., and E. O. Wilson. 1963. An equilibrium theory of insular zoogeography. *Evolution* 17: 373–387.
Margalef, R. 1958. Information theory in ecology. *General Systems* 3: 36–71.
Miller, G. T. 1993. *Environmental Science*. Belmont, CA: Wadsworth.
Myers, N. 1986. Mass extinction of species: A great creative challenge. Albright Lecture in Conservation, College of Natural Resources, University of California. Berkeley, California.
Myers, N. 1987. The impending extinction spasm: Synergisms at work. *Conservation Biology* 14: 15–22.
Odum, E. P. 1993. *Ecology and Our Endangered Life Support Systems*. 2nd ed. Sunderland, MA: Sinauer Associates.
Price, P. W. 1975. *Insect Ecology*. New York: John Wiley and Sons.
Purves, W. K., G. H. Orians, and H. C. Heller, 1992. *Life: The Science of Biology*. Sunderland, MA: Sinauer.

Raven, P. H., and G. B. Johnson. 1992. *Biology*. 3rd ed. St. Louis: Mosby–Year Book.

Raven, P. H., L. R. Berg, and G. B. Johnson. 1993. *Environment*. New York: Saunders, Harcourt Brace Jovanovich.

Ricklefs, R. E. 1991. *Ecology*. 3rd ed. New York: W. H. Freeman.

Ricklefs, R. E., and G. W. Cox. 1972. Taxon cycles in the West Indian avifauna. *Amer. Naturalist* 106: 195–219.

Sabrosky, C. W. 1952. *Insects: The Year of Agriculture, 1952*. Washington, DC, U.S. Department of Agriculture.

Soule, M. (ed.). 1986. *Conservation Biology: The Science of Scarcity and Diversity*. Sunderland, MA: Sinauer.

Southwick, C. H. 1976. *Ecology and the Quality of Our Environment*. 2nd ed. New York: D. Van Nostrand.

Stork, N. E. 1988. Insect diversity: facts, fiction and speculation. *Biological Journal of the Linnaean Soc.* 35: 321–327.

Systemmatics Agenda 2000: Charting the Biosphere. 1994. New York: American Museum of Natural History.

Wilson, E. O. 1984. *Biophilia*. Cambridge, MA: Harvard University Press.

Wilson, E. O. 1989. Threats to biodiversity. *Scientific American* 261(3): 108–116.

Wilson, E. O. 1992. *The Diversity of Life*. New York: W. W. Norton.

PART
IV

Human Prospects and the Quality of Life

Chapter 21

THE HUMAN CONDITION: ECONOMICS, DEMOGRAPHY, AND HEALTH

It is no secret that nearly 6 billion people are impacting planet earth in dramatic, often tragic, ways, but how is global change affecting us? What are human prospects as we enter the twenty-first century?

The human condition is an incredible mosaic of progress and defeat, prosperity and misery, peace and war, health and illness, happiness and despair, and all gradations between these extremes. The overall picture is not one to inspire pride or complacency.

Although the world has over 2 million millionaires (people with a net worth of over 1 million U.S. dollars), more than half the world's people live on a per capita annual income of less than $500. Most people in the world will earn less in their lifetime than a reasonable U.S. yearly income ($30,000). One hundred million people in the world are homeless, 400 million are so malnourished that "their minds and bodies are deteriorating," and over 1 billion human beings live in "absolute poverty," as defined by the UN (Durning, 1990; McMichael, 1993). Absolute poverty is a state of utter deprivation in which people do not have enough money to afford a sustainable diet, and do not know where their next meal is coming from.

The most tragic victims of poverty are children, who may suffer mental and physical retardation and often death. Even though early childhood deaths have been greatly reduced in the past 10 years (*UNICEF Report,* 1993), over 30,000 children still die every day from hunger and poverty (Anon, 1993)—a true Malthusian nightmare. The World Health Organization (WHO) and UNICEF estimate that more than 40 percent of the 14 million annual deaths among children could be prevented by vaccinations or other low-cost interventions such as oral rehydration (McMichael, 1993). Tragically, the result of high infant mortality is often an increased pregnancy rate; the highest birth rates typically occur in the poorest nations. We face again the fundamental question of whether poverty is the result of rapid population growth, or whether rapid population growth is the result of poverty. This highlights the common problem in ecology of distinguishing between cause and effect in complex interactions.

We do know that the highest rates of population growth occur in those nations of lowest economic status. The United Nations refers to this as the "PPE spiral, created in many developing countries by the interactions between poverty, rapid population growth and environmental degradation. Poverty leads to high rates of infant mortality, which encourages overpopulation. This in turn leads to unstable human pressure on land and other natural resources, bringing even greater poverty" (Lewis, 1993).

ECOLOGY AND WORLD HEALTH

Despite advancing medical science and increasing health care expenditures, the world health picture is very mixed. On one hand, life expectancy has increased dramatically in the latter half of the twentieth century (Weeks, 1992). In China, for example, life expectancy has increased from 51 in 1960 to 68 in 1995. In India, life expectancy has increased from 48 to 58.5 years in the same period. These increases offer clear evidence of better health care and better nutrition.

On the darker side of world health, some infectious diseases and many environmental illnesses are increasing markedly. Although the World Health Organization was once optimistic about eradicating malaria, over half the world's people still live in malarious areas, and 500 million are currently infected with malaria (Miller, 1992). An estimated 700 to 900 million people are infected with hookworm, and at least 200 million suffer from schistomiasis; both are debilitating parasitic diseases. Tuberculosis is increasing around the world at an estimated rate of 7 million new cases per year; at least 3 million of these confirmed annually in laboratory tests.

With the growth of food production slowing in many parts of the world, often slipping behind population growth rates in some developing nations, malnutrition and famine remain serious global problems. At least 1 billion people in the world are so poor they do not have adequate diets (Raven, Berg, and Johnson, 1995). There is often a virulent synergism between malnutrition (not eating the right kinds of food) or undernutrition (not eating enough food) and infectious disease. For example, chicken pox can be a fatal disease in malnourished children. Most of these infections and nutritional illnesses are related to ecology and environmental conditions (McMichael, 1993).

World health problems are greatly increased by the rise of the human immune viruses (HIVs) producing AIDS. These viruses attack us at one of our strongest yet most vulnerable points: our immune system. We have been able to survive and resist many pathogenic viruses, bacteria, protozoa, fungi, and helminths by virtue of our amazing immune competence. Human populations have survived devastating epidemics—because of our immune system. But HIVs cripple this system, making us vulnerable to many of history's scourges: influenza, tuberculosis, cholera, malaria, diphtheria, measles, pneumocystis, and others. Of the common infectious diseases of our past, only smallpox has been eradicated. This in itself is one of the great public health triumphs of the twentieth century, but it is not likely to be repeated with other pathogenic agents in the near future. The outcome of the global AIDS epidemic, which is affected by the propensity of viruses to mutate, the so-called "emerging viruses," is by no means predictable at this point (Stine, 1993).

Many chronic, noninfectious diseases, related to the environment, also have a great potential to increase. In both industrial and agrarian nations, chronic diseases such as asthma, emphysema, obstructive lung disease, cirrhosis, and many cancers, especially those related to environmental pollution, are on the increase. Global increases in lung and breast cancer, for example, while partially due to better diagnosis, are definitely real. Lung cancer is especially related to great increases in smoking in developing nations. We will have more to say on global patterns of health in the next chapter.

THE PROBLEM OF DISPARITY

We have seen that diversity in ecological systems is a desirable quality, but disparity in economic systems, the so-called "poverty gap," threatens the human future. Much of the violent turmoil in both domestic and international affairs can be interpreted as a product of economic disparity. Political power struggles, territorial incursions, terrorism, and violent crime have a common thread of anger and aggression over a lack of economic and political control. (This theme will be explored more thoroughly in Chapter 23.) Global violence usually has an economic base. Even the great political movements of communism and socialism originated in power struggles to deal ostensibly with the redistribution of wealth.

This places premium value on trying to understand the dimensions of economic and ecological disparity. Economic disparity refers to the differential distribution of wealth; ecological disparity refers to the differential distribution of life supports and successful environmental adaptations. The two are closely related but not exactly the same. A tribe of hunter-gatherers in an environment may have very little material wealth, but they may be successfully adapted to their life support system. It is not always easy to classify a given human condition as a problem of economic disparity or ecological disparity. For example, does famine in Somalia or Mozambique represent a failure of the environment or a failure of the economic and political system? The ecologist might say that the landscape is exhausted and incapable of supporting so many people, whereas the economist would say the famine is a result of human corruption, political chaos, and civil war. These disparities are obviously a result of both economics and ecology, but regardless of their origin, they have serious, sometimes deadly, implications.

In this chapter we will look at disparities in both economic and ecologic data for countries around the world—at economic factors such as GNP (gross national product) or GDP (gross domestic product) and energy consumption; at ecologic and demographic factors such as food consumption, life expectancy, and infant mortality; and at standard of living factors such as housing space, piped water, and electricity. While comparisons will be primarily between different nations, similar comparisons could be made within virtually every nation including our own.

Of approximately 192 nations in the world (184 in the United Nations as of January 1995), it has been common to classify them as industrial or agrarian; developed, developing, or lesser developed; first world or third world; or various other simplified categories. The facts are, of course, that nations differ in almost every conceivable quality: size, wealth, economics, politics, population, history, culture, language, topography, ecology, and climate. The comparisons presented here oversimplify very complicated human situations. Nonetheless, they illustrate some important facts about global disparities.

An additional caveat is that specific data sometimes vary between references. For example, the 1995 *World Almanac* lists the per capita GDP of the United States as $23,400 in 1992, whereas the 1994–95 *World Resources Guide* lists U.S. per capita GNP as $22,356 based on World Bank figures. Even with such discrepancies, however, broad trends can be discerned.

Another consideration is that these data, although collected internationally, are supplied by national sources within each country. There is no guarantee that the data were collected objectively. The accuracy of the data may sometimes be questioned because we do not know what criteria and sampling methods were used in each country. For example, Japan reports a literacy rate of 99 percent, Venezuela a literacy rate of 88 percent, and the United States a literacy rate of 95 percent. But each country may define literacy differently. Some nations consider an individual literate after one or two years of classroom schooling; others require an adult reading examination. A similar problem exists in governmental reports on the percentage of dwellings with piped water and electricity. Do these figures represent the total population? Do they adequately reflect both urban and rural populations? Are they based on actual household surveys or on questionnaires returned by mail? Virtually all national statistics involve sampling and survey methods which may differ considerably. Hence, it is important to recognize these problems, but despite their existence, the data at face value show many striking comparisons. Often disparities are so apparent that statistics merely add numbers to obvious conditions that anyone can observe.

ECONOMIC DISPARITIES

International data on GNP or GDP are highly variable and subject to many interpretations, but they still provide a general quantitative guide to the relative economic status of nations. They also provide a crude, but general indicator of per capita income. The richest nations of the world usually have per capita GNPs or GDPs between $15,000 and $30,000. Before the Persian Gulf War, Kuwait listed a per capita GDP of $19,700, a result of high oil prices and high oil production at the time. After the war, it reported a per capita GDP of $11,000, reflecting both lower oil prices and impaired oil production as a result of the war.

Most European nations fall in the GNP range of $5,000 to $20,000, and most of the more progressive developing nations in the range of $1,000 to $5,000 (Table 21.1). Some of the largest developing nations, such as China, India, and Indonesia, show less than $500 per capita GNP, but this may change rapidly with new trade policies and rapid economic growth in these countries.

Right at the bottom of the economic range are a number of poor nations, including Bangladesh, Ethiopia, Nepal, and Somalia, with per capita GNPs of less than $200. These countries have some of the most miserable living conditions in terms of inadequate housing, inadequate food supplies, and lack of health care,

TABLE 21.1
Comparative economic indicators (per capita) for various nations.

Nation	Annual GNP or GDP[1]	Energy Consumption[2] (gJl)	Daily Food Consumption[3] (cal)
United States	$23,400	280	3,641
Canada	19,600	288	3,340
Switzerland	22,300	108	3,449
Japan	19,800	111	2,852
France	18,900	114	3,529
United Kingdom	15,900	138	3,249
Germany	17,400	193	3,520
Spain	13,200	65	3,294
Israel	12,100	69	3,060
Kuwait	11,000	195	3,344
Czech Republic	7,300	183	3,393
Greece	8,200	66	3,668
South Korea	6,500	44	3,056
Portugal	9,000	38	3,204
Saudi Arabia	6,500	104	2,940
Mexico	3,600	50	2,870
Brazil	2,350	19	2,570
Malaysia	2,960	26	2,518
Venezuela	2,800	96	3,646
Iraq	1,940	15	2,146
Costa Rica	2,000	13	2,653
Indonesia	680	8	2,118
Haiti	340	2	1,905
China	360	19	2,426
India	270	7	2,056
Bangladesh	200	2	1,837
Somalia	170	4	1,986
Nepal	165		1,933
Ethiopia	130		1,793

[1] Data in U.S. dollars from *World Almanac*, 1995. GNP/GDP figures are variable in different data sources depending on several factors, including whether the "per capita" refers to adults only or all members of the population. The figures here refer to the later, and thus include children.
[2] Data in gigajoules (joules × 10^9, or one billion joules of heat energy). One gJl = $9.48 × 10^5$ BTU, or 278 kW-hr. One gJl is the equivalent of 0.1635 barrels of oil (approximately 6.9 gallons of oil). Data from Kurian (1991). Conversion formula from Gates (1985).
[3] In large Calories. Data from Kurian (1991).

education, and social services. Nations in this category often show a cluster of population characteristics including high infant death rates, high birth rates, high population growth rates, and low life expectancies.

Some analysts have argued that these nations operate on the barter system rather than by conventional economics, and thus GNP or GDP figures are meaningless. They contend that the cost of living in these nations is low. The facts are often the opposite; food, housing, and clothing expenses are often exorbitant, frequently higher than in the United States. It is difficult for us in the Western world to fully appreciate life for most people in countries with a GNP of less than $200. Usually they have an insecure existence in a mud or straw house with dirt floors, no windows, no electricity, no running water, and no heat in winter other than that provided by a fire of wood sticks or dung patties. They have no furniture other than a straw bed, rope charpoi (a wooden and rope cot), or reed mat on a hard floor. They have only one or two meals a day of rice or wheat cakes, occasional vegetables or fruit, rarely meat. They are frequently exposed to rain, wind, mosquitoes, and other biting insects, to bone-chilling cold in winter and to stifling heat and humidity in hot seasons. They are frequently ill; diarrhea, coughs, colds, fevers, and fatigue are almost daily facts of life. They have little or no schooling beyond a few years, and little or no prospect for a better job or a significant improvement in living standard. They have little or no access to health care and social services, not to mention entertainment and recreation beyond glimpses of satellite television.

Given these living conditions for a billion or more people, it is no wonder that the world has seen massive migrations of individuals from poor nations to Europe and North America. In the worst cases, economic disparities and political dislocations have brought whole societies to the point of dysfunction, collapse, terrorism, revolution, and war.

Some might argue that these problems are in the realm of the social sciences, not ecology, but they do involve the ecological relationships of the dominant population on earth, the human species. They involve competition and resource utilization. They literally involve a struggle for survival; individuals, families, and populations are leaving their homelands to find better living conditions. The world currently has more migrants and refugees than anytime in history, estimated in 1995 to number 23 million foreign refugees, 27 million refugees within their own country, and over 110 million migrants (Kane, 1995). Usually the factors stimulating these migrations are degraded environments and social conditions.

Nations can also be compared economically based on energy use. A commonly used index of energy use is the use of electricity, petroleum products, and other fossil fuels, and miscellaneous energy sources for both domestic and commercial purposes expressed as per capita consumption. The United States uses 280 gigajoules[1] of energy per person per year, whereas most European nations use less

---

1. A joule is a unit of energy equivalent to 0.239 calories, or 0.738 foot pounds, the latter being the amount of energy required to lift one pound approximately 9 inches. A joule is also equivalent to 9.48×10^{-4} BTU (British thermal units), or 2.278×10^{-4} watt hours. A gigajoule is 10^9 joules. What is important is not the absolute amount of energy used but the relative consumption patterns in different countries. The United States is an exorbitant user of energy, whereas many developing nations are very deficient in energy availability.

than 200, and most developing nations less than 100. The poorest nations, such as Bangladesh and Haiti, use only 2 gigajoules per person per year (see Table 21.1).

Energy plays a key role in both economic development (e.g., manufacturing, transportation, trade) and also the standard of living. Without electricity, for example, it is impossible to have modern homes, schools, and hospitals. Without electricity modern food storage is difficult, health care limited, and long-distance communication virtually impossible. Without motor vehicles, the speed of transportation is limited to a few miles per hour. The relative energy use of different nations varies by a factor of more than 100, as does the relative GNP or GDP. The United States, for example, has a GNP of more than 100 times greater than Bangladesh, and it also consumes more than 100 times more energy.

DISPARITIES IN THE QUALITY OF LIFE

The quality of life is an elusive trait that refers to the physical, psychological, and emotional aspects of life. A wealthy city-dweller in the United States or Europe may have many amenities of an abundant life, but may also be so overwhelmed by the tensions and stress of modern living that he or she dies of a heart attack at age 52, whereas a villager in China may have a very little material wealth but lead a peaceful and contemplative life until the age of 87. Which of them, we may wonder, has the better quality of life? "Standard of living" is a more objective term, but certainly does not give a full picture of human existence.

In physical terms, we can look at certain measures dealing with food consumption (see Table 21.1), housing, access to piped water, and electricity (Table 21.2). In terms of consequences or outcome, we can compare longevity, infant mortality, access to physicians, and literacy, as common measures of living standard (Table 21.3). At all times, however, it is important to keep these objective measures in perspective; they certainly give clues to the standard of living, but they do not completely reflect the quality of life.

Food consumption is an important attribute of both economic well-being and quality of life. Average adult food consumption in North America and Europe is generally over 3,200 calories per day. In terms of good health, this is often excessive, depending on the amount of physical activity. Our diet in the United States is too rich in calories, fats, and red meat proteins for ideal health. Japan is an exception among industrialized nations with a daily adult food intake of less than 3,000 calories (see Table 21.1), and also fewer calories in fats and red meat proteins. This diet may be one factor in the greater longevity of the Japanese (3.5 years longer on average) compared to Americans (see Table 21.3).

In per capita daily food consumption, developing nations typically average 2,100 to 2,800 calories, and the poorest nations drop below 2,000. Ethiopia and Bangladesh, for example, average around 1,800 calories per person per day. From 1988 to 1990 Ethiopia had only 73 percent of the calories per capita that were needed by its population, whereas the United States had 138 percent of its needed calories per capita (*World Resources,* 1994–95).

Caloric requirements vary greatly depending on body size, age, sex, activity,

TABLE 21.2
Comparative indicators of living standards (Kurian, 1991).

Nation	Living Space (rooms per dwelling)	Dwellings with Piped Water (%)	Dwellings with Electricity (%)	Physical Quality of Life
United States	5.1	97.6	100.00	98
Canada	5.7	99.5	100.00	98
Switzerland	3.6	100.00	100.00	99
Japan	4.7	94.00		99
France	3.7	99.2	89.9	100
United Kingdom	3.8			97
Germany	3.5	98.7	99.9	97
Spain	4.4	90.5	94.4	98
Israel	3.0	96.5	88.1	96
Kuwait	4.0	53.9	99.5	84
Czech Republic	3.5	91.6	100.00	93
Greece	3.5	81.3	88.1	97
South Korea	3.0	51.2	49.9	88
Portugal	3.9	73.4	77.6	91
Saudi Arabia				56
Mexico		66.3	74.6	84
Brazil	5.1	66.2	79.4	77
Malaysia	2.3	65.0	64.4	81
Venezuela		85.3		87
Iraq	2.4	20.8	17.1	62
Costa Rica	4.0	86.9	97.3	94
Indonesia	3.3	11.0	14.2	63
Haiti	2.2	12.0	1.1	48
China	2.2			
India	2.0	67.0	53.5	55
Bangladesh	2.0	56.8		43
Somalia				29
Nepal	3.7	47.7	30.2	36
Ethiopia				25

workload, climate (especially temperature), and other factors, but a generally accepted norm for typical adult energy is 2,100 to 2,700 calories (Goodheart and Shils, 1980). Intake below this level is unlikely to meet the basic maintenance requirements of an adult.

Calories are, of course, only one part of nutrition. The proper balance of carbohydrates, fats, lipids, and proteins is important, as are adequate amounts of vitamins and minerals. Many of the world's poor people have diets with adequate calories but inadequate protein, the cause of a severe deficiency disease known as kwashiorkor. Or they may have shortages of essential vitamins and minerals, which can lead to deficiency diseases such as Vitamin A blindness, beri-beri, scurvy, and rickets. Some of these deficiencies cause irreparable physical and mental handicaps in children.

Other quantitative measures of the standard of living include living space, dwellings with running water, dwellings with electricity, and and the physical quality of life index (PQLI), which is derived by averaging child mortality, literacy, and longevity.

Living space as measured by the number of rooms per dwelling gives a general idea of housing, but an overly conservative and incomplete one because it does not account for the size of the rooms or the numbers of people occupying them. The number of rooms per dwelling varies from an average of over five in the United States and Canada to only two in India and Bangladesh. If room sizes and family sizes are factored in, the discrepancies become even greater.

In regard to water and electricity in dwellings, most industrial nations, with a few exceptions, have well over 90 percent of houses and apartments with both water and electricity, whereas some poor nations have less than 15 percent with

TABLE 21.3

Comparative measure of health and education (The *World Almanac*, 1995).

Nation	Life Expectancy at Birth	Infant Mortality[a]	Population per Doctor	Adult Literacy[b] (%) Males	Females
United States	75.5	10	406	95.7	95.3
Canada	78.5	7	449	95.6	95.7
Switzerland	78.5	7	311	99.00	99.00
Japan	79	4	588	99.00	99.00
France	78	7	374	98.9	98.7
United Kingdom	77	7	611	99.00	99.00
Germany	76.5	7	319	95.9	89.9
Spain	77.5	7	257	95.9	89.9
Israel	78	9	345	95.0	88.7
Kuwait	75	13	515	80.5	73.1
Czech Republic	73	9	319	99.6	99.5
Greece	77.5	9	303	97.3	90.6
South Korea	70.5	22	1,007	97.5	87.9
Portugal	75.5	10	348	84.8	74.6
Saudi Arabia	68	53	523	58.0	34.6
Mexico	73	27	885		
Brazil	62	60	848	80.4	78.3
Malaysia	69	26	2,638	82.2	63.2
Venezuela	73	28	576	90.7	87.2
Iraq	66	67	1,922	65.9	26.0
Costa Rica	78	11	981	92.7	92.6
Indonesia	61	67	2,685	83.0	65.4
Haiti	45	109	6,083	37.1	32.5
China	68	52	648	83.5	61.2
India	58.5	78	2,337	54.8	25.7
Bangladesh	55	107	5,264	43.3	22.2
Somalia	54.5	126	19,071	18.4	6.5
Nepal	52.5	84	16,007	31.9	9.2
Ethiopia	52.5	106	36,660	93.0	0.5

[a] Infant mortality per 1,000 live births.
[b] Percent of adults who can read, based on criteria within each country and as reported by each country.

water, and Haiti has only 1 percent with electricity. Some of the numbers in Table 21.2 are suspect; for example it is unlikely that India has 67 percent of dwellings with piped water and 53.5 percent with electricity. Certainly the great majority of village India, which constitutes 70 percent of the total population, does not have water or electricity in their dwellings.

The PQLI, or physical quality of life index, varies from 99 or 100 to 25 among the nations listed in Table 21.2. This represents a combination of three of the traits in Table 21.3. In the industrial nations, life expectancies at birth generally average 74 to 77 years. In the poorest nations they average 45 to 55 years. Infant mortalities in the richest nations are usually less than 10 deaths per 1,000 live births; in the poorest nations they are over 100. Population per doctor in the richest nations averages around 400 persons; in the poorest nations, there are usually over 5,000 people per doctor, and as high as 36,660 in Ethiopia. In the latter case, if a doctor saw each person just once a year, the doctor would have to see more than 100 individuals per day, seven days a week! Obviously, adequate health care is impossible with such a ratio.

Finally, as an index of disparity in the quality of life, adult literacy varies widely, from 92 to 99 percent in the richest nations to less than 10 percent in the poorest. There are sharp differences between men and women, with women having the highest rates of illiteracy. Many desirable social goals—democracy, health care, education, family planning, and economic improvement—are largely unattainable if the majority of the population is illiterate.

Some of the problems of the data in the preceding tables have been mentioned, and, of course these data represent only a small sample of the data available. Even with such shortcomings, however, these figures show several important associations and trends.

First of all, it is apparent that economic status, energy use, and physical standards of living are positively related. The higher the GNP or GDP, the higher the energy use, and the higher the standard of living. Some nations, such as the United States, have extravagant energy use, and we could certainly maintain the same standard of living with much less energy if we conserved appropriately. Switzerland, for example, has a physical quality of life and longevity equal to or greater than the United States with less than one-half the energy use per person.

Secondly, it is apparent that food availability, medical care, life expectancy, and literacy all decline dramatically in the poorest nations. Infant mortality, on the other hand, increases markedly. There are some anomalies, however. China, with a very low per capita GNP, has a life expectancy greater than Saudi Arabia; it also has fewer people per doctor. The United States, one of the richest nations in the world, has an infant mortality greater than that of Spain even though the U.S. GNP per capita is $10,000 higher than that of Spain. In fact, despite the overall excellence of medical science in the United States, we usually rank nineteenth or twentieth in the world in infant mortality; that is, 19 or 20 nations have lower rates of infant mortality than we do. This statistic is sometimes explained by the diversity of ethnic backgrounds in this country, but it highlights a failure in our health care system.

Although there are some exceptions, the association between the economic status of a nation, its physical quality of life, and the health of its population is fairly consistent. Some of these associations are summarized in Table 21.4.

THE POVERTY GAP

The goal of many international organizations, governmental and nongovernmental, is to raise the economic standards of developing nations; that is, to narrow the poverty gap through grants, loans, technical advice, and cooperative assistance. This is the purpose of the U.S. Agency for International Development, the Peace Corps, the United Nations, the World Bank, the Asian Development Bank, the African Development Fund, the Inter-American Development Bank, the International Development Association, and many private organizations such as the Medical Assistance Program, Habitat for Humanity, various world relief agencies, and numerous philanthropic Foundations, to mention just a few.

International assistance to developing nations in economic development, agriculture, health, management planning, along with direct financial aid, has consistently been a multibillion dollar enterprise, not just from the United States but from many other high- and moderate-income nations as well.

TABLE 21.4
General associations between economic status, physical quality of life, and health in various nations.

Trait	Economically Wealthy Nations	Economically Poor Nations
GNP or GDP per capita	High $> \$10,000$	Low $< \$800$
Energy use per capita	High > 50 gJ1	Low < 20 gJ1
Daily calories per capita	High > 3000	Low < 2500
Dwellings with piped water (%)	High $> 90\%$	Low $< 50\%$
Dwellings with electricity (%)	High $> 90\%$	Low $< 50\%$
Physical quality of life index	High $> 95\%$	Low $< 65\%$
Life expectancy	High > 70 yrs.	Low < 60 yrs.
Adult literacy	High $> 95\%$	Low $< 70\%$
Infant mortality	Low $< 1\%$	High $> 10\%$
Birth rates	Low $< 20/1000$	High $> 30/1000$
Population growth rate	Low $< 1\%$	High $> 2.5\%$

Note: Some exceptions to these associations occur in individual nations, but these general trends prevail, and often the differences between wealthy and poor nations are more pronounced than shown here. Note that the first eight traits are all high in wealthy nations, low in poor ones, but the last three traits show the reverse pattern.

Many of these programs are genuine altruistic efforts to help people in low-income nations raise their standards of living. Other aid programs may have economic and political benefits to the grantor nations. These nations may be seeking to create markets, to win political hearts and minds, or to constrain and eventually defeat totalitarian regimes. The U.S. Foreign Aid budget in 1991 was 34.3 billion dollars; but unfortunately much of this spending took the form of military aid.

With massive amounts of foreign assistance, is the poverty gap narrowing or widening? Unfortunately, the old adage seems to be true: the rich are getting richer, and the poor are getting poorer, at least relatively so. Once again, there are many exceptions to this trend, but the overall pattern is one of a widening poverty gap. Figure 21.1 shows the average per capita GNP of industrial countries from 1960 to 1990, along with those of middle-income countries and low-income countries. GNP in the industrial nations has grown in this 30-year period from less than $4,000 to over $8,000, whereas in middle-income nations it has increased slightly from approximately $400 to $900, and in low-income nations it has remained less than $300. In 1960, the poorest 20 percent of the world's population earned 2.3 percent of global income, by 1991 this had fallen to 1.4 percent, whereas the richest 20 percent had increased their share from 70 to 85 percent (Kane, 1995).

In recent years, the United States per capita GDP has risen from $11,360 to $23,400, an increase in 10 years of 106 percent. In this same 10-year period, the per capita GDP of Mexico has risen from $2,130 in 1981 to $3,600 in 1991, an increase of 69 percent, and Brazil has risen from $2,050 to $2,350, an increase of only 14.6 percent.

Figure 21.1 Trends in per capita gross national product, 1960 projected to 1990 (from World Bank, 1979).

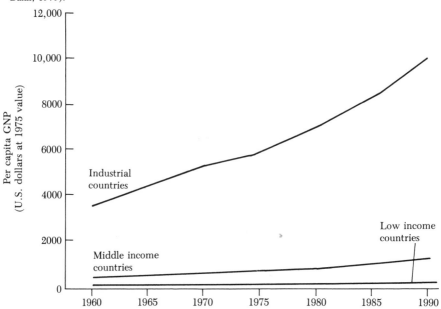

Among the low-income nations, India's per capita GNP has moved very slug-gishly from $240 in 1981 to $270 in 1993, an increase of 12.5 percent, and per capita GNP in Ethiopia actually declined from $140 in 1981 to $130 in 1993, a drop of 7.1 percent. Iraq's per capita GNP fell an even greater percentage, from $3,020 in 1981 to $1,950 in 1991, a drop of 35.4 percent. Could this have been a factor in Iraq's decision in 1991 to invade Kuwait, which at the time was en-joying a per capita GNP of $19,700, 10 times that of its belligerent neighbor?

Economic growth in the global economy is uneven and changeable, however, with some emerging nations showing remarkable development. Since 1960, sev-eral nations in East Asia have outdistanced Europe and the United States in eco-nomic growth rates. Starting with Japan, which increased per capita income four-fold from 1960 to 1985, other nations of the Pacific Rim—South Korea, Taiwan, Hong Kong, Thailand, Malaysia, and Singapore—have all shown similar in-creases.

The per capita GDP in Singapore, for example which became an independent nation on August 9, 1965, grew from $2,443 in 1969 to $16,500 in 1992, an increase of 575 percent in 23 years. This change in economic status has been accompanied by remarkable demographic changes. In less than 40 years, Singa-pore's birth rate declined from 42.7 per 1,000 in 1957, to 22.1 in 1969, to 17.0 in 1994. Infant mortality is now 6 deaths per 1,000, one of the lowest in the world and 40 percent lower than that of the United States. Life expectancy has increased from 64.4 years in 1969 to 76.0 in 1994. The population growth rate has declined from 4.3 percent per year in the 1950s, to 3.6 percent in the early 60s, to 1.1 percent in 1994.

Singapore's favorable economic and demographic changes have been accompa-nied by improvements in living standards. The physician to population ratio has improved from 1 doctor to every 1,623 people in 1969 to 1:757 in 1994. Virtually all measures of the physical quality of life—housing, food, education, electricity, communications, and transportation—have improved. Singapore has, of course, capitalized on its favorable trade location, but it has also shown ingenuity, hard work, and acceptance of change in its pursuit of economic betterment. It also has employed strong governmental inducements to change individual behavior, such as tax and housing incentives to limit family size and strong regulations against crime, including restrictions on individual freedoms.

The amazing economic success of smaller Pacific Rim nations shows signs of spreading to larger Asian nations such as China and Indonesia. Growth in East Asia was described in an editorial in U.S. News and World Report (Dec. 6, 1993) as follows (Zuckerman, 1993):

> By the end of this decade, there will be a billion Asians living in middle-class house-holds, virtually equal to the entire population of North America, South America and Europe. Four hundred million of them will have disposable incomes at least equal to those of the West today—able to afford houses, cars, international travel, holidays, edu-cation.
>
> . . . Asia is on the verge of an enormous consumer boom. Add the immense capital investment in such things as power plants, telephone switching, airplanes and air traffic control, and urban infrastructure, and it is easy to see that the modernization of Asia will be the most important international story for the next several decades.

China, with a population of 1.2 billion, is in a commercial frenzy. Deng Xiaoping has made many of the Chinese rich. One hundred million people may still remain in poverty in China, but 200 million people have moved out of it during the Deng era. The transformation between my first visit in 1977 and my visit this year is stunning. Construction cranes rise everywhere, markets overflow, department stores flourish.

If China can maintain the same growth rate over the next two decades that it has managed in the past 10 years, real per capita income will be equal to that of Japan in the '70's. China's gross domestic product would grow sixfold.

This picture of financial well-being should be tempered by several other considerations. First of all, probably more than 100 million of China's people remain in poverty; with a per capita GNP still under $500, the average Chinese citizen is a long way from enjoying an abundant and prosperous life. Secondly, both scientific reports and personal observations confirm that China's rapid industrial growth is producing massive pollution problems (Smil, 1993). Air and water quality in many of China's cities is totally unacceptable, and the living standard of most Chinese citizens do not begin to equal those of most Americans and Europeans.

SUMMARY

In terms of human economics and human prospects, the world represents great disparities of wealth and poverty, health and illness, happiness and despair, success and failure.

The industrial and high-income nations of the world, primarily those of the northern hemisphere, show high levels of GNP (Gross National Product) or GDP (Gross Domestic Product). Characteristically, the richest nations have the highest rates of energy use per capita and the largest caloric intakes per person, usually over 3,000 calories per day with abundant nutrients.

In contrast, the economically poorest nations of the world have very low GNPs or GDPs, as low as only 1 percent of that of the wealthiest nations. They have very low rates of energy consumption, and diets deficient in both calories and nutrients.

The richest nations of the world typically have larger housing units containing fewer people, and most dwellings have electricity and piped water. Their physical quality of life is quantitatively higher than that of low-income nations.

With regard to life expectancy, medical care, and literacy, the wealthy nations of the world are markedly better off than poor nations. The wealthy nations usually have much lower rates of infant mortality, lower birth rates, and lower rates of population growth.

The poverty gap between high-income and low-income nations widened between 1960 and 1990 and continues to show this trend. Some notable exceptions have occurred, but on average the industrial nations doubled their per capita GNP or GDP in this 30-year period, many middle-income nations increased only 25 percent, and some low-income nations increased less than 10 percent or actually declined.

International economic fortunes have undergone remarkable changes in the latter half of the 20th century, and show every indication of greater change ahead. Record economic growth has occurred since 1960 in several Pacific Rim nations, including Japan, South Korea, Taiwan, Hong Kong, Thailand, Malaysia, and Singapore. This economic growth is now extending to China and Indonesia, and hopefully it will come to other developing nations.

The prospects for human betterment are great, but environmental and population pressures threaten to counteract many of the gains. Rapid industrialization often leads to excessive pollution, urban sprawl, and overall environmental deterioration. The great challenge will be to achieve economic growth without destroying environmental and social values.

REFERENCES

Anon. 1993. Bread for the World Institute for Hunger and Development. Reported in *Worldwatch* 6(6): 37.

Durning, A. B. 1990. Ending poverty. In *State of the World, 1990,* ed. L. R. Brown, C. Flavin, S. Postel, and L. Starke. pp. 135–153. New York: W. W. Norton.

Gates, D. M. 1985. *Energy and Ecology.* Sunderland, MA: Sinauer Associates.

Goodhart, R. S., and M. E. Shils (eds.) 1980. *Modern Nutrition in Health and Disease.* Philadelphia: Lea and Febiger.

Kane, H. 1995. *The Hour of Departure: Forces that Create Refugees and Migrants.* Worldwatch Paper No. 125. Washington, DC: Worldwatch Institute.

Kurian, G. T. 1991. *The New Book of World Rankings.* 3rd ed. New York: Facts on File.

Lewis, P. 1993. Fatal diseases declining for children. Excerpted from N.Y. Times. *Boulder Camera* (Dec. 21, 1993).

McMichael, A. J. 1993. *Planetary Overload: Global Environmental Change and Health of the Human Species.* New York: Cambridge University Press.

Miller, G. T. 1992. *Living in the Environment.* 7th ed. Belmont, CA: Wadsworth.

Raven, P. H., L. R. Berg, and G. B. Johnson. 1995. *Environment.* Philadelphia: Saunders College Publishing.

Smil, V. 1993. *China's Environmental Crisis: An Inquiry into the Limits of National Development.* Armonk, NY: M. E. Sharp.

Stine, G. J. 1993. *Acquired Immune Deficiency Syndrome: Biological, Medical, Social and Legal Issues.* Englewood Cliffs, NJ: Prentice-Hall.

UNICEF Report on Child Health. 1993. New York: United Nations.

Weeks, J. R. 1992. *Population: An Introduction to Concepts and Issues.* 5th ed. Belmont, CA: Wadsworth.

World Almanac and Book of Facts. 1995. Mahwah, NJ: Funk and Wagnalls.

World Resources: 1994–95. A Guide to the Global Environment. New York: Oxford University Press.

Zuckerman, M. B. 1993. The wealth of nations. *U.S. News and World Report* (Dec. 6, 1993): 100.

Chapter 22

GLOBAL PATTERNS
OF HEALTH

International differences in longevity and infant mortality discussed in the preceding chapter are related to the types of disease in nations of different economic status. Human health patterns reflect, in part, the nature and quality of the human environment. At the level of individual health, there are many exceptions to this premise, but at the population level the correlation has considerable validity. The study of world health is generally called medical geography or epidemiology.

From the standpoint of the economic status of nations, there are two broadly discernible patterns of disease in high-income nations compared to low-income nations.

INDUSTRIAL HIGH-INCOME NATIONS

The primary health problems in the wealthiest nations of the world are noninfectious chronic diseases (Chivian et al., 1993). Heart disease and cancers lead the list in most industrialized countries. Major morbidity[1] and mortality[2] factors are cardiovascular diseases (diseases of the heart and blood vessels), malignant neoplasms (cancers), chronic respiratory diseases, and violent deaths through accidents, homicides, and suicides. In the United States, diseases of the heart and blood vessels and cancers account for over 60 percent of all deaths (Table 22.1). These plus chronic respiratory and digestive tract diseases, diabetes, and violent deaths account for nearly 80 percent of all deaths in the United States. In contrast,

---

1. Morbidity refers to disease or illness, specifically, the rate or prevalence of disease. Disease refers to any state of illness, whether it is infectious, nutritional, metabolic, congenital, developmental, or genetic in origin or etiology.
2. Mortality refers to death, specifically, the rate or prevalence of death.

TABLE 22.1
Estimated causes of death in the United States.

General Mortality Factor	Mortality Rate per 100,000 per Year	Percentage of All Deaths
Cardiovascular diseases (e.g., heart disease, hypertension, stroke, atherosclerosis, etc.)	345.0	39.2
Malignant neoplasms (all cancers, including lymphomas and leukemias)	205.8	23.4
Chronic noninfectious pulmonary diseases (including asthma, emphysema, chronic bronchitis)	39.2	4.5
Accidents (e.g., vehicular, home, workplace)	34.4	3.9
Influenza and pneumonia	31.7	3.6
HIV-AIDS	14.9	1.7
Other infectious and parasitic diseases[1]	23.4	2.7
Chronic digestive tract disease (e.g., cirrhosis, hernia, kidney disease)[1]	24.5	2.9
Homicide and suicide (including violent assault)	22.0	2.5
Diabetes (including related complications)	21.4	2.4
Congenital and perinatal deaths[1]	11.3	1.3
Miscellaneous and undetermined causes	105.7	11.9
All causes	879.3	100.0

Source: World Almanac (1995). Data from National Center for Health Statistics, U.S. Dept. of Health and Human Services, 1993, except as noted.
[1] Data from National Center for Health Statistics, U.S. Dept. of Health and Human Services, 1991. Not available in these categories for 1993–95.

infectious and parasitic diseases account for only 8 percent of all deaths in the U.S. (Table 22.1).

Other industrial nations have similar patterns of mortality. In industrialized nations in general, including those of Europe and Japan, cardiovascular diseases, related degenerative diseases, and cancers recently accounted for 74.5 percent of all deaths (Veil, 1992).

In the United States the major causes of death are closely linked to our physical and social environment. Medical experts concede that 70 to 80 percent of all cancers are "environmentally related," that is, triggered by environmental factors (Blumenthal, 1985; Chivian et al., 1993). This includes smoking as an environmental factor, and over 400,000 deaths per year in the United States are attributed to smoking. Smoking is also a major factor in cardiovascular diseases, especially hypertension and heart disease. The fast-paced and stressful social environment of industrialized nations also plays a role in chronic disease. In fact, the major causes of death in the United States are sometimes called "lifestyle diseases." Many of them are associated with our sedentary habits, excessively rich diets, stressful jobs, smoking and drinking, drug addiction, violent behaviors, and industrial pollution.

DEVELOPING LOW-INCOME NATIONS

Data on the causes of death in many developing nations are not available because medical determinations of death are not made routinely and when causes are assigned, national records are not always maintained.

Those reports that are available show that in nations with the lowest economic status the causes of morbidity and mortality are primarily infectious and nutritional diseases. A common designation of illness or death in medical records is FUO, standing for fever of unknown origin. Respiratory infections, such as flu, pneumonia, diphtheria, and tuberculosis, and gastrointestinal illnesses, such as viral diarrhea, amoebiasis, and bacillary dysentery, are the most common killers of both children and adults. Many childhood infections, such as measles, whooping cough, and chicken pox, become fatal in nutritionally deprived children. In famine conditions, the frequency of starvation deaths is often increased by infectious disease. Malnutrition weakens the immune system and increases both the likelihood and severity of infectious disease.

A recent report of the World Health Organization Commission on Health and Environment (Veil, 1992) concluded, "Interactions between nutrition and infection to produce the 'malnutrition/infection complex' create the greatest public health problem in the world." This report presented data from a group of developing nations showing that over 50 percent of all deaths were due to infectious and parasitic diseases (Table 22.2). Less than 15 percent of all deaths were attributable to cardiovascular diseases and less than 6 percent to cancers. Chronic diseases that claim approximately three-fourths of all deaths in the United States are responsible

TABLE 22.2
Estimated causes of death in developing nations (Veil, 1992).

General Mortality Factor	Mortality Rate per 100,000 per Year	Percentage of All Deaths
Cardiovascular diseases (e.g., heart disease, hypertension, stroke, atherosclerosis)	134.4	14.7
Malignant neoplasms (all cancers, including lymphomas and leukemias)	51.7	5.7
Chronic noninfectious pulmonary diseases (including asthma, emphysema, chronic bronchitis)	47.6	5.2
Accidents (vehicular, home, workplace; all external)	49.6	5.4
Respiratory infectious diseases (including influenza, pneumonia, tuberculosis, pertussis, diphtheria)	322.5	35.3
Other infectious and parasitic diseases (e.g., diarrheal disease, malaria, schistosomiasis)	159.2	17.4
Congenital and perinatal deaths	76.5	8.4
Miscellaneous and undetermined causes	72.3	7.9
Total	913.8	100.0

for only about one-fourth of deaths in these developing nations. These figures tell us a great deal about environmental conditions in different economic and ecologic situations, and they also have great implications for action plans and remedial programs.

Table 22.2 does not list data for malnutrition or starvation deaths. Such deaths are often not reported because most nations do not readily admit to starvation problems unless they have an absolute crisis such as Somalia in 1993. Many medical studies indicate, however, that malnutrition is an important factor in high perinatal mortality and in death and disease attributed to miscellaneous or undetermined causes.

In developing nations vector-borne parasitic infections are also a common cause of morbidity. The most prominent infectious diseases are malaria, schistosomiasis, trypanosomiasis (sleeping sickness in Africa, Chagas disease in South America), leishmaniasis, and filariasis. These diseases tend to be chronic and debilitating infections, but malaria is also a leading cause of childhood death in tropical countries. The others certainly shorten life span.

Malaria has increased dramatically in Africa and some other tropical areas. The causative organism, *Plasmodium,* has become resistant to commonly used antimalarial drugs such as chloroquine, and the mosquito vectors have become resistant to insecticides. The WHO has estimated that 2.1 billion people are now at risk of malaria in the world (McMichael, 1993). Schistosomiasis, or bilharzia, a helminthic infection, has spread with the increase in irrigation systems, with an estimated 600 million people now at risk (McMichael, 1993).

A recent report of the nongovernmental MAP International (originally the international Medical Assistance Program) showed that nearly 25,000 childhood deaths per day in the world are due to just seven infectious diseases: malaria, pneumonia, diarrhea, measles, whooping cough, tetanus, and tuberculosis. In many of these deaths, malnutrition is a predisposing factor.

Despite these tragic levels of infant mortality, there has been some encouraging progress in child health around the world. A recent UNICEF report, *The State of the World's Children,* indicated global reductions in infant deaths from pneumonia, diarrhea, measles, tetanus, and whooping cough, largely due to expanded vaccination programs in developing nations. In the last 10 years, UNICEF reported that deaths from measles have declined from over 2.5 million per year to just over 1 million per year. Deaths from neonatal tetanus dropped from just over a million to approximately 500,000, and deaths from diarrheal disease have declined from over 4 million to under 3 million. Polio has also dropped markedly from half a million cases of paralysis to an estimated 140,000 in 1992. UNICEF's primary goal is to reverse the damaging spiral of poverty, high mortality, high birth rate, rapid population growth, and environmental degradation (Grant, 1993).

IMPLICATIONS FOR PUBLIC HEALTH

In both developing and industrial nations health problems are intimately related to environmental factors. This point is emphasized by Dr. Anthony Cortese (1993):

We have known for centuries that a healthy environment is essential for human existence and health, and that contamination of the environment with heavy metals, microorganisms, physical agents, and certain organic compounds can cause serious illness and death.

Protection of the environment and preservation of ecosystems are, in public health terms, the most fundamental steps in preventing human illness.

In developing nations, lack of safe drinking water, inadequate food, poor housing, and poor sanitation are obvious factors in the high rates of disease. In urban slums people are crowded into miserable conditions in which cleanliness is practically impossible and airborne infections are virtually omnipresent. Vermin are also common in such habitats; flies, mosquitoes, cockroaches, fleas, lice, rats, and mice are all potential reservoirs or vectors of disease.

In rural environments as natural habitats are invaded by human agriculture, ecological conditions are created for the spread of zoonotic infections such as yellow fever, typhus fever, and Chagas disease. Often these zoonotic infections do not cause illness in their animal hosts. For example, the protozoan *Trypanosoma cruzi* is not seriously pathogenic in its wild mammalian host, the South American opossum, but when it is transmitted by a bloodsucking bug to a human being, it causes serious and ultimately fatal Chagas disease in the human host. The bug, also known as a kissing bug, or *Triatoma* bug, occurs in substandard housing in Latin America, especially in wooden dwellings with thatched roofs and dirt floors.

The ideal ecological situation for the occurrence of Chagas disease is tropical deforestation with the invasion of agricultural homesteading into forested areas. This situation promotes, even forces, ecological contact between the natural hosts, vectors, pathogens, and human hosts. Another tropical infection with the same ecological background is yellow fever, a virus infection transmitted by mosquitoes. Traditionally, epidemic yellow fever occurred in tropical cities of South America and Africa where *Aedes aegypti,* an urban mosquito, transmitted the virus. Antimosquito campaigns and yellow fever vaccination programs controlled the worst urban epidemics, but in more recent years, sylvan, or jungle, yellow fever has occurred at deforestation sites. Field studies have shown that the natural viral infection in monkeys, transmitted by a treetop mosquito, *Hemagoggus speggazini,* is carried to woodcutters, farmers, and other laborers working along the forest edge, especially those cutting trees. The mosquito, which normally lives in the forest canopy, is swept down to the ground as trees are cut. Their preferred hosts, the monkeys, flee, and as a secondary preference the mosquito bites a human being and transmits the yellow fever virus. Outbreaks of yellow fever from Brazil to Guatemala have had this type of epidemiology related to human invasion of natural habitats.

There are more than 150 infectious diseases of zoonotic origin. These diseases occur in both animals and humans, and many of them spread from animal to human populations as we invade or alter natural environments (Acha and Szyfres, 1991) or as animal hosts and vectors reach pest proportions. Some of the best known zoonotic infections are plague, rabies, encephalitis, giardiasis, tularemia, and Lyme disease.

For maximum public health impact in many developing nations, the greatest effort should be expended on improving environmental quality. This can be done by providing clean water, adequate food, better housing, better sanitation, less crowding, reduced population growth, control of insects and vertebrates that provide the vectors and reservoirs of disease organisms, and vaccination programs to eliminate diseases such as tetanus, diphtheria, whooping cough, measles, and polio (McMichael, 1993).

In high-income industrial nations, health statistics indicate that the greatest threats to human health come from the social and industrial environment. This is not to say that all problems of water quality, food supply, and housing have been solved, certainly not by any means, but the threat of infectious disease from impure water, contaminated food, poor housing, and vermin is less than in low-income nations. Poor health is more likely to be caused by stressful living, environmental pollution, overeating, drug use, alcoholism, and violence—all aspects of our social and cultural environment. The physical environment, of course, plays a prominent role in the transmission of toxic chemicals into our air and water. Some kinds of cancer, chronic respiratory disease, and perhaps certain metabolic diseases are related to industrial activities, including high energy use and high automobile use. Industry provides us with many conveniences and luxuries, of course, but it also contributes toxic elements to our environment, which may permanently alter the biosphere. Cortese (1993) summarized the effects of industry on public health and the environment as follows:

> The threatened destruction of this natural support system will be one of the most important international issues to be faced by society and by the public health community in the 21st century.
>
> The rate of industrialization, which is far outpacing the growth of population, is an even more powerful determinant of environmental transformation. In the past 100 years, the world's industrial production increased 100-fold. From 1950 to 1980 manufacturing increased by a factor or 7, the use of electricity by a factor of 8, the number of automobiles by several orders of magnitude. The impact on the global environment has been dramatic. In 150 years human activity has increased the atmospheric concentration of carbon dioxide by 24 percent, doubled the concentration of methane, and introduced long-lived ozone-destroying chlorofluorocarbons into the stratosphere. Man-made emissions of sulfur and nitrogen, which lead to acid rain, now equal or exceed the natural flux of these elements. And man-made emissions of lead, cadmium, and zinc exceed the natural flux by a factor of 18.
>
> It is time that [physicians, public health professions, and environmental specialists] joined forces to work on the inseparable goals of preserving the environment and promoting the health and well-being of the global population.

It is worth noting again there are similarities and differences in the public health needs in low-income and high-income nations. In both, the environment plays a major role, with different environmental components that are most important. In both, the need for education is strongly evident. In many developing nations, education is a necessary ingredient for improving water and food supplies, the quality of housing, the need for sanitation and infectious disease control. In industrial nations, education in necessary for improving diet and exercise, reducing stress and violence, and in helping people lead less consumptive, less polluting, and less destructive lives.

HUMAN IMMUNE VIRUSES AND AIDS

The World Health Organization statistics in Table 22.2 do not specifically list AIDS as a cause of death. Only recently has AIDS been listed as a mortality factor in the U.S. national health statistics. Perhaps this is because AIDS is a new and complicating factor in public health, or perhaps it is because technically HIV infections themselves are not a direct cause of death. AIDS contributes to death indirectly by causing a breakdown in the immune system, leading to death by either an opportunistic infection such as pneumonia or a malignant neoplasm such as Kaposi's sarcoma. This is now recognized as a technical point, and AIDS is in fact listed as a cause of death in the U.S. There is no doubt that AIDS is increasing in importance as a major mortality factor. In some age and sex classes, especially young men in the 25 to 45 age group, AIDS has become the major contributing cause of death.

Although not everyone with an HIV infection will develop AIDS, once the immune deficiency symptoms of AIDS occur, the disease is almost always fatal. Stine (1993) described the disease process as follows:

> Because of a suppressed and weakened immune system, viruses, bacteria, fungi and protozoa become pathogenic. [The result is] a demoralizing, debilitating, painful, helpless and unending struggle to stay alive. About 88 percent of deaths related to HIV infection and AIDS are caused by OIs [opportunistic infections] compared with 7 percent due to cancer and 5 percent due to other causes.

The mechanisms by which the human immune viruses cause AIDS are not well understood (Weiss, 1993), and there are actually some scientists who contend that AIDS is not caused by the HIVs (Duesberg, 1993). The latter opinion is based on the observation that not all HIV-positive individuals develop AIDS, although this is true of only a small minority and is entirely expected. It is also based on the observation that some AIDS patients show very low concentrations of viral antibodies in their blood or are HIV-negative. Both of these situations are true of other viral infections, however, and both are to be expected in the case of newly discovered viral infections with long latent periods and patterns of pathogenesis that are not well understood. In the case of AIDS and HIV infections, we do not fully understand if disease is caused primarily by the depletion of T-cell helper lymphocytes or by the stimulation of an autoimmune reaction (Weiss, 1993). The controversy over the role of human immune viruses in the causation of AIDS has been summarized by Cohen (1994).

Despite these questions, the overwhelming scientific evidence supports the view that HIVs are responsible for the clinical symptoms of AIDS. Stine (1993) summarizes this view:

> The virus causes few symptoms on its own; what makes HIV disease particularly horrible is that it leaves patients open to an endless series of infections that would not occur in people with healthy immune systems. *Pneumocystis carinii* pneumonia, toxoplasmosis, Kaposi's sarcoma, candidiasis, cytomegalovirus retinitis, cryptococcal meningitis, mycobacterium avium complex, herpes simplex, herpes zoster—these infections sicken, disfigure, and some eventually kill most people with AIDS.

These thoughts on the pathology of just one disease may seem a long way from the central issues of global ecology, but the fact is that AIDS is a unique ecological situation in the world. It is not just another potential epidemic like plague or cholera that can be controlled by medical technology. On a global basis, the human immune viruses are particularly insidious and dangerous, not only because of their devastating effect on human health, but also because HIV infections are often inapparent and undiagnosed for months and years. Antibodies do not normally appear until six to eight weeks after infection. Blood tests done during this period are usually falsely negative, yet this is the period of greatest infectivity. Clinical symptoms of disease in teenagers and adults may not appear for several years. In children, symptoms occur much sooner, but teens and adults may be unaware of their infection for years if they are not tested. In 1989 and 1990, the average time between HIV infection and the onset of AIDS was 9.8 years (Stine, 1993). During this time, infected individuals may spread the virus unknowingly to all sexual contacts.

There is thus a hidden backlog of infection in the population and considerable momentum for the silent spread of the epidemic. For every diagnosed case of AIDS, there are at least 10 people who are HIV-infected, and some estimates run as high as 50 or 100 people (Stine, 1993). HIV transmission is occurring throughout the world via heterosexual and homosexual intercourse, injection drug abuse, contaminated blood transfusions, childbirth, and breast-feeding. Not every HIV-infected individual will necessarily develop AIDS, but the great majority will, and for those that do the long-term outlook is "virtually 100% fatal" (Dr. C. Everett Koop, quoted in Stine, 1993).

AIDS was first recognized as a clinical entity in Los Angeles in 1981, although there is retrospective evidence from antibody surveys of frozen blood sera that the viruses responsible for AIDS may have existed since the 1950s. Various virus isolates capable of infecting lymphocytes were identified in the United States and France from 1978 to 1985. Different virus types and nomenclatures are now reduced to two types, HIV-1 and HIV-2, although these types are highly variable and capable of frequent mutation. It is very likely that more HIV types will be found.

The growth of AIDS in the United States is shown in Figure 22.1, and projected worldwide cases of HIV infections and AIDS are shown in Figure 22.2. Although there are various reports of the AIDS epidemic slowing in the United States, 1992 was notable by the addition of 2 million new AIDS cases in the world, and some projections indicate as many as 12 million AIDS cases worldwide by the year 2000. In 1994, four million people contracted HIV infections, including 400,000 newborn babies (Sachs, 1995).

AIDS is not only an expensive threat to the health-care systems of the United States and Europe, it is even more devastating to many developing nations. The U.S. government spent 1.3 billion dollars for AIDS research and prevention in 1994, and this did not include all the individual costs of treatment, lost productivity, and wasted lives. In the United States, hospital care, drug treatment, and blood screening are among the best that modern medicine can offer, but in many developing nations, medical care and health personnel are in short supply, blood

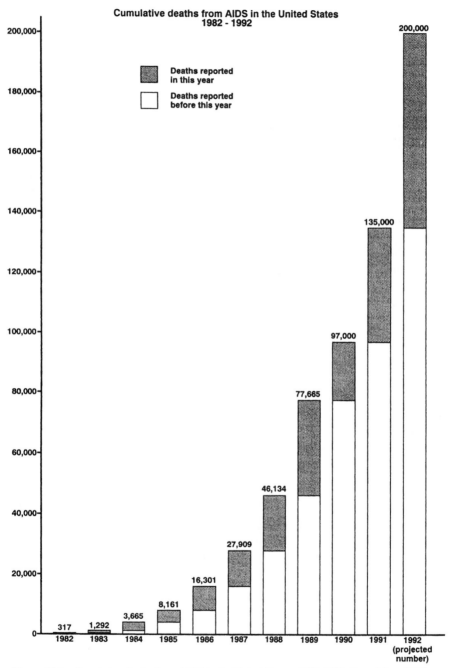

Figure 22.1   Cumulative deaths from AIDS in the United States, 1982–1992 (from Stine, 1993).

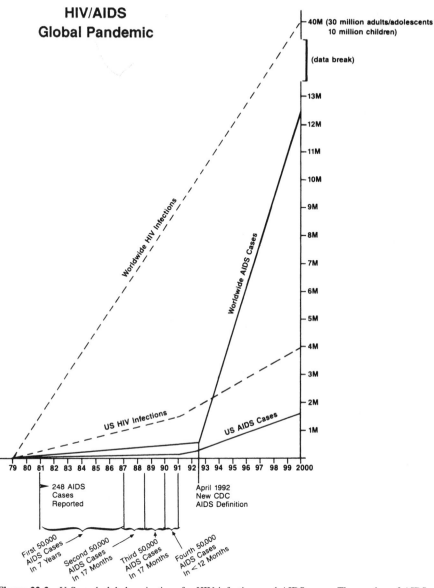

Figure 22.2   U.S. and global projections for HIV infections and AIDS cases. The number of AIDS cases in the United States by the year 2000 is projected to be about 1.6 million, worldwide about 12.4 million cases (from Stine, 1993).

banks are often contaminated, and education, literacy, and public health awareness are acutely limited. Thus the AIDS epidemic is having a much greater impact on human society. Adult populations in the villages of some African nations are being decimated, and a generation of orphans is arising. The AIDS epidemic in developing nations contributes to poverty, to high birth and death rates, to environmental destruction, and to only slight reductions in the rates of population growth.

In 15 nations in Africa that have the highest prevalence of HIV infections, people are increasingly susceptible to infectious diseases. The United Nations estimates that AIDS will contribute to at least 13 million excess deaths in these nations by the year 2015, and that 7 million fewer children will be born due to AIDS death of women. Nonetheless, Africa's population is expected to double by the year 2015 (Holden, 1992). Under these circumstances, it will be difficult if not impossible for these nations to make progress in economic development, agriculture, education, family planning, or proper environmental management. The ecological costs are immense.

The many organizations and individuals dedicated to solving the AIDS epidemic are working on many fronts. Medical research is focusing on antiviral agents as a possible cure and on vaccines for prevention. No immediate breakthroughs are in sight, as of 1995, but several promising leads are being pursued. One antiviral approach involves ribozymes that have the potential of slicing up the viral RNA of HIVs and thereby inactivating the viruses. As of 1995, this is only a theoretical approach, but it offers great potential. In vaccine development, the greatest need is to understand the basis of retroviral[1] immunity and the nature of retroviral immunogens. As of 1993, expectations for a 100 percent effective vaccine had been dampened (Haynes, 1993). A major problem with both antiviral agents and vaccines is the propensity of HIVs to mutate and evolve rapidly, so the race for both a cure and a vaccine is really a duel between a labile viral pathogen and human scientific ingenuity. The other side of prevention is education: encouraging safe sex or abstinence, preventing drug abuse, ensuring 100 percent safety of blood supplies. All of these involve practical difficulties.

CONVERGENCE IN WORLD HEALTH PATTERNS

Ecological trends in world health make it likely that some convergence in the global health patterns of industrial and developing nations will occur.

First of all, as longevity increases in developing nations, more people will develop chronic diseases associated with aging, namely, cardiovascular diseases and cancers. In nations with life expectancies below 60 years of age, heart problems and cancers are masked by early deaths from infectious diseases. Many nations are making dramatic advances in longevity due to improvements in medical care and food supplies. As their populations age, cardiovascular diseases and cancers will become more evident.

Secondly, many developing nations are industrializing at a furious pace, often without reasonable controls on pollution. Air and water pollution are becoming acute problems in the cities of China and India, for example (Smil, 1993). As these nations develop heavy industries that produce metal products, petrochemicals, plastics, electronics, pesticides, rubber, paint, cement, and so on, air and

---

1. The human immune viruses that cause AIDS belong to a type known as retroviruses because they can reverse the flow of genetic information. Normally in living cells, genetic information is copied from DNA to RNA to protein synthesis. In the retroviruses, RNA is copied into DNA by means of an enzyme known as reverse transcriptase. This involves a high error rate, which creates difficulties in vaccine development.

water pollution problems grow at an alarming rate. Also, automobile, truck, and bus traffic in many nations is growing at an exponential rate, and many vehicles lack any semblance of emission controls. The air quality in cities such as Beijing, Shanghai, Chengdu, Calcutta, Bombay, and New Delhi is atrocious. A gray haze usually hangs over these cities, even on sunny days, without any particular weather-related cause. The midday air stings the eyes, irritates the throat, and obscures visibility to just a couple of city blocks.

The problem of air pollution, coupled with high rates of smoking, will inevitably lead to high rates of lung cancer and other environmentally induced cancers. Whereas the smoking rate for males in the United States has declined below 30 percent in the past 25 years, the rate in developing nations is usually over 75 percent. Smoking rates are exceptionally high in Asia. U.S. tobacco companies are strongly promoting smoking in developing nations around the world.

Another factor that will bring mortality patterns in developing nations more in line with those of industrial nations is the increasing accident rate. As vehicular traffic increases, automobile, truck, and bus accidents are soaring in developing nations. In India now, major highway accidents are common experiences in daily life; in fact, it is unusual to drive more than 200 miles on major roadways in India without seeing a serious accident. Vehicular traffic has increased more rapidly than basic improvements to roads and bridges, and yet many roads are still crowded with bicycles, motor scooters, bullock carts, horse-drawn carts, and pedestrians. All are competing for narrow roadway spaces with trucks and buses traveling 40 to 50 miles per hour, high rates of speed for the existing conditions. At many hospitals in India and other developing nations, accident trauma has become a new primary cause of emergency care.

At the same time, infectious disease is likely to become a more prominent aspect of medical care in the industrialized nations. At least four factors contribute to this. First of all, HIV infections continue to expand, and no successful cure or preventive vaccine is in sight. AIDS-related infectious disease will undoubtedly become an increasingly important cause of death. In the United States in 1993, AIDS was the primary cause of death in young men in the age range from 25 to 44 (National Center for Health Statistics, 1993).

Secondly, newly evolving viruses, or "emerging viruses," can occur spontaneously with virulent pathogenicity. In recent years emerging viruses have been the cause of Lassa fever and Ebola fever in Africa and four-corners disease in the southwestern United States. These are rodent-borne or primate-borne viruses that can be fatal in human beings. Four-corners disease is apparently a new mutation of a Hanta virus, normally associated with Korean or Argentinean hemorrhagic fever, a serious but usually nonfatal illness. This disease, which appeared in the Four Corners area of Arizona, New Mexico, Utah, and Colorado in the spring of 1993, is unusual in that it causes an acute, sometimes fatal respiratory disease rather than nonfatal renal or hepatic pathology. A major ecologic concern for world health is the possibility that a mutant influenza virus with a high fatality rate could arise in any country in the world and be transported globally in 24 hours via normal jet travel. There are no biological reasons why this cannot occur, and medical science would be hard-pressed to cope with such an event. Influenza viruses are likely candidates for causing such a global epidemic because they are highly variable, have a high mutation rate, are airborne, and move readily between animal and human infections with an increased probability of mutation in so doing (Garrett, 1994). There has been, however, considerable exaggeration and emotionalism in the popular press and entertainment industry regarding the worldwide threat of viruses such as Ebola. While emerging viruses are serious medical and public health concerns, they should not be the basis of panic.

The prevalence of infectious disease in the United States is also likely to increase due to declining vaccination rates in infants and children. For example, cases of whooping cough rose significantly in 1993, an illness readily prevented by DPT shots in early childhood.

Finally, infectious bacteria remain a potential, even increasing, threat to human health because of their ability to mutate toward antibiotic resistance. Staphylococcal, pneumococcal, and gonococcal bacteria have long been famous for antibiotic resistance, and newly emerging strains of the bacteria causing tuberculosis and dysentery are also showing antibiotic resistance. The mutations are not stimulated by antibiotic use—that is, they do not occur originally in response to antibiotics—but the widespread use of antibiotics selects for resistant strains. The phenomenon is largely one of artificial selection, in which only resistant bacteria survive, giving rise to new and dangerous populations of bacteria.

SUMMARY

Human health is a reflection of the environment. Many scientists and physicians point out that the preservation and maintenance of environmental quality is the first step in the prevention of human illness.

Human health problems in wealthy nations are dominated by chronic noninfectious diseases, especially cardiovascular diseases and cancers, with chronic lung disease, accidents, homicide, and suicide also playing major roles. As mortality factors, these problems are often related to lifestyle and socio-behavioral aspects of the human environment.

In developing nations, health problems are dominated by infectious, parasitic, and nutritional diseases. These problems are closely linked to the biotic and physical environment, especially food and water supplies, housing, and sanitation.

Some convergence is occurring in world health problems, however. In developing nations, at least four factors are contributing to higher rates of chronic disease: (1) increasing life expectancies, (2) greater pollution, (3) higher rates of smoking, and (4) more crowding. Accident rates are also rising sharply in developing nations.

In industrial nations, even the wealthiest, higher rates of infectious disease are occurring due to: (1) AIDS-related infectious diseases, (2) the threat of emerging viruses, (3) antibiotic-resistant bacteria, and (4) declines in vaccination rates.

If health and illness are viewed as responses of individuals and populations to their environments, the conditions of the world's population is improving as evidenced by increased longevities, but as evidenced by child mortality, we are still in bad shape. In both cases we are seriously threatened by pollution, poverty, stress, and microbial mutation. The human immune viruses and AIDS represent an unprecedented threat to human health and progress. By damaging the immune system, HIVs contribute to a wide range of infectious disease problems, and they exacerbate the vicious cycle of poverty and environmental degradation.

REFERENCES

Acha, P., and B. Szyfres. 1991. *Zoonoses and Communicable Diseases Common to Man and Animals.* 2nd ed. Scientific Publication 503. Washington, DC: Pan American Health Organization.

Blumenthal, D. S. 1985. *Introduction to Environmental Health.* New York: Springer.

Chivian, E., et al. 1993. *Critical Condition: Human Health and the Environment.* Cambridge, MA: MIT Press.

Cohen, J. 1994. The Duesberg phenomenon. *Science* 266: 1642–1644.

Cortese, A. 1993. Human health risk and the environment. In *Critical Condition: Human Health and the Environment,* ed. E. Chivian et al. Cambridge, MA: MIT Press.

Duesberg, P. 1993. HIV and the Aetiology of AIDS. *Lancet* 341: 957–958.

Garrett, L. 1994. *The Coming Plague: Newly Emerging Viruses in a World out of Balance.* New York: Farrar, Straus, and Giroux.

Grant, J. P. 1993. *The State of the World's Children: UNICEF Report.* New York: United Nations.

Haynes, B. F. 1993. Scientific and social issues of human immuno-deficiency virus vaccine development. *Science* 260: 1279–1286.

Holden, C. 1992. African AIDS toll. *Science* 257: 1627.

MAP International. 1993. *The Age Without Innocence: Report on Child Health.* Report 17 (6). Brunswick, GA: MAP International.

McMichael, A. J. 1993. *Planetary Overload: Global Environmental Change and the Health of the Human Species.* New York: Cambridge University Press.

National Center for Health Statistics. 1993. Bethesda, MD: United States Public Health Service.

Sachs, A. 1995. HIV/AIDS cases rise at record rates. In *Vital Signs: The Trends That Are Shaping Our Future.* (eds.) L. R. Brown, N. Lenssen, and H. Kane. New York: W. W. Norton.

Smil, V. 1993. *China's Environmental Crisis: An Inquiry into the Limits of National Development.* Armonk, NY: M. E. Sharp.

Stine, G. J. 1993. *Acquired Immune Deficiency Syndrome: Biological, Medical, Social and Legal Issues.* Englewood Cliffs, NJ: Prentice-Hall.

Veil, S. 1992. *Our Planet, Our Health: Report of the WHO Commission on Health and Environment.* Geneva: World Health Organization.

Weiss, R. S. 1993. How does HIV cause AIDS? *Science* 260: 1273–1279.

WHO. 1992. World Health Organization. See Veil, 1992.

Chapter 23

COMPETITION AND CONFLICT

Competition is a fact of life for all organisms. Ecologists define competition as "the use or defense of a resource by one individual that reduces the availability of that resource to other individuals" (Ricklefs, 1990). Odum (1993) noted that the word *competition*

> denotes a striving for the same thing, as in the familiar case of two businesses striving for the same market. At the ecological level, competition becomes important when two organisms strive for something that is not in adequate supply.

Competition can occur between two individuals of the same species (intraspecific), for example, two mule deer consuming a limited supply of forage, or between individuals of different species (interspecific), for example, cattle, pronghorn antelope, and grasshoppers all consuming the same grassland.

An additional note on competition is necessary before we discuss its ecological aspects. The people of many nations, the United States included, are fascinated by competition, in sports, in science, in business, and in education. The thrill of competition is in winning. As Jimmy Johnson, former coach of the Dallas Cowboys, phrased it: "We play for just three reasons: to win, to win, to win." Our entire sports enterprise, amateur and professional, and a good part of our economic system is geared toward winning. No one wants to be a loser. The goal is to be number one. Capitalism is based on competition. Marketing and manufacturing success often depend on the best combination of quality and price. In virtually all endeavors, competition makes products better and prices lower. In human terms, competition provides a motivation for achievement. All of these are true, and our society is, in fact, based on competition.

This chapter does not attempt to refute any of these ideas. The purpose here is not to take away from any of the positive aspects of competition in sports, economics, education, science, and human advancements. Rather, this chapter is based on the ecological side of competition. Competition does involve a loser, and it involves conflict. Conflict is not inherently bad but it can lead to violence. Before getting into this issue, we will first look at some biological aspects of competition.

COMPETITION AND EVOLUTION

Charles Darwin was keenly aware of competition in nature and considered it a building block in his theory of evolution. Darwin had been influenced by Thomas Malthus's "Essay on the Principle of Population," which proposed that populations of living organisms tend to increase geometrically (as in the series 2, 6, 18, 54, in which each number is three times the former), whereas the means of their subsistence increases only arithmetically (as in the series 2, 5, 8, 11, in which each number is greater than the former by 3). Darwin felt that this principle of population inevitably leads to competition and survival of the fittest.

Darwin also noted individual variation in members of a species and presumed that some of this variation was inherited. He theorized that competition would favor those individuals best adapted to the environment and best able to survive and reproduce. Thus, Darwin's ideas followed a logical sequence with high reproductive potential leading to competition, and competition leading to survival of the fittest. Furthermore, given variations in the traits of individuals and inheritance of some of these variations, over time there would be a natural selection of those individuals most fit to survive and reproduce. The result would be gradual change in the traits of a given species and the evolution of new species.

Darwin's ideas have been attacked, vilified, modified, and refined on many fronts, but they are largely intact today as the core of modern biology. We now recognize different mechanisms of evolutionary change, including mutation, geographic isolation, recombination, and genetic drift, and we also understand that survival of the fittest refers to reproductive fitness rather than physical fitness per se. Darwin's concept of fitness as the strongest is now replaced by the concept of reproductive fitness in terms of surviving offspring. Thus, the important indicator of evolutionary success is the contribution of genes to the next generation. Darwin's ideas about genetics were incomplete; *Origin of Species* went through three editions before Mendel's research and theories on genetics were published in 1865. Nonetheless, Darwin was fundamentally right in his principle of natural selection. The living world makes virtually no sense without his concept of evolution, and "the evidence supporting [Darwin's] theory has grown progressively stronger" (Raven and Johnson, 1992).

FORMS OF COMPETITION

Competition can involve any resource and can take any form from the subtle and mildly detrimental to the violent and lethal. It can be direct or indirect. Direct competition involves two individuals interacting over a limited supply of food, water, space, nesting sites, escape routes, safety zones, or some other finite resource. Indirect competition is competition for social rank, which in turn influences access to resources. It is indirect because the competition does not directly involve a physical resource, but is over status which secondarily determines access to resources. Many animals do not interact directly over food or mates but conduct elaborate displays and fights that establish dominance, or pecking order. Quite a lot of animal behavior, and human behavior as well, is unintelligible without understanding its origins in competition and evolution.

INTERSPECIFIC COMPETITION

The science and industry of agricultural pest control is an attempt by humans to out compete other species: the bugs, weeds, worms, and vermin that eat our crops and destroy our food. All organisms live in a network of competitors, often very diverse in taxonomic status. Few organisms could be further apart than wheat rust, a grasshopper, and human beings, but all are competing for the stored food supplies in domestic wheat. The destruction of field crops and stored food products throughout the world by unwanted pests is awesome. In India, it has been estimated that one-third of all food production is lost to nematodes, insects, and rodents. In Africa, entire annual crops have been devastated in some years by locust plagues. Pest problems are dramatic examples of interspecific competition between humans and other species. Most animals experience such competition. Some examples are parrots and monkeys in a tropical forest, prairie dogs and bison in a temperate prairie, and white-tailed deer and gypsy moths in a New England forest; these diverse animals compete directly for limited food supplies. Animals of different species compete for other resources as well. Woodpeckers and flying squirrels compete over limited tree holes. Octopus and tilefish vie for a limited number of crevices in a coral reef.

For closely related species, interspecific competition is often subtle but ultimately unsustainable. Gause (1934) was one of the first ecologists to demonstrate that closely related species with nearly identical resource requirements cannot coexist. If strong competition exists between such species, one tends to displace or eliminate the other. This principle of competitive exclusion was demonstrated by Gause's classic study of *Paramecium* (Fig. 23.1).

Closely related species do coexist in ecological communities, but the secret of their success is niche differentiation. MacArthur and MacArthur (1961) demonstrated that closely related species of warblers all feed on insects, and may even feed in the same tree, but they do so by specializing on certain parts of the tree and on certain species of insects. Thus the Cape May warbler feeds on insects at the very top of a conifer tree, the bay-breasted warbler feeds in the middle, and the Blackburnian warbler feeds at the base of the same tree.

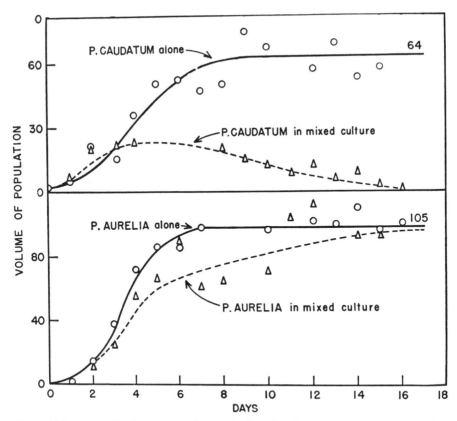

Figure 23.1 Competition between two closely related species of protozoa in the genus *Paramecium*. When separate, both species grow well and exhibit logistic growth in controlled cultures with constant food supply; when together, *P. caudatum* is eliminated (from Odum, 1971).

COMPETITION AND COOPERATION

Our emphasis on competition should not obscure the fact that cooperation is also a powerful force in the living world. In fact, cooperation and competition might be considered as two sides of the same coin, the coin of interspecific relationships. Cooperation, deliberate or inadvertent, is a common biological reality and can take many forms: altruism, where one individual assists another, often at the risk of life or limb; commensalism, where one species benefits and the other is unaffected; or mutualism, where both species benefit.

Cooperation between individuals or between species readily reverts to competition, however, when resources are in short supply. This sets the stage for conflict.

Two kinds of forces tend to create resource shortages: (1) large numbers of consumers and (2) unanticipated resource shortages. The two factors are obviously related; the problem revolves around the ratio between consumer demand and resource supply. A shortage of gasoline like that in the United States in the early 1970s, which caused economic dislocations, would barely have been noticed in

China since there were so few vehicles in China in the 1970s. Now, in the 1990s the situation in China is very different, and a petrol shortage there would have serious economic and social repercussions.

THE SEVERITY OF COMPETITION

The severity of competition is greatest between individuals that have the closest ecological requirements. In most cases this means intraspecific competition between members of the same species. When lemmings are at population highs, they compete most severely with other lemmings for burrow space and food. A point of population density is reached where they either fight or flee, with either process leading to extensive mortality. In famine conditions, people compete most directly with other people. Since members of the same species share the same ecological requirements, competition within any given species is usually the most frequent and intense.

When competition becomes severe because of large numbers, high densities, or resource shortages, conflict is an inevitable result. Displays of social dominance become acts of overt aggression. Overt aggression turns into violence.

COMMUNICATION OR VIOLENCE?

Animals are often accused of violent savagery. In the 1980s, a popular U.S. President responded to examples of violent human behavior by saying, "They act like animals." Not only was his metaphor inappropriate, his biology was inaccurate. Intraspecific violence leading to death in animals is relatively rare. Although it does occur—humans are not the only animals that kill their own species—violent aggression, infanticide, and fights to the death among animals are less common than popularly thought. What is often portrayed as violence in the animal kingdom is actually predation, or food getting. A wolf pack viciously attacks a white-tailed deer, a leopard stalks a gazelle, or an alligator devours a water bird. The overwhelming majority of intraspecific animal aggression involves communication, which precedes and normally prevents violent behavior.

HIERARCHIES AND TERRITORIES

Two behavioral systems, mediated by communication, both stimulate and reduce violence in animal societies: dominance hierarchies and territorialism. Dominance hierarchies are systems of social rank that determine relationships between individuals and influence access to resources. These hierarchies may originally be established through aggressive, though not necessarily injurious or violent, behavior. Once established, dominance hierarchies are usually maintained by ritualized displays. Communication maintains law and order, so to speak, without violence, and stable hierarchies serve to reduce the amount and severity of injurious aggression and violence.

Territorialism provides a system of partitioning spatial resources. The individual or group holding a territory originally may have had to fight for possession, but such fights are usually not fatal to the loser. Once territories are established, the victor can maintain his, her, or their territory by display communication.

The communication systems by which hierarchies and territories are maintained may be visual, auditory, or chemical. For example, mountain gorillas, falsely depicted in the past as ferocious animals, are remarkably gentle and peaceful. Their stable social life is maintained by dominant male "silverbacks" who assert their authority by chest-beating or by intimidating postures and lunges.

Howler monkeys have famous territorial vocalizations that maintain the spatial separation of groups. The predawn howling of males identifies the location of each group in the forest, as if they are proclaiming, "This is our space!" Group movements are often accompanied by grunting or lesser vocalizations which indicate the whereabouts of each group, enabling them to avoid competing groups in their normal daily activities. This system helps neighboring groups to respect each other's territorial rights.

If populations are dense, however, or if certain groups are looking for new resources, howling may actually attract neighboring groups, bringing them to territorial boundaries where "howling battles" ensue. Howling battles can also occur during severe thunderstorms when groups accidentally come together because the noise of rain and thunder has muffled or distorted their communications. Under such circumstances, the animals, upon seeing members of another group, become highly agitated and upset, but normally they do not fight. Even under these accidental circumstances, violence does not occur. The fascinating thing is that the behavioral evolution of the howler monkey societies has moved beyond the point of violence between groups. They have apparently achieved a behavioral inhibition that prevents violence in most cases. This is not to deny published accounts of infanticide in howler monkeys. Infanticide does occur but it is not common.

Gibbons have a similar system of vocal communication to maintain group territories and reduce intergroup conflict. Gibbon groups are normally much smaller than howlers; usually a small family unit consists of an adult male, adult female, and one or two offspring, instead of the 10 to 40 individuals in a typical howler group. But the early morning territorial calls of the gibbon, one of the most beautiful sounds in the primate world, send the same message as the raucous howling of the howler: "This is our territory!"

The howler monkey and the gibbon have seemingly achieved levels of social evolution in the control of aggressive behavior that is above our own. Humans do avoid much violent behavior through communication, diplomatic posturing, and threat behavior. For example, World War III was avoided by the massive expenditures of the Cold War. Basically, the United States and the former Soviet Union threatened each other with destruction, and these threats successfully substituted for actually going to war. A policy of nuclear weapons buildup, called MAD, for Mutual Assured Destruction, succeeded in preventing nuclear holocaust. Now a similar policy to deter violence must be achieved at regional and local levels, where we have yet to succeed in controlling injurious aggression and violence.

ANIMAL VIOLENCE

While most animal species regulate aggression and avoid or reduce violence through communication, exceptions do occur. Exceptions occur most commonly in species that do not have elaborate or highly developed display communication, or that experience degraded or disturbed ecological conditions.

Two examples of animals that show violent aggression are domestic rodents, especially house mice and Norway rats, and a group of monkeys known as the macaques, especially the rhesus monkey. Neither of these groups have elaborate patterns of ritualized display, neither are strictly territorial, and both are capable of living in and exploiting degraded environments.

In the 1950s a series of studies at the University of Wisconsin on the population ecology of wild house mice in old buildings showed that populations supplied with unlimited food became more aggressive and violent as populations increased. If dispersal routes were available, individual mice moved into new areas with lower population densities. When space was limited and dispersal prevented, the violence became more injurious, and the mice inflicted serious, often fatal, wounds. Mortality rates soared. The populations were ultimately controlled when the levels of violent aggression reached approximately one fight per hour per mouse (Fig. 23.2). Reproduction and birth rates remained high in these populations, but most young litters of mice died as a result of parental abandonment or infanticide.

The basic cause of increased violence and mortality seemed to be a breakdown of the normal social organization. This societal breakdown disrupted parental behavior. Nests were not properly built and maintained, nursing was disrupted, and normal male-female relationships deteriorated, often to the point of fatal violence, especially for infants and juveniles. Normal peaceful social interactions such as huddling and grooming were replaced by fighting.

An important aspect of this research is that food and water were in constant overabundance, and space was also available. Over half of the nest boxes were unused, and the large pens in which the mice lived still had a considerable amount of empty space. In fact, some of the six replicate populations were limited at densities as low as 1 mouse per 10 square feet. No population even achieved 1 mouse per square foot, whereas in laboratory cages mice are happy and healthy in densities of 8 to 10 mice per square foot.

These studies showed that density per se is not the critical factor in producing violent behavior. Crowding was a factor, but crowding occurred at different densities according to individual differences within each population. In other words, a house mouse is not simply a house mouse like all other house mice. Individual personalities, dominance relations, and social structure vary between populations. Some individuals are more aggressive than others, some are more tolerant of crowding than others. In all cases, however, crowding at whatever level it oc-

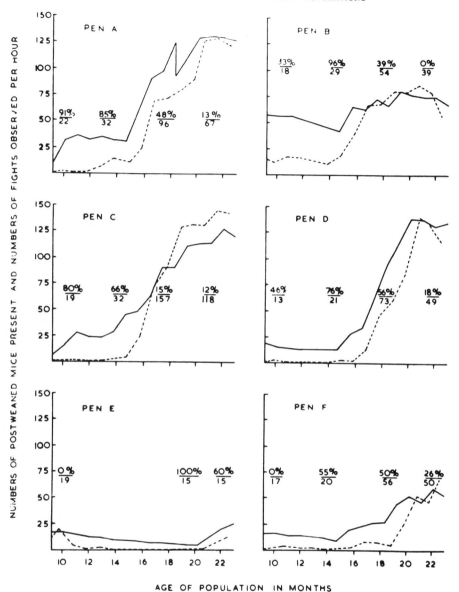

Fig. 1. Relationships of population growth and litter survival to amount of aggressive activity observed.

———————   Numbers of postweaned mice

- - - - - -   Numbers of fights observed per hour

$$\frac{91\%}{22}$$ Per cent survival to weaning
of young born in a two-month period
Number of young
born in a two-month period

Figure 23.2   Relationships between population growth and aggression in seminatural populations of house mice. Aggression was not related to population density per se but to social structure. In all populations, as aggression increased, infanticide increased and litter survival decreased (from Southwick, 1970).

curred, not definable as a numerical density, had a disruptive effect on social organization and normal behavior patterns. Social pathologies cropped up and ruled the day.

Variations on this theme have been found in many other studies of house mice and Norway rats. In a number of research projects in England, crowded populations with surplus food sometimes showed reduced reproduction and sometimes showed increased mortality (Crowcroft, 1966). In all cases, the key factor was disturbed social structures.

Many investigations of Norway rat populations showed related phenomena. Some crowded populations developed excessive violence. Others developed a strange phenomena of "pathological togetherness" in which countless individuals huddled in a deteriorating environment, much to the detriment of their own health. Calhoun (1962) found these pathological aggregations self-stimulating and called them "behavioral sinks," referring to the maladaptive nature of such groups. Individuals in these aggregations did not show violent aggression; rather, they showed passive, nonfunctional behaviors. From the standpoint of normal animal societies and individual health, the two extreme conditions of violent aggression on one hand and passivity on the other are both ultimately destructive of individuals and society.

Among the primates, which are our closest biological relatives, macaques, especially rhesus monkeys, share some of the ecological and behavioral traits of house mice and Norway rats. They are highly successful animals—adaptive, intelligent, and capable of exploiting a variety of habitats. They are also aggressive and at times violent.

Rhesus monkeys are notable for several reasons. They have the widest geographic and ecological distributions of any primate other than human beings, occurring naturally from the mountains of Afghanistan and northern China, to the tropical forests of central India, Burma, Thailand, and Vietnam. They live in the slums of crowded cities such as Calcutta, the mangrove forests of Bengal and Bangladesh, the agricultural villages of the Gangetic plains, and the pine forests of the Himalaya Mountains in Pakistan and Nepal. In these different habitats, their social behavior varies considerably, especially with regard to aggressive behavior. Rhesus are also notable as a mainstay of biomedical and behavioral research. They are the prime subjects for a great deal of research on learning, behavioral development, physiology, immunology, and virtually all of the applied biomedical sciences.

Field studies of rhesus in India have shown that those living under the competitive pressure of cities, where both space and food is limited, are four to five times more aggressive than rhesus living in forests. Their home ranges in cities are one-tenth to one-hundredth the size of comparable groups in forested habitats. When these aspects of aggression were analyzed in greater detail, and also examined experimentally in captive groups, it was found once again that the primary variable was not density per se but social structure and organization. Aggressive and violent behavior increased much more sharply with changes in the social composition of groups than with changes in density or food supply (Southwick, 1970). For example, crowding an established group of 20 monkeys into one-half of its normal space (500 square feet instead of 1,000 square feet), which was equivalent to doubling of the population density, resulted in an approximate doubling of the aggression rate. In contrast, replacing two females in the group with two new females of the same age and size resulted in a 10-fold increase in aggression rate. It was obvious that rhesus monkeys are acutely aware of social structure and individual identities in a social group. They are strongly xenophobic, that is, suspicious of and aggressive toward strangers.

This same phenomenon has been documented in chimpanzees, the species of primate most like ourselves. Normally chimps live in relative peace, but they can be violently aggressive to other chimps outside the established social groups. Jane Goodall (1986) has observed the equivalent of chimpanzee murder in small roving groups that seek and find outsiders, individuals that do not belong in the group, on the edge of their home ranges. Frans de Waal (1989) has shown that chimps in captivity and nature can be intensely "biopolitical," that is, very much aware of group identity and social rank. These perceptions can lead to social acceptance and friendships or to murder.

The propensity for competition and aggression in many animals is so great that most animals have developed communication and reconciliation systems to substitute for violence. They have gestures and signals that help to ease aggressive tensions. Signals of subordination, which effectively promote reconciliation, are usually recognized and respected (de Waal, 1989). When these signals become thwarted or overwhelmed by competitive pressures or ecological circumstances, violence can break out.

The leap from animals to humans is usually thought to be a large one, but in the case of competition and conflict the similarities are striking and the analogies are strong. If we analyze and contemplate the extensive literature on animal aggression and on human aggression, we can find remarkable similarities. In many ways, the house mouse and Norway rat studies of the 1950s are predictive of human behavior. They showed that the integrity or disintegration of a society is indeed related to population growth, that competition and aggression are closely linked, that crowding can produce a breakdown of normal societal structure, and that this breakdown can lead to a variety of social pathologies: aggression, violence, and illness.

HUMAN AGGRESSION

Human aggression, like all things human, is exceedingly complex, and no one in his or her right mind would claim that it is as simple as animal aggression, which is complicated in its own right. In the study of human behavior, it is important at the outset to distinguish between individual behavior and collective behavior. Individual aggression involves such behaviors as abuse, assault, rape, suicide, homicide, and individual acts of terrorism. Collective aggression includes riots, organized crime, group terrorism, revolution, and war. From this vast spectrum of human behavior, we shall consider just two major types: individual violence, especially homicides, and the most organized of all collective violence, revolution and war.

There is an extensive literature and continuing debate on the sources of human aggression and violence. Why are we so aggressively violent, both individually and collectively? Is this part of our instinctive behavior or our environmental experience? These questions are so complicated and have been discussed in so many different disciplines, that we can do no more than touch upon the nature of the controversy here. Disciplines involved in the study of human aggression range from the natural sciences of physiology, neurology, and genetics to the social sciences of psychology, sociology, and political science. Ecology attempts to bridge these sciences by looking at the roles of aggression and violence in interactions between a population and its environment.

One group of scientists, represented by ethologist and physician Konrad Lorenz and anthropologist Robert Ardrey, believed that human beings are innately aggressive. According to this view, discussed in Lorenz's 1966 book *On Aggression* and Ardrey's *The Territorial Imperative*, our aggressive tendencies are inherited; they are part of our genetic makeup from millions of years of evolution. Lorenz and others maintained that aggressive behaviors in animals are adaptive and a necessary part of survival in natural selection; only those species and individuals with sufficient aggressive drive are able to survive. They further maintained that human beings share this genetic inheritance, and our problem now in the modern world is to control and channel aggressive tendencies into productive pursuits.

Another group of scholars, represented by the anthropologist Ashley Montagu and psychologist J. P. Scott, believed that attributing our aggressive behavior to genetics and evolution was ridiculous, unwarranted, and dangerous. Instead, they believed that human aggression is a product of environmental pressures and social problems experienced by individuals and social groups. They found no evidence for an innate drive to fight. Montagu (1968) cited many examples of cooperative and peaceful human societies, some with virtually no record of fighting and no words of aggression or violence. Scott (1958) recognized genetic differences in aggression among animals but concluded, "The notion that man is a 'killer ape' and motivated to destroy his own kind because he is a hunter is without foundation." Both scholars denied the existence of an aggressive instinct in human beings.

Scientists in both camps have mustered impressive evidence to support their positions, and the debates in the scientific and popular literature have been lengthy and heavily documented. Lorenz pointed to the frequency of human violence, both individual and collective, throughout history and also referred to tribal groups such as the Yanomamo Indians of Venezuela and certain tribes in the New Guinea Highlands for whom warfare was an annual event. Montagu, on the other hand, cited innumerable examples of peaceful and cooperative human behavior, and referred to tribes such as the Mbuti pygmies of the Congo Basin and the Hazda of South Africa who were gentle, peace-loving people.

The question arises, of course, Why consider these complicated aspects of cultural anthropology and social psychology in the field of global ecology? The answer is that human aggression has had and continues to have such a drastic impact on the environment and on human resources that the subject inevitably falls within the realm of ecology.

At the same time, the environment impacts human behavior. In the case of hunter-gatherer peoples, this was dramatically shown by the field studies of anthropologist Colin Turnbull (1972). Turnbull studied a peaceful people of northern Kenya who turned to vicious aggression and violence when their environment deteriorated through drought and population pressures. These hunter-gatherers, known as the Ik people, were driven from their traditional hunting grounds, through the creation of a wildlife reserve, into a sparse mountain habitat bordering Kenya, Uganda, and Sudan. Faced with an overcrowded and inadequate environment, these people experienced "a social disintegration [that] parallels in many ways that of our own society." They tried to become farmers, but

in less than three generations, they have deteriorated from a group of prosperous and daring hunters to scattered bands of hostile people whose only goal is individual survival, and who have learned that the price of survival is to give up compassion, love, affection, kindness and concern—even for their own children. The Ik can and do steal food from the mouths of their parents, throw infants out to fend for themselves, abandon the old, the sick and crippled to die without a backward glance. Walled into their compounds, living in fear of each neighbor, indifferent to anything but their own individual welfare, the Ik have evolved a society whose destructive qualities mirror the cold and lonely selfishness of our own—though we do not have their excuse of starvation and exile.

In short, Turnbull documented the decay and collapse of a people formerly known for their prosperous and peaceful lifestyle. The turning point was competition induced by a harsh environment. The prominent anthropologist Carleton Coon (1972) noted:

Turnbull comes to the sad conclusion that the same kinds of societal breakdown so vividly portrayed by his meticulous study of the Ik are beginning to produce the same results among ourselves . . .

and Ashley Montagu observed, "The parallel with our own society is deadly." These comments were made in 1972, following the turbulent decade of the 1960s, but we have now come to recognize this was still an early stage in the American experience with individual violence.

VIOLENCE IN AMERICA

At the level of individual violence, the United States is an example of a nation gripped in a wave of crime and assault. Between 1960 and 1992, the frequency of violent crimes (murders, rapes, and violent assaults) climbed from a rate of 161 per 100,000 population to 758 per 100,000 population, more than a four-fold increase (Gest et al., 1994). Many cities exceeded this national rate. Washington, DC, for example, reported a violent crime rate of 2,832 per 100,000 population in 1992.

The total number of violent crimes reported in the United States increased from just around 300,000 in 1960 to 1.9 million in 1992 (Fig. 23.3). A U.S. Justice Department survey considers this an underreporting, with the true number of violent crimes exceeding 6 million per year (Gest et al., 1994).

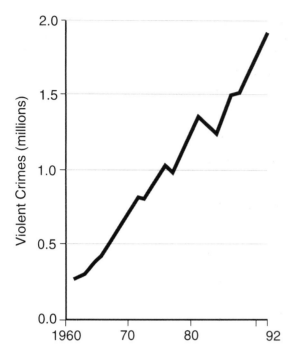

Figure 23.3 Reported violent crimes in the United States, 1960 to 1992 (from Gest et al., 1994).

Although some cities have recently reported declines in violent crimes, violent crime in the United States is at an all-time historic peak. Homicide has become the leading cause of death in certain groups of teenagers, especially black youth age 16 to 19.

The public response to increasing violent crime is fear and outrage. A 1994 cover story on crime in *Time* magazine noted, "As crime spreads into small towns, many Americans barricade themselves." In many states, the only public tax increases approved by voters in recent years have been for more police officers and more prisons. Tax measures for education, basic research, environmental restoration, and conservation go down to defeat. In 1994, the new presidential anticrime bill calls for $3.4 billion in spending and 50,000 new police officers. This spending allocation is greater than the budget of the National Science Foundation, which supports basic research and education in the natural sciences, and greater than the budget for agricultural and environmental research.

Why is the United States such a violent nation? This is indeed a very complicated question, involving our cultural history as well as current sociology and economics. Ecology, often considered a science of just plants, animals, and microorganisms, can offer some insights.

In the 1960s, national anguish over the assassinations of President John F. Kennedy, Attorney General Robert Kennedy, and Civil Rights leader Martin Luther King Jr., led to efforts to come to grips with our violent society. Two select national committee reports, the *Kerner Report* (U.S. National Advisory Commission on Civil Disorders, 1968) and the Eisenhower report (Graham and Gurr, 1969) emphasized ecologic conditions and social change as underlying factors in violence. The final section of the Eisenhower report on ecological perspectives considered the relationships between overcrowding, aggression, and environmental and social factors. The anthropologist B. J. Siegel further developed this theme, noting that violence is "one among several strategies of response to environmental threat" (Siegel, 1969), and the medical psychologist G. M. Carstairs concluded:

> In summary, it seems that overpopulation only aggravates the widespread threat to social stability presented by masses of our population who are basically unsure of their personal future, who have lost confidence in their chance of ever attaining a secure place in their community (Graham and Gurr, 1969).

The Kerner Report (1968) emphasized depressing environmental and social conditions in inner-city ghettos as a prelude to violence:

> The ghettos too often mean men and women without jobs, and schools where children are processed instead of educated, until they return to the street—to crime, to narcotics, to dependency on welfare, and to bitterness against society.

None of these studies, or those of animal ecologists, claim that high population density per se is the cause of increased aggression. Rather, a combination of factors, including population crowding, competition, inadequate access to resources, frustration, and social disintegration, constitute a set of circumstances conducive to violence.

The Eisenhower report concluded that the most important cause of increased violence is the widespread frustration of individuals and groups who feel socially deprived in relation to the goods and services, which they believe are theirs by right. Extremes of scarcities and abundance contribute to these frustrations, which are further exaccerbated by the media. Competition and unemployment generate bitterness in the have-nots. Many acts of violence have the common thread of extreme bitterness against society, often fueled by failure in a competitive economic world.

These conclusions were reached almost 30 years ago, with the hope and intention that we could do something about violence, yet violence has continued to increase. A number of changes in the United States from the 1960s to the 1990s support this trend: (1) an increase in population of over 75 million, from 179.3 million in 1960 to 262 million in 1995, (2) an increase in unemployed workers, from 3.8 million in 1960 (5.5% unemployment) to 8.7 million in 1993 (6.8% unemployment), despite a 10-fold increase in our national economy (from a GNP of $515.3 billion in 1960 to $5,950 billion in 1993), (3) an increasing discrepancy between rich and poor, (4) an increase in media violence, especially on television and in movies, (5) an increase in narcotics use and drug trafficking, (6) a disintegration of family structure, especially in inner cities, with an increasing number of single-parent families and more births out of wedlock, and (7) greater accessibility and use of lethal weapons, especially guns.

This is not intended to be a comprehensive or definitive list, but it is one that shows environmental and social changes occurring during the same years as our rising violence. These comparisons come down in their basic forms to issues of competition, deprivation, early experiences, and social organization. Dr. William Bennett has recently pointed out that the period from 1960 to 1990 saw "a 560 percent increase in violent crime, a 400 percent increase in illegitimate births, a quadrupling of divorces, a tripling in the percentage of children living in single parent homes, and a 200 percent increase in teenage suicide rate". . ., all evidence of societal breakdown (Bennett, 1995).

As complex as the issues of individual violence are, even more complex are those of collective violence. The most complicated of all is war, a highly organized social behavior that springs from the aggressive nature of human beings but is modified by group dynamics, political action, tribalism or nationalism, religious fervor, international competition, and so on. In the next chapter, we will sort through some of the ecological factors contributing to war, as well as the impact of war on the global environment.

SUMMARY

Competition and conflict are ecological facts of life for all organisms. Darwin recognized that competition provides the driving force of natural selection and evolution.

Competition may be interspecific, between different species, or intraspecific, between members of the same species. Both types of competition can be intense and limiting at times, but intraspecific competition is usually the most severe. Closely related species avoid competition by niche specialization. Gause's principle, or the principle of interspecific exclusion, states that two closely related species with identical ecological requirements cannot coexist. One will ultimately displace the other.

Competition often leads to aggressive interaction. Animals are popularly accused of violent aggression, but what is portrayed as violence in animals is usually display behavior, a form of communication in which a threat or dominance is asserted without direct fighting. Intraspecific violence leading to death does occur in animals, but it is relatively rare. Ritualized display is much more frequent. Dominance hierarchies and territorialism are patterns of animal social behavior that involve aggressive display but reduce violent and injurious behavior.

Ecological factors that contribute to violence in animal populations include rapid population growth and crowding, which lead to social disintegration. The breakdown of dominance hierarchies and stable territorial boundaries often leads to more violent aggression.

Human violence can be considered in terms of individual violence (e.g., murder, assault, rape) or collective violence (e.g., riots, group terrorism, revolution, and war). These forms of violence often overlap.

Extensive debate has occurred in both the natural and social sciences as to whether human beings are innately aggressive by virtue of our evolutionary history or whether aggression and violence are entirely learned through competition and strife in early experiences. Substantial evidence can be mustered to support both points of view. Their common ground is the recognition that humans do have strong aggressive tendencies which are further provoked by ecological pressures and environmental circumstances.

Individual aggression has been increasing with alarming frequency and severity in the United States over the past 30 years. The rate of homicide has risen four-fold in the population at large, and many times that in certain locations and specific age groups.

Several national commissions and many independent scientists studying the problem of violence have come to the conclusion that social disintegration, economic frustration, and ecological factors such as population growth and over-crowding have all contributed to the rise of violence. Cultural factors such as the media, the widespread availability of weapons, the use and distribution of narcotics, and the loss of stable family life have also played a part.

Although the study of aggression and violence involves many disciplines and is a topic of great complexity, ecological factors are strongly implicated. Competition and conflict influence global ecology in many ways, and will have profound effects on the survival of the world's people as well.

REFERENCES

Allee, W. C., et al. 1949. *Principles of Animal Biology*. Philadelphia: Saunders.

Ardrey, P. 1966. *The Territorial Imperative*. New York: Atheneum.

Bennett, W. J. 1995. Redeeming our time. Imprimis 24 (11): 1–8.

Calhoun, J. 1962. *The Ecology and Sociology of the Norway Rat*. Bethesda, MD: U.S. Public Health Service Publication 1008.

Coon, C. 1972. Review comment on dust jacket of book by Colin Turnbull. *The Mountain People*. New York: Simon and Schuster.

Crowcroft, W. P. 1966. *Mice All Over*. Brookfield, IL: Chicago Zoological Society.

de Waal, F. 1989. *Peacemaking Among Primates*. Cambridge, MA: Harvard University Press.

Gause, G. F. 1934. *The Struggle for Existence*. Baltimore, MD: Williams and Wilkins.

Gest, T., et al. 1994. Violence in America. *U.S. News and World Report* 116(2): 22–33 (Jan. 17, 1994).

Goodall, J. 1986. *The Chimpanzees of Gombe: Patterns of Behavior*. Cambridge, MA: Harvard University Press, Belknap Press.

Graham, H. D., and T. R. Gurr. (eds.). 1969. *Violence in America*. New York: New American Library.

Kerner Report. 1968. U.S. National Advisory Commission on Civil Disorders. New York: Bantam Books.

Lorenz, K. 1966. *On Aggression*. New York: Harcourt, Brace and World.

MacArthur, R. H., and J. W. MacArthur. 1961. On bird species diversity. *Ecology* 42: 594–598.

Montagu, M. F. A. 1968. *Man and Aggression*. New York: Oxford University Press.

Odum, E. P. 1971. Fundamentals of Ecology. 3rd ed. Philadelphia: Saunders.

Odum, E. P. 1993. *Ecology and Our Endangered Life Support Systems*. 2nd ed. Sunderland, MA: Sinauer Associates.

Raven, P. H., and G. B. Johnson. 1992. *Biology*. 3rd ed. St. Louis: Mosby–Year Book.

Ricklefs, R. E. 1990 *Ecology*. 3rd ed. New York: W. H. Freeman.

Scott, J. P. 1958. *Aggression*. Chicago: University of Chicago Press.

Siegel, B. J. 1969. Defensive cultural adaptation. In *Violence in America,* ed. H. D. Graham and T. R. Gurr. New York: New American Library.

Smith, R. L. 1992. *Elements of Ecology*. 3rd ed. New York: Harper Collins.

Southwick, C. H. (ed.). 1970. *Animal Aggression*. New York: Van Nostrand Reinhold.

Turnbull, C. 1972. *The Mountain People*. New York: Simon and Schuster.

Chapter 24

THE ECOLOGY
OF WAR

In any decade since World War II there have been more than 30 wars in progress. As recently as 1990, 100 wars were being fought in more than 40 different countries (Nietschmann, 1990).

The specific number of wars underway depends, of course, on one's definition of war. When does an armed conflict become a war? In *Webster's Dictionary, war* is "the state or fact of exerting violence or force against a state or other politically organized body." This is a broad definition; in this chapter we use the term to refer to armed conflicts of this type that have considerable political organization. Human beings are perhaps unique in the extent to which they have organized group conflict.

What does war have to do with ecology? Revolution and war have always been considered within the realms of history, political science, sociology, economics, international relations, and perhaps psychology, but usually not in the domain of ecology. Ecology and war, however, are closely related in at least two ways. First of all, environmental factors—competition, space, resource requirements, and social interactions—are important causes of war. Nations compete for land, resources, power, and domination. Secondly, wars impact the environment directly, ravaging the earth, often with longlasting effect, and indirectly, by diverting, consuming, and destroying valuable resources that could be used for productive purposes. As Nietschmann succinctly states (1990): "The environment has been both a military target and a casualty of war."

Many wars have arisen from ecological pressures. Political turmoil is often associated with a burgeoning population, poor living conditions, a stagnant economy, and shortages of food, space, and jobs, as well as runaway inflation resulting from resource shortages. These problems, accompanied by rising expectations, have added to problems of social and political instability. As Eric Hoffer noted in his insightful book *The Ordeal of Change* (1963);

> A population subject to drastic change is a population of misfits—unbalanced, explosive, and hungry for action. In other words, drastic change, under certain conditions, creates a proclivity for fanatical attitudes, united action, and spectacular manifestations of flouting and defiance; it creates an atmosphere of revolution.

THE CAUSES OF WAR

A classic study of war was undertaken between World Wars I and II by a group of social scientists at the University of Chicago (Wright, 1942). Wright and his colleagues analyzed 278 major wars from 1480 to 1940, averaging 6 major wars per decade, with an average duration of 3.6 years per war. Wright's study found that 48 percent of these wars were "balance of power" wars, where nations were competing for domination and power; 28 percent were "civil" wars, hierarchical power disputes within nations; 16 percent were "imperialistic" wars of territorial expansion; and 8 percent were "defensive" wars, where both parties felt threatened or about to be attacked.

These categories suggest that humans fight for basically the same reasons that animals do: dominance, territory, and protection. However, the extent of violence in human conflicts far surpasses that of the much-maligned animal kingdom. Aggression in most animal societies takes the form of display and ritual. Relatively little of this behavior is injurious or fatal. Furthermore, animal fighting lacks the elaborate political organization of human warfare. Although there are coalitions of aggressive behavior in primates (Noe, 1994) and there are voracious group attacks in social insects such as army ants, such behaviors are not the norm in most species of animals.

Among the unique features of human warfare are the elaborate nature of the conflicts and the complex social organization and political ideologies that underlie them. Also unique is the use of military tools, which have become increasingly powerful. The aggressive behavior of no other species involves so much planning, organization, and material preparation. No other species uses force with such lethal intent against its own kind.

Another feature of human warfare is loyalty, the loyalty of individuals to a belief, an ideal, a leader. The violence of war is accentuated by subjugation of individual will to the goals of the group. The commander is to be obeyed, even at the risk of death. Soldiers blindly charging into battle and modern jet pilots unleashing their missiles are simply "doing their job." The individual soldier follows the dictates of his or her commander, who follows the dictates of his or her nation. Thus, faith and obedience are an essential part of human warfare, far more so than individual survival, which is the primary drive in animal aggression.

Recent civil wars in Angola, Bosnia, Serbia, Croatia, Cambodia, Rwanda, Somalia, and Chechnya demonstrate the ferocity of human conflict when different social groups are involved and central control is lost. In each situation above, one societal group has set out to subjugate or destroy another. Competition for limited resources sets the stage, while ethnic rivalries trigger the aggression and fan the violence. The result is the disintegration of society at large. Human atrocities are committed in the name of political loyalties.

Robert Kaplan, in explaining the recent rise of violence and wars in Africa and the Middle East, has emphasized the role of resource scarcities, overpopulation, and tribalism (1994). In his analysis, ecological factors play a causative role in the social disintegration that leads to violent conflict. The disastrous civil war in Rwanda had not happened at the time his article was published, but it is a clear example of his thesis; tribal conflict in one of the most crowded and impoverished nations of Africa led to a human disaster of mammoth proportions. Within a few short months, Rwanda experienced over a million deaths, largely of innocent civilians caught up in uncontrolled violence, and several million more refugees fled into Zaire and Burundi. Kaplan saw such conflicts as "The Coming Anarchy"—a response to environmental and social dysfunction. In this "revenge of the poor," individual violence coalesces along racial, religious, tribal, or political lines, causing the breakdown of rational society. Kaplan concluded, "Future wars will be those of communal survival, aggravated or, in many cases, caused by environmental scarcity" (Kaplan, 1994).

These national wars also have ecological consequences for conservation and endangered species. In Rwanda, for example, the war has added a new threat to the survival of the mountain gorilla. The Karisoke Reserve, home of 300 mountain gorillas, which represent 50 percent of the total population in the world of these magnificent animals, was invaded by thousands of Rwandan refugees fleeing the war. The refugees trampled and cut new escape routes to Zaire through the forested mountains of Karisoke, and they set up encampments in the forests that destroyed habitat and increased the likelihood of transmitting tuberculosis, dysentery, cholera, and flu to the gorillas. As of early 1995, we have no reports of serious losses of the gorilla population, but the danger of extinction is substantially increased by the magnitude of this human tragedy. The Karisoke gorilla population became famous through the work of George Schaller and Dian Fossey, especially Dian Fossey's book *Gorillas in the Mist* and the movie of the same title about her life.

A BRIEF ENVIRONMENTAL HISTORY OF WAR

The violence of warfare has impacted and brutalized the environment as well as human beings. Three thousand years ago (1,000 B.C.), the Jordanian wars provided one of the early recorded cases of chemical warfare that damaged the environment. The army of Ambimelech spread salt on agricultural fields to deprive the Jordanian cities of Shechem and Nablus of food. In 512 B.C., in the Persian-Scythian war, as the Scythians retreated, they practiced a deliberate scorched-earth policy to lay waste to landscape and hinder the Persian advance into what is now modern-day Iran. All of these conflicts were accompanied by loss of human life through starvation and disease as well as battle injuries.

In the Peloponnesian Wars of 431 to 404 B.C., the Spartans destroyed Athenian grain crops with the same intent of environmental destruction. Three centuries later, between 149 and 146 B.C., in the Roman attacks on Carthage, the Romans spread salt on the fields in a deliberate attempt to destroy the agricultural environment.

This theme continued in Europe in the Middle Ages. In 405 A.D., the Vandals sacked Rome, inflicting terrible damage on the environment as well as the human population, and from 1213 to 1224 A.D., the Mongols invaded and pillaged eastern Europe, destroying the environmental life-support system. In the Thirty Years

War from 1618 to 1648, when the Protestant armies of Germany and Sweden fought against the Catholic armies of the Holy Roman Empire, 40 percent of all agricultural fields were destroyed, and the human population of Bohemia was reduced 75 percent, with civilians suffering as much as soldiers.

The Napoleonic wars, the U.S. Civil War, and many other conflicts have involved deliberate attacks on the environment, including scorched-earth campaigns intended to deprive the enemy of resources.

Despite the frequency and intensity of war-related destruction prior to the twentieth century, most environments have been able to recover in 20 to 50 years. Forests, grasslands, and agricultural fields have generally returned to productive states. Looking at some of the Civil War battlefields in the United States, we find it hard to imagine the violence they sustained 130 years ago.

There are exceptions, of course. Some environments in the Middle East and North Africa have never recovered from the effects of war, but usually other factors have been involved, such as bad agricultural practices, prolonged overgrazing, and related forms of exploitative land use. As discussed in Chapter 13, some areas of the Middle East today have environmental problems that originated centuries ago, even before the time of Christ. These environmental problems have led to recurrent economic and political problems. Warfare has often worsened environmental conditions, but it has not been the sole cause of irreversible environmental change.

ENVIRONMENTAL ASPECTS OF WAR IN THE TWENTIETH CENTURY

The history of war has essentially been one of increased firepower and destruction. During World Wars I and II armies laid waste to the cities and farms of Europe and parts of eastern Asia. Most of these sites have recovered; cities have been rebuilt, farms returned to productive states, and forests regenerated. As with some Civil War battlegrounds, it is a leap of imagination to recall the destruction that occurred just over 40 years ago (Figure 24.1).

With the Korean and Vietnam wars, however, new elements were introduced. The military policy of interdiction became a practiced art. Interdiction was basically the policy of using heavy munitions against suspected enemy positions or enemy resources. Massive firepower was directed against landscape where the enemy was thought to be hiding or entrenched, and against croplands, water supplies, and transportation routes that the enemy used or was thought to use. Thus, the environment became the target of destruction.

In the Korean War of the 1950s, 75 percent of U.S. munitions expenditures were used for interdiction, and in the Vietnam War of the 1960s and 1970s, 85 percent of munitions expenditures were used for interdiction. This involved the blanket use of scatter bombs, firebombs, napalm, and patterned saturation bombing. This type of bombing was generally used when the enemy could not be seen but was thought to be hiding in forests or underground. The idea was to harass the enemy, destroy resources, or kill the enemy by overwhelming firepower.

A new dimension of warfare during the Vietnam War was defoliation. Herbi-

Figure 24.1   The brutalities of war affect all components of an ecosystem in many direct and indirect ways. In this photo, U.S. Marines recapture a Japanese position on Okinawa in 1945. Local environments can show remarkable ability to recover from conventional warfare, but as firepower and munitions devoted to interdiction increase, both human and environmental costs escalate. Recovery becomes difficult or impossible. (U.S. Marine Corps photo by Thomas D. Bartlett, Jr., courtesy National Archives)

cides were used to deprive the enemy of cover or food crops. The herbicides 2,4-D and 2,4,5-T were used extensively in a mixture known as Agent Orange to defoliate forests and poison crops. They were sprayed from aircraft on over 8 million acres of forest and 3.8 million acres of agricultural cropland (Orians and Pfeiffer, 1985). Single exposures caused trees to lose their leaves and crops to wither; repeated applications actually killed trees and poisoned agricultural fields for many years. Forests were targeted in both the central highlands of Vietnam and mangrove coastal zones. Land destruction in Vietnam involved 80 percent of all forestland and over 50 percent of coastal mangrove habitats (Pfeiffer, 1990).

Other forms of environmental destruction in Vietnam included cratering, bulldozing, and land mines. Fourteen million tons of bombs were rained on Vietnam and Cambodia in less than 20 years of saturation bombing. In Vietnam 500-pound bombs produced some 20 million craters 20 to 50 feet across and 10 to 20 feet deep. Many of these have permanently pockmarked the landscape, and those in lowland areas filled with stagnant water, creating breeding grounds for mosquitoes. This has accentuated the transmission of mosquito-borne diseases, including malaria, filariasis, Dengue fever, and encephalitis.

Bulldozing involved the intentional leveling and razing of forestland to allow better patterns of artillery fire and easier movements of troops and mobile armor. Bulldozers of 25 tons with 11-foot blades, known as Roman ploughs, were capable of knocking over small to moderate-sized trees, walls, houses, and other obstructions that could provide cover for the enemy. Over 1 million acres were bulldozed by Roman ploughs.

Land mines also remain a problem today, not only in Vietnam, but in Cambodia, Afghanistan, Angola, Kuwait, and other countries with recent or current wars. Deadly land mines are estimated to number 100 million throughout the world (Chelminski, 1994). Tens of thousands of civilians and children have been killed and maimed, and such tragedies are still daily occurrences in many areas. Cambodia has an estimated 35,000 land-mine amputees; Angola has at least 20,000. More than 600 people are killed or injured by land mines every month, and as many as 1 million land-mine deaths have occurred in the last 40 years (Chelminski, 1994).

Another environmental problem in old battlefields is metal fragments from exploded bombs, artillery, and land mines that have been imbedded in trees. Such fragments are hazards for foresters engaged in the timber industry.

While most of the environmental scars of warfare through World War II could usually be restored in 40 years, those of the Vietnam War will continue much longer, and some damage may prove to be irreversible. Much of the tropical hardwood forest that was repeatedly sprayed with Agent Orange converted to weedy plants such as imperator grass, Lantana, or bamboo, all graze-resistant and commercially of little or no value. Tens of thousands of acres of valuable forests were essentially destroyed, including the wildlife in these forests. Vietnam had 11 species of primates, including gibbons, the rare and endangered Douc langur, and several species of macaques. Many local populations of these valuable animals have disappeared, and the few that remain are in scattered, usually isolated, populations.

As a result of long years of warfare, a weakened communist government, a severely damaged environment, and a large population in a small area (in 1994 73 million people in an area less than one-half the size of Texas), the Vietnamese economy has been poverty-stricken in the 20 years since the end of the war. The per capita GNP of Vietnam was $180 in 1987 and only $230 in 1993, making it one of the poorest nations in the world. Vietnam is struggling to create a productive economy, but the process will be a long and difficult one hampered by a severely damaged environment.

THE PERSIAN GULF

Modern warfare reached a new level of automated destruction in the Gulf War of 1990. Although nuclear weapons were not used, the extensive use of guided missiles from land-based stations, aircraft and naval vessels made this a high-tech conflict in which both the people and environment of Kuwait and Iraq suffered mightily. More than 500 oil-well fires burned for weeks, and in some cases for several months, creating clouds of dense smoke across the Middle East and carbon-particle fallout as far as the Himalaya Mountains of India and Nepal (Canby, 1991). In Kuwait, day was turned into night. Oil spills blanketed the western coast of the Persian Gulf for 300 miles, as far south as the island nation of Bahrein. The "tide of destruction" described by Canby smothered coral reefs, fish-spawning areas, shrimp nurseries, pearl oyster beds, waterfowl feeding and breeding sites, sea turtle nesting beaches, and even some mangrove areas. In addition, in the desert landscape of Kuwait and southern Iraq, a new semipermanent feature is the debris of thousands of vehicles, civilian and military, and the discarded hardware of war. Although the people of Kuwait and Iraq suffered tragic losses, the science advisor of Jordan, Abdullah Toukan, noted that the most senseless casualty was the environment. Although the global effects of the oil-fire clouds were not as great or as persistent as originally feared, the ecological as well as the social scars of this conflict will remain for many decades to come.

NUCLEAR WAR

History has shown us the increasing firepower and environmental devastation which comes with each advance in military technology. From swords, arrows, spears, horses, gunpowder, cannons, and bombs to napalm, herbicides, and modern aircraft and missiles, the potentials for human and environmental devastation increase. The ultimate weapon at the present time is the nuclear bomb, including atomic, hydrogen (thermonuclear), and neutron bombs. We had a glimpse of their destructive power when the first atomic bombs were dropped on Hiroshima and Nagasaki in August 1945, but these bombs were small in destructive effect compared to the hydrogen bombs developed since then.

To provide a general idea of the increasing power of nuclear bombs, the largest bombs used in Europe in World War II were the "blockbusters," bombs capable of leveling a city block with the equivalent of 10 tons of TNT. The atomic bomb dropped on Hiroshima on August 6, 1945, the first nuclear weapon used in war, was the equivalent of 12.5 kilotons of TNT (12,500 tons), more than 1,000 times more powerful than the famous blockbuster. This bomb devastated the city of Hiroshima and caused between 80,000 and 150,000 human deaths. The Nagasaki bomb, dropped three days later, was almost twice as powerful (22.0 kilotons) and killed over 200,000 people. The justification for using these nuclear weapons was that they saved lives by bringing the war to a speedy end. Japan surrendered six days after the Nagasaki bomb.

After World War II, the nuclear arms race advanced rapidly; scientists developed bombs with greater destructive power and designed more efficient ways of delivering nuclear bombs to the enemy. The Soviet Union exploded its first test nuclear bomb in 1949 (Table 24.1). In 1954, the United States developed the first hydrogen bomb, 50 times more powerful than an A-bomb. We tested it on Bikini atoll in the Marshall Islands of the central Pacific. Thirty years later, Bikini was

TABLE 24.1
A brief outline of the nuclear arms race.

July 16, 1945	First nuclear bomb detonated by the United States in the New Mexico desert
August 6, 1945	First nuclear bomb used in war on Hiroshima
August 9, 1945	Second nuclear bomb used in war on Nagasaki
1948	United States deploys first intercontinental bomber with nuclear bombs
1949	Soviet Union explodes its first nuclear bomb
1954	United States develops hydrogen bomb, 50 times more powerful than the atomic bomb
1955	Soviet Union develops hydrogen bomb
1960	United States develops submarine-launched ballistic missile
1966	United States develops multiple nuclear warhead
1968	Soviet Union develops submarine-launched ballistic missile and their first multiple nuclear warhead
1968	United States develops anti-ballistic missile (ABM)
1970	United States develops multiple, independently targeted reentry vehicle (MIRV)
1972	Soviet Union develops ABM
1975	Soviet Union develops MIRV
1982	United States develops long-range cruise missile
1983	United States develops neutron bomb for operational use; i.e., actual use in wartime if needed.
1991	United States and Soviet Union sign Strategic Arms Reduction Treaty (START I) to reduce strategic nuclear warheads from 19,165 to 8,180 by the years 2000 to 2003. The U.K., France, and China are not party to START and not subject to any of its limits.
1992	The breakup of the Soviet Union dramatically reduces the threat of global nuclear war, but the presence of nuclear weapons in Russia, Ukraine, Belarus, Kazakhstan and their spread to other nations including Iraq, Iran, Israel, South Africa, India, Pakistan, and North Korea pose regional nuclear threats. The long-range outlook for nuclear proliferation remains troubling (Nuckolls, 1995).

still too radioactive to be habitable. By the 1960s, we had developed H-bombs, also called fusion bombs or thermonuclear bombs, with the destructive power of millions of tons of TNT.

A one megaton bomb (1 megaton equals 1 million tons) would destroy an area of 80 square miles with its blast effect, heat, and radiation pulse. There would be a blinding fireball and total human destruction within a two-mile radius. The intense heat would char flesh within 7 miles. All buildings would be destroyed within 5 miles. There would be dangerous radioactivity 200 miles downwind, potentially fatal to people 50 miles away. All this from one megaton!

In the 1970s the United States tested a single bomb of 28 megatons, and Soviets had tested one of 58 megatons. By the 1980s both superpowers were talking about 100-megaton bombs. (One megaton is equivalent to the amount of TNT it would take to fill a freight train 300 miles long.)

The new delivery systems were equally awesome. One outdated B-52 bomber became capable of carrying twice the destructive power of all the bombs used in World War II. One Poseidon submarine could carry nine megatons, equivalent to 720 Hiroshima bombs.

By the mid-1980s the world had stockpiled 20,000 megatons of nuclear bombs. These were deployed in intercontinental ballistic missiles (ICBMs), intermediate- and short-range missiles, short- and long-range bombers, submarines, and guided missiles on naval ships. In 1980 the United States had 30,000 strategic and tactical nuclear missiles the Soviet Union over 21,000. The United Kingdom, France, and China had over 1,000 together. The global arsenal of nuclear weapons was approximately 55,000 in the late 1980s (Renner, 1989). The use of even one-quarter of this nuclear arsenal could result in the immediate deaths of an estimated 30 million Americans and Soviets (Chandler, 1988).

NUCLEAR WINTER

One of the dire predictions coming from an analysis of the human and environmental costs of nuclear war was the concept of nuclear winter. A group of prominent American and European scientists, primarily in the field of atmospheric science, calculated that a nuclear war would ignite fires around the northern hemisphere. Forests, farms, grasslands, cities, oil refineries, and chemical plants would ignite and produce circumpolar clouds of smoke (Turco et al., 1990). These clouds would shut out sunlight, impair photosynthesis, and drastically reduce surface temperatures of the earth. Both natural and agricultural plants would wither and die. Life would be unsupportable within six months if the fires and smoke persisted. Ironically, the earth would freeze as a result of the devastating blasts and heat of nuclear war. Whereas the direct effects of a limited superpower nuclear war could kill 1 billion people, nuclear winter would kill the remaining 4 or 5 billion in the following months (Miller, 1992).

These conclusions were based on computer modeling of fire, smoke, and surface temperatures. Other scientists challenged the results; they granted that tragic consequences would ensue but calculated that the fires and smoke would not persist long enough to extinguish life on earth as the Turco model predicted (Schneider and Thompson, 1988).

The world was provided with a glimpse of the cooling effects of smoke on a small scale in the Yellowstone fires of 1988 in the United States and the Gulf War oil fires of 1990. In the Yellowstone forest fires, clouds of smoke moved southeast and caused a minor but measurable lowering of temperature for hundreds of miles, as far away as Nebraska and eastern Colorado. Temperature declines were only a few degrees, however. In the Gulf War, more than 600 Kuwaiti oil-well fires sent dense billowing clouds of smoke thousands of feet into the air for several months. Oil smoke blackened the snow in Kashmir, 1,500 miles to the east. Daytime air temperatures dropped significantly in Kuwait City and were 7.5° F below normal as late as May 1991 (the war began on August 2, 1990) in Bahrein, 200 miles to the south (Williams, Heckman, and Schneeberger, 1991). No global climatic effects were observed, however, and in general the atmospheric consequences of the Gulf War oil fires were less than expected.

Other natural experiments on the climatic effects of smoke have occurred following massive volcanic eruptions. In 1815, following the Tambora eruption in Indonesia, one of the largest on record, weather and temperature effects were noted in the United States. The summer of 1816 was exceptionally cold in New England, with freezing temperatures in July, making 1816 the "year without a summer" (de Blij and Muller, 1993). The eruption of Mount Pinatubo in the Philippines in 1991 also created dust clouds that altered global weather in 1991 and 1992.

Thus, the atmospheric and climatic threats of nuclear winter are real, but perhaps not so drastic as originally modeled.

THE COLD WAR

Depending on one's viewpoint, the nuclear arms race was either a triumph or a tragedy. It was a triumph in that we have not had World War III. The threat of mutual destruction deterred the United States and the Soviet Union from launching a nuclear attack on each other. In this view, the Cold War prevented the ultimate global tragedy—nuclear destruction. All of the world's peoples can be thankful for this, and if the Cold War was indeed responsible for avoiding a Hot one, we can consider it a triumph.

We will never know, however, if the Cold War was actually necessary to avoid World War III, or if we could have found a more reasonable way had the world not been under such a military tension following World War II. The Cold War was tremendously expensive, and its costs will be with us for a long time. In this sense, it was a financial tragedy that we could not find a more rational approach to avoiding war.

Both superpowers, the United States and the former Soviet Union, virtually bankrupted themselves with military expenditures. During the 1980s, when both nations were suffering from numerous social and environmental problems, the costs of the Cold War helped to create massive national deficits in both nations. Military budgets exceeded $300 billion dollars per year in each nation. The U.S. deficit deepened as military costs of the Cold War expanded in the 1980s (Fig. 24.2).

Many nations other than the United States and USSR were involved in the upward spiral of military expenditures. Worldwide, military costs totaled more the $940 billion dollars in 1985, more than the combined GNP of China, India, and all of Africa south of the Sahara (Corson, 1990). In the mid-1980s the world was spending $163 per person per year on military costs, compared to only $140 per person per year on education. The world is still spending well over $10,000 per year to train each soldier, less than $100 per year to educate each child. In some developing countries, military expenditures dominate all other government spending. For example, in 1984 Iraq spent 50 percent of its gross national product on military costs. While the people of many nations are ill-fed, poorly housed, and

Billion
Dollars
(1987)

Figure 24.2   U.S. military expenditures and economic deficits, 1950–1987 (from Renner, 1989).

lacking in education and medical care, their governments are buying missiles and bombers.

Military expenditures, made in the name of security, imply that virtually all people feel threatened by international competition and attack. Rare, indeed, are nations that do not maintain substantial military forces. Costa Rica, in Central America, is one such nation; it has devoted significantly larger shares of its national income to conservation, education, and health care. It has also had more stable political affairs than its neighbors Nicaragua, Panama, El Salvador, Honduras, and Guatemala.

INDIRECT COST OF CONFLICTS

Part of the tragedy of high military spending in most nations, including the superpowers, is the diversion of financial and natural resources to destructive weapons rather than productive uses. Two days of global military spending, estimated in 1989 to be $4.8 billion dollars, equaled the annual cost of the proposed United Nations action plan to halt desertification in the developing world over a 20-year period (Table 24.2). Three days of global military spending, an estimated $6.5 billion dollars, would fund the tropical forest conservation action plan for five years. The cost of one Trident nuclear submarine, with its awesome destructive

TABLE 24.2
Examples of what could be done if military expenditures were used for environmental and social purposes (Renner, 1989).

Military Priority	Cost	Social/Environmental Priority
Two days of global military spending	$4.8 billion	Annual cost of proposed UN action plan to halt desertification in developing nations over 20 years
Three days of global military spending	$6.5 billion	Five-year funding for tropical forest conservation action plan
Two weeks of global military spending	$30 billion	Annual cost of proposed UN Water and Sanitation Decade
One Trident submarine	$1.4 billion	Global four-year child immunization program against six deadly diseases
Trident submarine and F-18 jet fighter programs	$100 billion	Estimated cost of cleaning up 10,000 worst hazardous waste dumps in the United States
Stealth bomber program	$68 billion	Two-thirds of the estimated cost to meet U.S. clean water goals by the year 2000
3 B-1 bombers	$680 million	U.S. government spending on renewable energy, 1983–85
One-hour operating cost, B-1 bomber	$21 thousand	Community-based maternal health care in 10 African villages
Two months of Ethiopian military spending	$50 million	Annual cost of proposed UN antidesertification plan for Ethiopia

power, at $1.4 billion dollars, would fund a worldwide five-year immunization program against six infectious diseases.

The costs of military "security" are incredible, and these comparisons suggest that if we could only control our aggressive and violent behaviors, we would have vast resources to devote to productive enterprises. We could improve the quality of life and the quality of the environment for everyone.

Those holding the security-minded point of view ask, What good are environmental and social priorities if the world disintegrates in nuclear war? This would certainly be one's viewpoint if one believed that the arms race of the Cold War prevented nuclear attack. According to this viewpoint, disarmament has never really worked. The only effective deterrent is the threat of force, and we can be thankful that we had the military strength to prevent World War III.

On the other hand, those who believe that neglecting social and environmental causes, along with military buildups, actually increases the likelihood of war insist that human beings must use rational behavior to control human violence. They point out that the exaggerated buildup of military hardware adds to the probability that it will be used to resolve conflicts and fear that if an obituary is ever written on the human race, it would read, "Out of fear and aggression, they spent their money on the wrong things!"

In addition to the costs of building military armaments, there are substantial costs associated with storing them, dismantling them, and cleaning up after use or storage. Even outside of war, direct land use for military bases and storage has been estimated at 750,000 to 1.5 million square kilometers (Renner, 1991), an area equal to that of Indonesia, the world's fourth most populous nation. In a world with shortages of agricultural land this is a problem. In the United States, the military holds about 100,000 square kilometers, an area about the size of Virginia.

In some cases, military land has been protected from undesirable sprawl or development. In other cases, this land has been degraded and polluted by military activities. Desert lands in southern California still bear the damage of tank-training maneuvers dating back to the early 1940s. Bombing ranges in Nevada have been permanently closed to human use because of unexploded bombs.

Pollution problems have also occurred on military bases. In recent years, our military establishments have produced more toxic pollution annually than the top five chemical industries combined. Otis Air Force base in Massachusetts has contaminated ground water with trichloroethylene, a known cancer-causing agent. In adjacent towns, lung cancer and leukemia rates are 80 percent above the state average (Table 24.3). Rocky Mountain Arsenal outside of Denver, Colorado, was a depository for 125 toxic chemical residues from nerve gas production. The projected cost of decontamination exceeds $1 billion.

Radioactive pollution at nuclear weapons plants such as Hanford, Washington; Rocky Flats, Colorado; Fernald, Ohio; and Savannah River, Georgia, represent our nation's most serious cases of long-term environmental radioactivity. The costs of cleanup run to hundreds of billions of dollars.

We also find that the threat of nuclear war has not disappeared entirely, though it is now greatly reduced. More nations than ever have the capability of building and delivering nuclear bombs, and constant international vigilance is necessary to prevent illegal trade in enriched plutonium. Nuclear terrorism, and even nuclear war, is still a frightful possibility despite the breakup of the Soviet Union.

SIGNS OF PROGRESS

The Strategic Arms Limitation Treaty (SALT) represents a move away from the threat of nuclear annihilation. If this agreement proceeds satisfactorily, if nuclear proliferation can be reduced or eliminated, and if regional wars can be stopped, human prospects are much brighter. If we could take most or even a good part of the $900 billion or more the world now spends on military competition and devote it to education, health, agriculture, renewable energy sources, better housing, cleaner and safer transportation, and environmental restoration, the possibilities for human betterment are immense. At the same time, economic growth could flourish. The potential for economic growth would be immense if everyone in the world had decent housing, good food, education, health care, and recreational opportunities, and the American Dream, "life, liberty, and the pursuit of happiness."

Obviously, this would require a massive reordering of the world's industrial complex, a daunting and extremely difficult task. It would require a shift toward environmentally acceptable lifestyles based on energy efficiencies not yet available

TABLE 24.3
Examples of environmental pollution problems at U.S. military bases
(Renner, 1991).

Location	Observation
Otis Air Force Base, MA	Ground-water contaminated with trichloroethylene (TCE), a known carcinogen, and other toxins. In adjacent towns, lung cancer and leukemia rates 80 percent above state average.
Picatinny Arsenal, NJ	Ground water at the site shows TCE levels at 5,000 times Environmental Protection Agency (EPA) standards; it is also polluted with lead, cadmium, polychlorinated biphenyls (used in radar installations and to insulate electrical equipment), phenols, furans, chromium, selenium, toluene, and cyanide. Region's major aquifer is contaminated.
Aberdeen Proving Ground, MD	Water pollution could threaten a national wildlife refuge and habitats critical to endangered species.
Norfolk Naval Shipyard, VA	High levels of copper, zinc, and chromium discharges. Contamination of Elizabeth River and of Willoughby and Chesapeake Bays.
Tinker Air Force Base, OK	Concentrations of tetrachloroethylene and methylene chloride in drinking water far exceed EPA limits. TCE concentration is highest ever recorded in U.S. surface waters.
Rocky Mountain Arsenal, CO	Some 125 chemicals from nerve gas and pesticide production dumped over 30 years. The largest of all contaminated sites is called "the most contaminated square mile on earth" by the Army Corps of Engineers.
Hill Air Force Base, UT	Heavy on-site ground-water contamination, including volatile organic compounds up to 27,000 parts per billion (ppb); TCE up to 1.7 million ppb; chromium up to 1,900 ppb; lead up to 3,000 ppb.
McClellan Air Force Base, CA	Unacceptable levels of TCE, arsenic, barium, cadmium, chromium, and lead found in municipal well system serving 23,000 people.
McChord Air Force Base, WA	Benzene, a carcinogen, found on-site in concentrations as high as 503 ppb, nearly 1,000 times the state's limit of 0.6 ppb.

and cleaner and more reasonable patterns of consumerism. The entire world certainly cannot afford to adopt the present American lifestyle, which is too destructive of environmental resources. Part of the industrial and economic challenge of the future is to achieve the good life with less energy use, less pollution, less waste, and less destruction. The conversion of the world's industrial prowess from systems of waste and destruction to systems of human and environmental improvement could indeed lead to the global prosperity and peace we desire.

Ultimately, we must come to grips with these challenges. Whether we can respond to them satisfactorily still remains to be seen. Few of us would deny that the human potential for good works and positive values is great, but at the same time our propensity for violence, both individual and collective, is frightening.

The point of discussing warfare in our consideration of global ecology is that conflict is a part of competition, and competition is basically an ecological issue.

Many destructive ecological trends today are directly related to increased competition, and these must be reversed or resolved in rational ways.

SUMMARY

Ecology is relevant to the study of war for at least two reasons: (1) ecological factors, including competition for resources, territorial ambitions, ethnic rivalries, and struggles for dominance, or power, are frequently involved in causing war, and (2) warfare and preparation for war impacts the environment in many ways, both directly and indirectly.

A major study of war conducted between World Wars I and II by a group of social scientists at the University of Chicago (Wright, 1942) analyzed 278 major wars from 1480 to 1940 and found that 48 percent of these were "balance of power" wars where nations were competing for domination, 28 percent were civil wars, or hierarchical disputes within nations, 16 percent were imperialistic wars of territorial expansion, and 8 percent were defensive wars, where both parties felt threatened or attacked. Ecological factors played a prominent role in virtually all of these conflicts.

Although the world has successfully avoided World War III in the past 50 years, numerous smaller wars have occurred throughout the world, recently numbering at least 30 at any given time. Most of these have involved power disputes, such as recent wars in Afghanistan, Somalia, Rwanda, Bosnia, Serbia, and Chechnya, or territorial expansion, as in the Persian Gulf War.

Military history over the past 3,000 years has shown that the environment has often been a target in warfare through deliberate scorched-earth policies to deprive the enemy of food or other resources.

The environmental impacts of warfare have become increasingly severe and long-lasting throughout history with great advances in firepower and destructive capabilities. Atomic bombs dropped on Hiroshima and Nagasaki in August 1945 introduced the awesome reality of nuclear warfare. Relatively small atom bombs devastated both cities and produced over 300,000 human casualties. The nuclear arms race from 1945 to 1991 produced increasingly powerful hydrogen and neutron bombs and more efficient methods of delivery.

The Cold War may be seen as a triumph, if one views it as the main reason the world has avoided World War III, or as a tragedy, if one considers the financial, natural, and human resources that were devoted to the arms race, and the fact that we felt we could not avoid World War III by more rational means.

The Korean and Vietnam wars reached new levels of environmental destruction through the increased use of conventional firepower in the policy of interdiction. This involved saturation bombing of the environment to deprive the enemy of cover and resources. In Vietnam, interdiction also involved the use of herbicides to defoliate forests and croplands. Ecological studies have shown that some of the environmental changes produced in Vietnam are virtually irreversible. Whereas the battlefields of World Wars I and II have recovered, many of those in southeast Asia have not and, apparently, will not.

The indirect costs of preparation for war are often economically devastating to both large and small nations. In the United States, the federal deficit deepened as military expenditures of the 1980s increased. Many developing nations exhaust their resources to maintain military forces, and many tear themselves apart socially and environmentally in regional conflicts. The world spends more on military preparedness than on education, health, basic research, and environmental conservation combined. If the more than $900 billion dollars expended annually on military activities could be shifted to positive and productive enterprises, the potential for human and environmental betterment would be immense. The opportunities are great if the human propensity for aggression and violence can be directed toward beneficial activities.

The end of the Cold War and international agreements on the Strategic Arms Reduction Treaty are positive steps forward, but other ecological trends are leading to increased competition and potential conflict.

REFERENCES

Canby, T. Y. 1991. After the storm. *National Geographic Magazine* 180(3): 2–33.

Chandler, W. U. 1988. Assessing SDI [Strategic Defense Initiative]. In *State of the World, 1988: A Worldwatch Institute Report*, ed. L. R. Brown et al. New York: W. W. Norton.

Chelminski, R. 1994. The new killing fields. *Reader's Digest* (Mar. 1994): 107–112.

Corson, W. H. 1990. *The Global Ecology Handbook*. Boston, MA: Beacon Press.

de Blij, H., and P. O. Muller. 1993. *Physical Geography of the Global Environment*. New York: John Wiley and Sons.

Hoffer, E. 1963. *The Ordeal of Change*. New York: Harper and Row.

Kaplan, R. D. 1994. The coming anarchy. *Atlantic Monthly* 273(2): 44–76.

Miller, G. T. 1992. *Living in the Environment*. 7th ed. Belmont, CA: Wadsworth.

Nietschmann, B. 1990. The ecology of war: Battlefields of ashes and mud. *Natural History* (Nov. 1990): 34–37.

Noe, R. 1994. A model of coalition formation among male baboons with fighting ability as the crucial parameter. *Animal Behavior* 47: 211–213.

Nuckolls, J. 1995. Post-cold war nuclear dangers: proliferation and terrorism. *Science* 267: 1112–1114.

Orians, G., and E. W. Pfeiffer. 1985. Ecological effects of the war in Vietnam. In *Global Ecology*, ed. C. H. Southwick, pp. 279–292. Sunderland, MA: Sinauer Associates.

Pfeiffer, E. W. 1990. The ecology of war: Degreening Vietnam. *Natural History* (Nov. 1990): 37–40.

Renner, M. 1989. Enhancing global security. In *State of the World, 1989: A Worldwatch Institute Report*, ed. L. R. Brown et al., New York: W. W. Norton.

Renner, M. 1991. Assessing the military's war on the environment. In *State of the World, 1991: A Worldwatch Institute Report*, L. L. Brown et al. New York: W. W. Norton.

Schneider, S. H., and S. L. Thompson. 1988. Simulating the effects of nuclear war. *Nature* 333: 221–227.

Turco, R. P., O. B. Toon, T. P. Ackerman, J. B. Pollack, and C. Sagan. 1990. Climate and smoke: An appraisal of nuclear winter. *Science* 247: 166–176.

Williams, R. S. Jr., J. Heckman, and J. Schneeberger (eds.). 1991. *Environmental Consequences of the Persian Gulf War, 1990–1991*. Washington, DC: National Geographic Society.

Wright, Q. 1942. *The Study of War*. Chicago: University of Chicago Press.

Chapter 25

SUSTAINABILITY

"Sustainable" is the buzzword of the 1990s. Appropriately so, for we are all interested in sustaining life. Most individuals and organizations do not want to think of their own demise, and we certainly don't want to think about the end of the civilized world. Yet there is much concern about the survivability of planet earth, and even speculation about whether the world as we know it will go out with a bang or a whimper. The threat of nuclear war has not been eliminated, nor has creeping deterioration of the earth's environment.

The inexorable growth of populations, the persistence of poverty, the breakdown of society, the continuation of conflict, the geographic expansion of pollution, the loss of soil—all of these trends must somehow be controlled.

Numerous books have highlighted these problems over the past 50 years, including *Our Plundered Planet* (Osborn, 1948), *The Limits to Growth* (Meadows, Meadows, and Randers, 1972), *Planet Under Stress* (Mungall and McLaren, 1990), *Beyond the Limits: Confronting Global Collapse, Envisioning a Sustainable Future* (Meadows, Meadows, and Randers, 1992), and *Living Within Limits* (Hardin, 1993).

The solution is sustainability. But what does this mean? Politicians and business leaders talk about sustainable growth. Economists and planners talk about sustainable development. Ecologists talk about a sustainable environment. These are all quite different uses of the word, with very different consequences. The purpose of this chapter is to consider the meaning and validity of these different uses.

HISTORY AND DEFINITIONS

Environmental books and ecology texts of the 1970s and early 1980s rarely, if ever, used the word "sustainable." Comparable terms, such as "survivability," "ecological balance," "stability," and "equilibrium," were common, but not "sustainable" or "sustainability."

Webster's New International Dictionary does not actually define the word "sustainable" except to list it as the adjective of the verb "sustain" which means . . . "to provide support, to maintain, to cause to continue, to keep up, to prolong. . . ." Sustainability is not even listed in the 2nd edition of Webster's New International Dictionary.

The word "sustainable" leapt into prominence following the publication of *Our Common Future* by the United Nations World Commission on Environment and Development (WCED). This publication, known as the Brundtland report (1987), since the commission was chaired by Ms. Gro Brundtland, the Prime Minister of Norway, vividly portrayed the growing problems of poverty, environmental degradation, and conflict experienced by many people in developing nations. The report recognized that much environmental damage was caused by poor populations exploiting the environment merely to survive. It recognized other ravages caused by wealthy nations exploiting the environment to support extravagant lifestyles. To deal with this double onslaught, the Brundtland report emphasized the need for sustainable development:

> What is needed now is a new era of economic growth—growth that is forceful and at the same time socially and environmentally sustainable.

To the ecologist, this simultaneous call for vigorous economic growth and growth that is environmentally sustainable is an internal contradiction. Ecological principles emphasize that planet earth is finite and that all growth has limits. Ecological growth, in the sense of population growth and consumption of nonrenewable resources, must ultimately come to an end, despite the American dream expressed by President Franklin Delano Roosevelt in the 1930's, "We shall expand indefinitely."

The Brundtland Report did recognize that population growth cannot proceed without limits:

> Thus sustainable development can only be pursued if population size and growth are in harmony with the productive part of the ecosystem.

The use of the word "harmony" implies ecological balance, but it does not come to grips with the issue of ultimate limits. In fact, subsequent passages in the Brundtland report suggest that some population growth is desirable, so long as rapid growth can be curtailed:

> The issue is not just numbers of people, but how those numbers relate to available resources. Urgent steps are needed to limit extreme rates of population growth.

The Brundtland report supported 12 major priorities that are necessary to "sustain human progress into the distant future": (1) slow the rate of population growth, (2) reduce poverty, inequality, and debt in the developing world, (3) make agriculture sustainable, (4) protect forests and habitats, and curb loss of species, (5) protect ocean and coastal resources, (6) protect freshwater quality and improve water efficiency, (7) increase energy efficiency, (8) develop renewable energy resources, (9) limit greenhouse gases and other air pollutants, (10) protect the stratospheric ozone layer, (11) reduce wastes, and (12) shift military spending to sustainable development.

This list refers to both sustainable agriculture and sustainable development. The Brundtland report offered a definition of sustainable development that has become broadly accepted:

> Sustainable development is development that meets the needs of the present without compromising the ability of future generations to meet their own needs.

Clearly, this is an admirable goal, but a very general one. Are we coming close to this now? Are we meeting the needs of the present generation? Obviously not, when the United Nations classifies 23 percent of the world's population as living in abject poverty. Obviously not, when nearly one child in five in the United States lives in what is legally defined as a state of poverty. And certainly not, when deserts spread, famines occur, species disappear, and devastating wars erupt around the world. The Brundtland report continued its focus on sustainable development and sounded a cautiously optimistic note:

> Our global future depends upon sustainable development. It depends upon our willingness and ability to dedicate our intelligence, ingenuity, and adaptability—and our energy—to our common future. This is a choice we can make.

Here, the responsibility for the future world is placed directly on our own shoulders. We can achieve a better world if we make the right choices.

BOX 25.1
Development: For the People and the Environment

Given the problems faced by humanity at the turn of this century, notably those concerning world population and the environment, it cannot be doubted that development is the key word to human salvation. And whereas it is indisputable that economic growth is vital for development, it is no less important to ascertain that the social and environmental dimensions are not sacrificed in the effort to achieve it. It must be remembered that the connotation of the term "development," which was traditionally perceived merely as economic growth, has gradually undergone a change in the past few decades to include—and currently even centre upon—the wellbeing of people. The present worldwide situation shows that efforts in this direction have not only failed to improve the social and economic condition of the major part of humanity, or the state of the natural environment, but that they have not even succeeded in containing the downward slide, which in certain cases has acccelerated markedly. The increasing interdependence of all the nations of the world, in everything from economic markets and population movements to environmental issues, makes the adoption of a global approach to these problems imperative.

Excerpted verbatim from *Connect, UNESCO-UNEP Environmental Education Newsletter*. Vol. XX(1): 1–2. 1995. UN Secretariat of the World Summit for Social Development, New York, NY 10017.

FACING THE FACTS

The Brundtland report was clearly a step forward in global environmental awareness. It sparked a host of publications, including *Toward a Sustainable World* (Ruckleshaus, 1989), *Strategies for Sustainable Economic Development* (MacNeill, 1989), and *Toward a Sustainable Future* (Corson, 1990), but it still left many critical issues in a state of limbo. Most of these publications avoided the difficult questions of limits.

No one knows what the ultimate limits of development are, but their existence must be more widely recognized. That recognition seems to be political poison, judging by its dearth of support. In fact, the 1992 Earth Summit in Rio de Janeiro, which was the largest assembly of world leaders in history to address problems of the environment, avoided the hardest issues of population growth and ecological limits. The 1992 Earth Summit noted the dangers of population growth and excessive consumption as follows:

> The spiraling growth of world population fuels the growth of global production and consumption. Rapidly increasing demands for natural resources, employment, education and social services make any attempts to protect natural resources and improve living standards very difficult. There is an immediate need to develop strategies aimed at controlling world population growth (Sitarz, 1993).

This statement certainly provides recognition of the problem, but reaches only a vague conclusion. The term "controlling," unlike "stopping," is subject to various interpretations. It could mean controlling population growth to 1 percent per annum, which would be a favorable reduction from the present 1.7 percent increase per year, but even with a controlled 1 percent increase, world population would still grow 270 percent in the next century. The Rio Earth Summit clearly avoided the issue of stopping population growth completely because it was too controversial, too hot to handle in economic and political terms.

Elsewhere, the Rio Earth Summit considered world population growth inevitable:

> The long-term consequences of human population growth must be fully grasped by all nations. They must rapidly formulate and implement appropriate programs to cope with the inevitable increase in population numbers (Sitarz, 1993).

The use of the term "inevitable" implies resignation; nothing can be done, so we must prepare for inevitable growth.

The failure of governments, scientists, writers, and the public at large to face the world population problem was recently noted by May (1993, quoted in Bartlett [1994]). In reviewing a new book on biological diversity, Robert May, Royal Society Research Professor at Oxford University, noted that the book:

> says relatively little about the continuing growth of human populations. But this is the engine that drives everything. Patterns of accelerating resource use, and their variation among regions, are important but secondary: problems of wasteful consumption can be solved if population growth is halted, but such solutions are essentially irrelevant if

populations continue to proliferate. Every day the planet sees a net increase (births less deaths) of about one quarter of a million people. Such numbers defy intuitive appreciation. Yet many religious leaders seem to welcome these trends, seemingly motivated by calculations about their market share. And governments, most notably that of the United States, keep the issue off the international agenda; witness the Earth Summit meeting in Rio de Janeiro. Until this changes, I see little hope.

IS SUSTAINABILITY POSSIBLE?

The only kind of sustainability that makes long-term sense is ecological sustainability. Sustainable growth is not possible because all growth must come to an end at some point. Sustainable development has long-term possibilities, depending upon one's expectations, but it also has limits. The message of ecology is essentially one of the limits—planet earth is finite.

Real sustainability involves some kind of equilibrium. This does not necessarily mean stability—natural ecosystems have great variability, and are often inherently unstable—but they tend toward equilibrium; that is, they have internal checks and balances. They have the capacity of self-regulation and self-restoration within limits. They are cybernetic systems.

The capacity of self-regulation in natural ecosystems can be illustrated with several examples that highlight the basic ecological principles discussed in Chapters 4 through 7. If human-dominated ecosystems are to become sustainable, we must also observe these principles. Perhaps our greatest educational need is to realize that we are not exempt from the laws of nature, however far we may bend and stretch them.

SUSTAINABILITY IN NATURAL ECOSYSTEMS

Wilderness, wherever it may occur, provides examples of sustainable ecosystems. All types of natural communities—forests, grasslands, estuaries, coral reefs—are sustainable. They are not necessarily stable. All natural systems change from disturbances such as storms, fires, climatic fluctuations, and even internal processes, but they respond to these changes in adaptive ways, either returning to former states or succeeding to other community patterns. They remain viable and productive as long as the disturbances are not life-threatening to the entire ecosystem. An ecologically life-threatening disturbance could be a massive climatic change such as an ice sheet, a total drought for centuries, or a volcanic eruption that covered the entire substrate with impenetrably hard lava.

Natural systems are sustainable because they have the capacity to restore equilibrium in the basic ecological functions of production, consumption, and decomposition. Organic matter is organized, produced, and recycled through the biogeochemical cycles (outlined in Chapter 6). Water, carbon, nitrogen, phosphorus, and other essential chemicals recycle through biotic and abiotic components of the system. Energy flows naturally from sunlight into plants, animals, and microorganisms. Oxygen is produced and carbon dioxide utilized through photosynthesis.

Organic materials pass through food chains, and biodiversity is maintained in the natural mosaics of the environment.

None of this requires or even implies stability, but it does mean sustainability.

In natural ecosystems disturbance is often part of sustainability. An aspen forest in the Rocky Mountains or a prairie grassland in the midwestern United States, for example, is maintained by periodic fires. Natural fires from lightning strikes sweep through the ecosystems, preventing the excessive buildup of dead organic matter that is slow to decompose, opening up natural recycling patterns, initiating new growth, and revitalizing the ecosystem. New aspen growth and new forb and grass growth in the prairie depend on periodic removal of old debris.

This pattern of perturbation, common in some ecosystems but not universally shared by all, is often used as the justification for cutting old-growth forests, but here different ecological principles apply. Not all ecosystems are adapted to drastic change. So-called climax communities are more stable. They have achieved a sustainable balance in a more consistent physical and climatic environment. Old-growth redwood and Douglas fir forests of the Pacific Northwest, for example, survive in a consistently wet temperate climate in which trees live to great age. Fire is not a regular part of the ecosystem. Disturbance is limited to occasional tree falls creating gaps for renewed growth. The great age and stability of the system reduces productivity but enhances biodiversity. Innumerable organisms depend on stability and diversity for survival in such communities. The promotion of clear-cutting destroys the system entirely, and even selective cutting, depending on how destructively it is done, may reduce survival chances for some species.

The important lesson here is that we must have some understanding of natural ecological processes in different ecosystems. We should not try to restrain natural fires in ecosystems that are fire-adapted, nor should we promote timber harvest in old-growth forests, which depend on stability for their survival and high biodiversity.

Different ecosystems vary considerably in their patterns of floral and faunal stability and the frequency of disturbances to which they are typically subject (Table 25.1). The Arctic tundra, for example, has relatively stable plant communities but usually unstable animal populations of species such as lemmings, snowshoe hares, Arctic ground squirrels, snowy owls, and caribou. Plant and animal populations of boreal forests usually show more stability, with the exception of snowshoe hares. Both tundra and boreal forests are subject to extreme seasonal fluctuations in daylight and temperature, but the biota are adapted to these extremes.

Temperate forest populations have more moderate seasonal extremes, but only moderate stability since fire and precipitation are often quite variable. Many exceptions to the broad generalizations listed in Table 25.1 can be found but the table does illustrate the point that natural ecosystems vary in both disturbances and stability. Their common feature is the capability of self-recovery and restoration. They can return to either the former community or move on to the next natural successional stage. In either case, viability is maintained.

With sufficient knowledge and proper management we may be able to harvest natural products from certain ecosystems without destroying their basic nature. In

TABLE 25.1
Examples of natural ecosystems showing sustainability with different degrees of perturbation and stability.

Ecosystem	Degree of Stability		Examples of the Types of Disturbance or Perturbation
	Flora	Fauna	
Arctic or alpine tundra	High	Low	Temperature and daylight extremes
Boreal forest	High	High	Temperature and daylight extremes
Temperate deciduous forest	Moderate	Moderate	Occasional fire, seasonal heat and cold, leaf fall
Temperate coniferous forest	Moderate	Moderate	Fire, seasonal temperature and precipitation
Temperate grasslands, savannah	Moderate	Low	Fire, seasonal temperature and precipitation
Tropical monsoon forest	Moderate	Low	Fire, seasonal drought
Tropical rain forest	High	High	Gap dynamics, natural tree fall
Oligotrophic lakes	High	High	Seasonal turnover
Eutrophic lakes	Low	Low	Seasonal turnover, nutrient pulses
Estuaries and coastal zones	Low	Low	Tidal action, storm action
Coral reefs	High	High	Wave action, hurricanes

tropical rain forests, for example, a wealth of natural products exist to provide new biochemical compounds, medicinals, foods, fibers, and genetic resources (de Onis, 1992; Reid, et al., 1993). Several economic studies have shown that sustainable harvesting of these products can produce more income over the long run than deforestation and marginal agriculture (Muul, 1993; Plotkin and Famolare, 1992).

HUMAN-DOMINATED ECOSYSTEMS

Over 30 years ago, the ecologist Lawrence Slobodkin noted that human populations have so modified the surface of the earth that we must now consider the world a combined factory and garden (Slobodkin, 1961). Perhaps this view was overly pessimistic at the time, and it may even be contested today, but it emphasizes the point that human influence permeates virtually every part of planet earth. Antarctic biologists discovered 30 years ago that residues of DDT occurred in penguins hundreds of miles from any known use of the pesticide. We know that airborne lead particles have been deposited in the Greenland icecap since the beginning of the industrial revolution. At the same time there are vast stretches of wilderness where little noticeable human impact can be discerned, unless one is using refined measures of atmospheric gases or trace levels of persistent pesticides. Slobodkin's point is well taken, however, when we consider the vast portions of the earth's surface that have been modified by human action.

Some ecosystems modified by human action have been so drastically changed that it is not at all likely they will ever return to their former state. Manhattan Island in New York City is essentially a human labyrinth of concrete and steel. It is hard to picture the forested landscape that existed on Manhattan just 500 years ago; only Central Park and the New York Botanical Garden can give us a hint of its former nature. Cities are highly unstable ecological systems. They constantly require imports of fresh air, water, and food and daily exports of gaseous, liquid, and solid wastes. If they suffer so much as a few days of blockage in the transport of any of these imports or exports, critical conditions arise very quickly. Odum (1993) diagrammed the basic ecological activities of cities as heterotrophic systems, that is, systems that depend on the support of surrounding ecosystems.

Agricultural ecosystems represent a less drastic environmental modification than cities and suburbs do, but they are still highly unstable ecosystems. The modification is less drastic in that they remain plant-dominated systems, but the natural plant community has been totally changed. Prairie habitats in the midwestern United States, for example, were once luxuriant grasslands with over 120 species of native grasses and forbs existing on natural precipitation. These communities have been simplified to monocultures of wheat or corn that are now totally dependent on artificial planting and cultivation, fertilization, weed control, insect control, harvesting, and often the addition of irrigation water. If any one of these exogenous inputs is omitted, the system can break down. Even with massive use of pesticides, insect outbreaks can sometimes occur, decimating the productivity of the system. Usually such outbreaks are avoided by advance warning and proper management. Still, the agricultural ecosystem is inherently unstable.

Many agroecosystems are also basically unsustainable. Modern agriculture as practiced in the midwestern United States is utilizing soil and water resources faster than they are being renewed. Ground-water resources in the Ogalala aquifer of the central United States are being depleted, sometimes at the rate of several feet per year. Soil losses exceed replenishment by several inches per year. U.S. agriculture has remained productive because of the vast original wealth of both soil and water and the fact that we have been using these resources for less than 150 years. But we have maintained and enhanced our productivity only by digging deeper wells, using more fertilizers, and applying more pesticides, creating a system that is both unstable and unsustainable.

For several years we have had warnings that this modus operandi cannot continue (Eckholm, 1976), and now evidence is mounting that agricultural yields are leveling off or even declining in many areas. The agricultural economist Lester Brown noted (1994):

> As the nineties unfold, the world is facing a day of reckoning. Many knew that this time would eventually come, that at some point the cumulative effects of environmental degradation and the limits of the earth's natural systems would start to restrict economic expansion. But no one knew exactly when or how these effects would show up. Now we can see that they are slowing growth in food production—the most basic of economic activities and the one on which all others depend.

Brown's statement is based on recent data that show corn yields in the U.S. (Fig. 25.1) and rice yields in Asia leveling off (Fig. 25.2). Projections into the

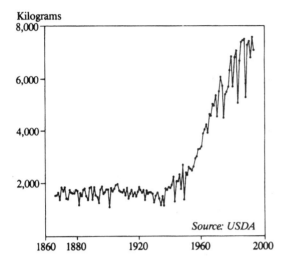

Source: USDA

Figure 25.1 U.S. corn yield per hectare, 1866–1993 (from Brown, 1994).

twenty-first century, now only a few years away, show actual declines in both yield and per capita output. These projections are based on the trends of the last few years. The technological optimists believe that these are temporary setbacks and that we can resume continued growth in agricultural productivity by increasing use of modern agricultural science. We have barely begun to use the wonders of genetic engineering in crop production, and the possibilities are endless, so they argue. On the other hand, ecologists believe that our agroecosystems are showing signs of stress and may be very near their carrying capacities; that is, yields are already pushing the limits of soil, water, and nutrient resources. In any case, it is apparent that we have a long way to go in developing sustainable ecosystems in any of our domestic environments.

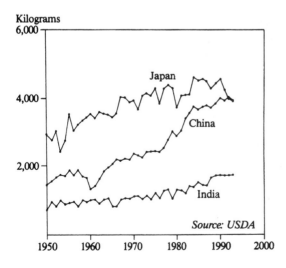

Source: USDA

Figure 25.2 Rice yield per hectare in China, India, and Japan (from Brown, 1994).

CRITERIA FOR SUSTAINABILITY

Sustainability is not a mysterious concept. True, the term has been misused, and bantered about too freely, but it is basically a sound idea. Ecological sustainability means that all fundamental ecological relationships are in balance or capable of being restored to a steady state. The idea of balance applies not only to production, consumption, and decomposition but also to the basic ecological cycles involving soil, water, and air quality. We must apply these principles to economic planning and land use.

In agricultural development we need to operate with no net loss of soil, nutrients, or water and no buildup of toxic wastes on any part of the land. In the long run, energy sources must be renewable. These tough standards are necessary if sustainability is to last more than a few decades, or at best a century.

In urban, suburban, and industrial development, sustainability requires much greater attention to land use and restoration. We can no longer afford to let cities, highways, airports, and similar developments expand like a malignancy into the surrounding countryside. Greenbelts and open space must be maintained. Some of our philanthropic benefactors in previous centuries realized this when they set aside such areas as Central Park in New York, Richmond Park in London, and the Maidan in Calcutta. The growth mania of the twentieth century seldom, if ever, matched this insight.

In some ways, ecological sustainability is like financial sustainability for an individual or an institution. Proper financial management involves attention to two goals: (1) cash flow, or operational budget, and (2) capital preservation. Most financial planners also address the element of growth to allow for inflation, but let us assume for the time being that inflation is under control and the total economic environment is in a steady state. Under these conditions, sustainable financial management requires that income and expenditures be equal over time. In other words, the budget should balance. Furthermore, capital is necessary for long-term expenses, illness, and unforseen exigencies.

An individual who does not maintain a balanced operational budget over time gets into trouble, especially if capital is depleted in the process. Deficit spending can last only so long. Most governments, including our own, are past masters at deficit spending, supported only by the faith and good will of lenders who continue to buy notes and bonds issued by the government. A day of reckoning is always just over the horizon, and servicing the debt takes money from current needs. If the U.S. government was suddenly called for all or most of its outstanding debt, it would indeed close down, and most Americans would suffer grievous losses.

The political hue and cry in the United States in 1995 was for a balanced budget amendment, as part of the Republican "Contract with America." A plan to balance the budget is long overdue, but not immediately realistic given the gap between the spending habits and tax revenues of our government at the present time.

Just as the United States and most other governments are living dangerously

with financial deficits, so the world's populations are living dangerously with ecological deficits. Global data on poverty and health show that we are not maintaining a "balanced budget" between population growth and resource use, nor are
we preserving "capital," that is, resources. We are consuming ecological resources—soil, water, forests, grasslands, wetlands, and species diversity—more
rapidly than they are being replaced. Neither the developed nor the developing
world is being managed sustainably. In one way or another, we are depleting the
natural heritage of our planet, and this will assuredly catch up with us if it has not
done so already. A few examples of nonsustainable trends in resource use are
listed in Table 25.2.

RESTORATION ECOLOGY

A balanced budget, whether financial or ecological, implies that all is well. But if
that balance is achieved at the poverty level, there is obviously room for improvement. Not only are many of the world's peoples (approximately 1 billion individuals) living in poverty, so are many of the world's environments. Many environments have been stripped of their vegetation, their soil, their clean water, their
natural species, and even their clean air. Simply balancing the budget on these
landscapes is not adequate. They need to be restored to some semblance of health.
Functional natural communities, even if they have fewer species in them than in

TABLE 25.2
Examples of nonsustainable trends in human-dominated environments.

Subject or Unit	Current Use, Trends, or Consequences
Agricultural soil in the midwestern United States	Being lost at the rate of 5 to 40 tons per acre per year
Development of saline soils in areas of extensive irrigation	Decline of agricultural production; loss of food supplies in developing nations
Cattle and sheep populations in Africa and parts of Asia	Increasing numbers, with severe overgrazing contributing to desertification
Ground water in the Ogalala aquifer	Being depleted faster than it is being replenished; water table dropping in many areas
Worldwide production of carbon dioxide	Accumulating in the atmosphere, with potential climatic effects
Worldwide production of 40 million cars and motor vehicles annually	Serious deterioration of air quality, with unfavorable consequences for natural ecosystems, agroecosystems, and human health
Population growth in India exceeding 1 million people per month as net increase	Dehumanizing urban crowding, severe unemployment problems, pollution problems, and potential food shortages
Unprecedented world population growth with net increase globally of 1.7 million people per week	All of the above, plus excessive competition and conflict over space and global resources, threatening the security of all people

their original form, can be built up only with biodiversity. Restoration ecology is that subdivision of ecology devoted to restoring damaged landscapes.

Restoration ecology has been recognized as an ecological speciality for many years (Cairns, Dickson, and Herricks, 1977), and it is coming into greater prominence as the need for ecological repair becomes more and more evident. The field is vast, involving activities ranging from pollution abatement and cleaning up toxic waste dumps to the restoration of vegetation and species on ravaged landscapes. A few examples will emphasize the extent and potentials of restoration ecology.

In the early part of the twentieth century, the Thames River flowing through London was a public sewer contaminated by many kinds of domestic and industrial waste. Its once abundant fish populations were reduced to just two or three species that could survive in spite of heavy pollution. In the 1950s, a regional Thames River authority began work to remedy this situation, developing and enforcing controls on sewage, urban runoff, and factory waste. By the late 1970s, a remarkable resurgence of fish species and populations had occurred. Salmon and trout returned to the river, and over a hundred species of freshwater fish could again survive in the Thames (Fig. 25.3).

In the United States, Lake Erie was declared "dead" from pollution in the 1960s, and one of its tributary rivers, the Cuyahoga, gained notoriety by catching fire from the petroleum pollution on its surface. Through admirable regional efforts, Lake Erie has returned to a pleasant and productive body of fresh water. Lake Washington, east of Seattle, faced a problem of domestic pollution and excessive eutrophication, but local pollution enforcement brought its nutrient levels down and greatly improved water quality. There are many other examples of pollution abatement and restoration of aquatic ecosystems. Such efforts require ecological knowledge and a local or regional commitment to take the right action.

In terrestrial environments, notable efforts have been made to restore strip-mined areas. Success has been mixed because such lands have not only lost much of their soil, but they are often highly eroded and polluted as well. Surface mines

Figure 25.3 Cumulative number of fish species found in the River Thames since 1964, when treatment of sewage entering the river first began (redrawn from M. J. Andrews and D. G. Rickard, 1980).

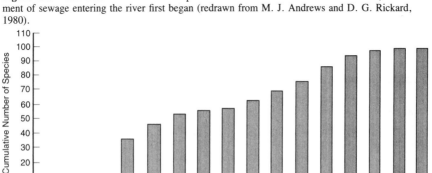

typically have acidic runoff from rainstorms. Nonetheless, in many states laws require that strip-mined areas must be returned to some approximation of their original land contours and revegetated. Apart from the problem of enforcing such legislation is the problem of getting plants to grow again on such devastated landscapes. Some hardy species of grasses (including various strains of ryegrass) and legumes (such as native vetches) are being used as vegetative ground covers to begin the process of soil rebuilding (Crowder, et al., 1995).

Surface-mine areas can become permanently damaged if toxic wastes are involved in the mining process. In gold mines in the western United States, for example, cyanide is used to leach gold from mine spoil, and this cyanide poisons both the remaining soil and surface stream drainage. Gold mining has also added to the environmental deterioration of Amazonia, along with deforestation and rampant homesteading (de Onis, 1992).

Ecological restoration involves both team effort and public commitment. The most severe problems often require the expertise of engineers, environmental toxicologists, botanists, soil experts, and hydrologists. As important as restoration is, it is usually much more expensive than prevention would have been. As in human health, curing a medical problem, such as heart disease, is often more difficult and costly than preventing it in the first place.

GOALS OF RESTORATION

One of the first tasks in any restoration project is to determine whether the goal is to restore the former habitat, enhance the former habitat, or convert the damaged environment into a different, worthwhile habitat. Conversion might be the plan of choice where a virgin stand of timber had been destroyed by strip-mining or clear-cutting. The virgin forest cannot be replaced, but a secondary replanting of trees would be preferable to leaving the area denuded.

In the case of enhancement, particular species of vegetation might be planted to provide better wildlife food and cover, or a stream might be enhanced to provide better trout habitat. Proper wildlife management affords many opportunities for environmental enhancement for wildlife or human use, ideally for both.

Perhaps the most common goal of all in restoration ecology is to return a damaged habitat to something approaching its former condition. This is often called remediation or reclamation. In the Alaskan oil spill of 1989 when the tanker *Exxon Valdez* hit a submerged rock and spilled 11 million gallons of crude oil, the main goal was to clean up the mess so it would not permanently pollute Glacier Bay, which had rich populations of seabirds, otters, whales, and other wildlife. The oil slick quickly spread and killed an estimated 580,000 birds, over 5,000 otters, 30 seals, 22 whales, and unknown numbers of fish (Miller, 1992). Initial cleanup costs ran $2 billion, and additional cleanup and legal fees may have doubled this to $4 billion. Remediation efforts involved measures to remove as much oil as possible from water and beaches and to treat as many sick and handicapped animals as possible. Oil removal required heroic physical efforts as well as bioremediation in which special strains of bacteria capable of breaking down crude oil into less harmful byproducts were sprayed on beaches. Bioremediation offers some hope as a means of coping with other toxic chemical spills, but we have a long way to go to make these methods completely effective. It is almost always less expensive and much less damaging to prevent environmental damage in the first place than to restore habitats once serious damage has occurred. The *Exxon Valdez* oil spill could have been avoided entirely by a more alert crew keeping the vessel on course or by building the ship with a double hull. The latter would have cost the Exxon Corporation $22.5 million, but it would have saved Exxon over $2 billion in cleanup costs. The Secretary of the Interior at the time, Rogers Morton, told the American people that oil tankers using Alaskan waters would be required to have double hulls, but this requirement was dropped under pressure from the oil companies (Miller, 1992).

Another type of restoration ecology involves restoring species to former habitats and ranges where they have been extirpated. For example, by the beginning of the twentieth century, the peregrine falcon, wild turkey, river otter, and desert bighorn sheep had all been hunted or trapped to extinction in the Dolores Canyon in southwestern Colorado. Since the habitat was still satisfactory, the Colorado Division of Wildlife began a program of reintroduction in the 1970s. Stocks of each species were obtained from other areas and released in the Dolores. Now each of the species has developed successful breeding populations.

In China, a successful reintroduction program in recent years involved a species of deer native to north-central China known as Pere David's deer. Formerly abundant over vast stretches of China, Pere David's deer was hunted to extinction in the nineteenth century. A small population remained in England, however, on the Duke of Marlborough's estate in Oxfordshire. This population originated from a few live specimens taken to England in the early 1800s. The English herd had prospered, and in the 1980s an international conservation effort was undertaken to return Pere David's deer to China. Suitable habitat was found, a sanctuary was secured, and a small herd of deer was reintroduced. This herd is now breeding, and shows promise of successfully reestablishing the species in its original home.

Numerous other reintroductions are being undertaken to restore species in imminent threat of extinction to their former habitats. Three in the United States that have received widespread publicity in the 1990s are the black-footed ferret in Wyoming, the peregrine falcon throughout the country, and the gray wolf in Yellowstone National Park. This is an area of restoration ecology where zoos can play a vital role in saving and restoring species. As important as it is ecologically to restore charismatic species, such as the gray wolf and peregrine falcon, it is enormously expensive to do so. Such restorations involve detailed scientific research, captive breeding or capture and transport of wild stock, and careful monitoring after release. Accurate assessments and protection of habitat are also necessary. Preventing the loss of these animals would have been much less costly than the cure after the loss has occurred.

The importance of restoration ecology is growing daily as more environments and species are endangered by human action. This importance was highlighted in a recent book on rehabilitating damaged ecosystems (Cairns, 1995), with applications to both aquatic and terrestrial environments. Restoration is not simply an ideal; or just a nice goal—we will find sooner or later that it is a necessity for our own survival.

THE BOTTOM LINE

It should be apparent by now that sustainable growth is not possible if by that we mean growth in numbers and the use of nonrenewable resources. If global ecology teaches us anything, it teaches us that numerical growth must ultimately be limited. Does accepting limits eliminate the pipe dream of sustainable development?

Not necessarily. Development can mean growth in quality as well as quantity, and quality is less limited ecologically than sheer numbers. The opportunities for quality development are immense. Instead of accepting the prospect of more and more people on earth competing for fewer and fewer goods, why not find ways to create a better life for those people? Why not create better-quality food, housing, education, jobs, medical care, and social and recreational opportunities? Such a plan would not necessarily encourage rampant consumerism or materialism. It could encourage healthy living, personal development, and creative pursuits. When we look at the world today, we see vast opportunities for human and environmental enhancement. So many damaged lives and so many degraded habitats are in need of restoration. Business and economics can find unlimited opportunities if they focus on quality rather than quantity when assessing potential markets.

If any one doubts the need for development, he or she should spend some time in any one of millions of rural villages in the developing world, villages without electricity, without clean or consistent water supplies, without schools, health facilities, or transportation other than walking. Food and fuel supplies are often marginal and unreliable. If someone becomes seriously ill or injured, it may take days to get assistance, if it is available at all. Life is harsh and unrewarding, and the demands of survival are relentless. This is the fate of over a billion people in the world; it is no wonder that they flock to urban slums in preference to rural poverty.

The bottom line is that sustainable development is possible if our objectives can be reoriented. We must find ways to reduce numerical growth and achieve quality development in both human and environmental terms.

SUMMARY

Sustainability is the key to the human future. The word "sustainable" is used in several contexts, however: political, ecological, and economic. The widely used definition of sustainable development is "development that meets the needs of the present without compromising the ability of future generations to meet their own needs."

Natural ecosystems are sustainable through the basic ecological processes of biogeochemical cycling and energy flow. Sustainability does not mean stability, since natural ecosystems are subject to many changes. These systems, however, have the ability to return to a former state or change to the next successional stage following disturbances. They are resilient, adaptive, and capable of self-renewal.

Many human-dominated systems are not sustainable, in that they deplete nonrenewable resources or exceed recycling rates of renewable resources. Thus, current human populations are overusing water, air, soil, forests, grasslands, and other fundamental resources at rates faster than they can be replaced.

Evidence of the nonsustainable nature of many current ecological trends can be found in the rates of human population growth and worldwide environmental deterioration.

Sustainability must be based on ecological principles. We need to understand the importance of maintaining sustainable ecosystems in natural reserves, agriculture, urban areas, and consumer patterns. Achieving sustainability will require greater emphasis on conservation, recycling, environmental restoration, and the rational limitation of growth.

Restoration ecology is an increasingly important field with the purpose of restoring damaged habitats to productive ecosystems. Goals may vary from attempting to return deteriorated areas to their original landscape type to mitigating acute environmental problems such as oil spills and toxic waste sites. The restoration of extirpated species and the conversion of highly degraded areas such as landfills to worthwhile environments such as public parks are also valid concerns of restoration ecology.

The commonly used term "sustainable growth" is a self-contradiction since unlimited numerical growth on a finite planet is impossible. Sustainable development has validity if we include the enhancement of quality as a goal of development.

Anyone who has worked or traveled extensively in developing nations recognizes the vast need for human and environmental improvement. We do not even have to go outside our own country to see this, or even much further than our own television set if we watch the daily news.

In both rural and urban areas there is an urgent need for the development of better housing, education, food, social and recreational opportunities, medical care and health services, environmental restoration, and conflict resolution. The emphasis must be on improvements in the quality of life rather than the quantity of life. Achieving sustainability should not be viewed as economic stagnation. It should be viewed as a reorientation of business from numerical growth to quality enhancement. The opportunities are enormous.

REFERENCES

Andrews, M.J. and D.G. Rickard,1980. *Marine Pollution Bulletin* 11: 327.

Bartlett, A. A. 1994. Reflections on sustainability, population growth, and the environment. *Population and Environment* 16: 5–35.

Brown, L. R. 1994. Facing food insecurity. In *State of the World, 1994,* ed. L. R. Brown. New York: W. W. Norton.

Brundtland, G. H. 1987. *Our Common Future.* World Commission on Environment and Development. New York: Oxford University Press.

Cairns, J., K. L. Dickson, and E. E. Herricks (eds.). 1977. *Recovery and Restoration of Damaged Ecosystems.* Charlottesville, VA: University Press of Virginia.

Cairns, J. (ed.) 1995. *Rehabilitating Damaged Ecosystems*. Boca Raton, FL: Lewis.

Corson, W. H. 1990. *The Global Ecology Handbook*. Boston, MA: Beacon Press.

Crowder, A. A., E. A. Ripley, R. E. Redman. 1995. *Environmental Effects of Mining*. Boca Raton, FL.: Lewis.

de Onis, J. 1992. *The Green Cathedral: Sustainable Development of Amazonia*. New York: Oxford University Press.

Eckholm, E. P. 1976. *Losing Ground: Environmental Stress and World Food Prospects*. New York: W. W. Norton.

Hardin, G. 1993. *Living Within Limits: Ecology, Economics and Population Taboos*. New York: Oxford University Press.

MacNeill, J. 1989. Strategies for sustainable economic development. *Scientific American* 261(3): 154–165. (Sept. 1989).

May, R. M. 1993. The end of biological history? *Scientific American* 268: 146–149 (Mar. 1993).

Meadows, D. H., D. L. Meadows, and J. Randers. 1972. *The Limits to Growth*. New York: Universe Books.

Meadows, D. H., D. L. Meadows, and J. Randers. 1992. *Beyond the Limits: Confronting Global Collapse and Envisioning a Sustainable Future*. Post Mills, VT: Chelsea Green.

Miller, G. T., Jr. 1992. *Living in the Environment*. 7th ed. Belmont, CA: Wadsworth.

Mungall, C., and D. J. McLaren (eds.). 1990. *Planet Under Stress: The Challenge of Global Change*. New York: Oxford University Press.

Muul, I. 1993. *Tropical Forests, Integrated Conservation Strategies and the Concept of Critical Mass*. Man and the Biosphere Digest 15. Paris: UNESCO.

Odum, E. P. 1993. *Ecology and Our Endangered Life Support Systems*. 2nd ed. Sunderland, MA: Sinauer Associates.

Osborn, F. 1948. *Our Plundered Planet*. Boston, MA: Little, Brown.

Plotkin, M., and L. Famolare (eds.). 1992. *Sustainable Harvest and Marketing of Rain Forest Products*. Washington, DC: Center for Resource Economics, Island Press.

Reid, W. V., et al. 1993. *Biodiversity Prospecting: Using Genetic Resources for Sustainable Development*. Washington, DC: World Resources Institute.

Ruckleshaus, W. D. 1989. Toward a sustainable world. *Scientific American* 261(3): 166–172 (Sept. 1989).

Sitarz, D. (ed.). 1993. *Agenda 21: The Earth Summit Strategy to Save Our Planet*. Boulder, CO: Earth Press.

Slobodkin, L. B. 1961. *Growth and Regulation of Animal Populations*. New York: Holt, Rinehart and Winston.

Chapter 26

ASSESSMENT AND
AGENDA

There is no doubt that the world faces unprecedented problems and opportunities. Ecologists tend to focus on problems, and they have been warning of environmental and social disasters for many years. Many of these disasters have come to pass, affecting innumerable individual lives.

Businesspeople or entrepreneurs tend to focus on opportunities and foresee unlimited opportunities for growth and human development. Ecologists might agree that every problem is an opportunity, but the two groups differ radically in the types of solutions they think are feasible and beneficial.

We have come full circle in this book, and it is time now to look again at the controversies and challenges of global ecology. One way to do this is to view the world through the eyes of an ecologist and the eyes of a businessperson or entrepreneur.

THE WORLD ACCORDING TO ECOLOGY

A realistic appraisal of environmental and social trends shows the heart and soul of planet earth slipping away. Those qualities we value the most—beauty, peace, security, diversity, and freedom—are in grave jeopardy. Not this year or next, but in the next 20, 50, or 100 years, gradual degradation of the global environment will catch up with us, our children, and our grandchildren. For many of the world's peoples, it has done so already. The problem is one of creeping deterioration in our physical, biotic, and social environments. The results will be insecurity, violence, and sickness. Dr. A. J. McMichael, professor of community medicine at the University of Adelaide, Australia, has pointed out, "Today we face the health consequences of disruption of the world's natural ecosystems."

The basic causes of this deterioration are unprecedented population growth and rampant consumerism. Neither of these processes is slowing down enough to prevent serious global damage. Since we exhibit few signs of controlling these trends rationally, external controls will be imposed by either unwanted governmental regulations or more drastic laws of nature. In either case, our freedoms will be diminished, our choices limited.

In southern California, the public did not curtail home construction on steep hillsides in a chaparral environment or on floodplains in lowland areas, nor did governmental regulations prevent such construction. Consequently, many homeowners have tragically seen their homes destroyed in windswept fires typical of chaparral habitats or in unrelenting mudslides and floods when winter rains begin. Public protests would have been loud and long, however, if governmental agencies had prevented construction in such areas.

The American public in the western United States stews and fumes about environmental controls that limit ranching, logging, and mining. Many of these vested interests defiantly claim that the National Environmental Protection Act, the Clean Air and Clean Water Acts, and Mine Reclamation Acts are taking away their freedoms and their livelihood.

Yet without reasonable enforcement of these acts, western rangelands, water supplies, and forests are in a sad state. The American West is strewn with degraded lands and remnants of exploitation. Some economic interests wish to push this deterioration even further. They have a frontier mentality that says, "Nobody, especially in Washington, DC, is going to tell me how many cattle I can run on my ranch or how many trees I can cut!" The Government is taking away my livelihood and my God-given freedoms! Indeed, political movements are underway to return national forests and national parks to private interests, opening them for exploitation.

There are, however, many examples where the government itself is the problem. Nuclear test blasts in Nevada, toxic wastes and plutonium storage sites in Colorado, unregulated population growth in southern California and Arizona following expansive federal water projects—all of these government-related activities have planted the seeds of environmental and financial problems.

This is not to say the government is always wrong or always right—both harmful actions and protective regulations can be attributed to government agencies. The point is that without rational self-regulation of human activities, external forces come into play, and either the government or natural forces are capable of causing even more devastation. In the American West the desire for personal and corporate gain has ruled the day, and the common heritage of all Americans has suffered as a consequence (Hardin, 1968).

Are things really this bad? Or are these exaggerated claims of environmentalists? What is the evidence of ecological disasters on a global scale?

A BRIEF REVIEW OF THE ECOLOGICAL EVIDENCE

Many of the trends emphasized by *The Limits to Growth* in 1972 (Meadows, Meadows, and Randers) and *The Global 2000 Report* in 1980 (Barney) are still with us, and some are stronger than ever.

In the mid-1990s, world population is increasing at the rate of 1.7 million people per week. This means an unprecedented increase of between 800 million and 1 billion people per decade (Bongaarts, 1994). Although many countries have made significant strides in population planning, contraceptive use, and fertility reduction, the world's population as a whole remains out of control. At the same time, 23 percent of the world's peoples are living in poverty, worldwide unemployment is dangerously high, and regional military conflicts are numerous and widespread. The irrefutable fact is that we are not taking adequate care of the 5.8 billion people now living in the world, and are not providing satisfactorily for the net increase of 1.7 million more every week. Where will the jobs, space, housing, and food come from to meet the basic needs of 1.7 million more people every week?

We are not providing adequate education, health care, recreation, and personal security for large numbers of the world's people, especially children. Even in the United States, one of the world's more prosperous countries, one child out of five under the age of six lives in poverty.

Nor are we providing adequately for environmental quality. We are sapping the world's natural resources at an unprecedented rate. Seventy-five billion tons of soil are lost annually, and to date one-third of the world's arable land has been lost (Pimentel et al., 1995) In the last 50 years, nearly 2 billion hectares of the earth's land surface has been seriously degraded by human action (Postel, 1994). This is more than twice the size of the United States. Living in the United States, most of us are largely unaware of the vast areas of degraded lands of Africa, Asia, and Latin America. Much of this land has been rendered permanently barren and economically worthless.

Although we have made great advancements in air and water pollution control in the richest nations, pollution is rapidly getting out of hand in developing nations which are industrializing at a furious pace without pollution controls. Air and water quality in cities such as Mexico City, Cairo, New Delhi, Bombay, Calcutta, Bangkok, Jakarta, Beijing, and Shanghai are atrocious—far worse than human beings should have to tolerate.

Not only water-quality problems but outright water shortages are increasing in many regions. Twenty-six nations, home to 230 million people, now fall into the scarce-water category (Postel, 1993). Nine of 14 countries in the Middle East face significant water shortages. Depletion of both surface and ground water is a serious environmental problem in every continent except Antarctica.

The interactions of the health of the environment and the health of human populations, which we discussed in Chapters 21 and 22, are central in ecological views of the world today. Deteriorating air and water quality in developing nations, expanding industrialization, increasing use of toxic chemicals, the spread of personal afflictions such as smoking, alcoholism and drug addiction, and widespread sexual promiscuity are all creating chronic health problems that are damaging the quality of life and limiting future human prospects. Whatever weakens human life damages the environment and vice versa through the medium of wasted lives and wasted resources.

The loss of forests, the spread of deserts, the permanent alteration of many habitats, and the onset of more variable and violent weather related to anthropogenic forcing will undoubtedly make life more tenuous and unstable for many of the world's people.

The populations of many important organisms are declining markedly, indicating environmental problems. These organisms include songbirds, waterfowl, amphibians, several species of marine mammals, and forest-dwelling primates. The causes .of these declines are generally well known: loss of habitat, pesticide poisoning and other forms of pollution, overhunting, and even, in the case of amphibians, atmospheric changes such as increased ultraviolet radiation.

The facts are evident, according to the ecological view, that we are destroying the life-support system on which we depend. Many of our individual and collective behaviors are life-threatening. While there is little doubt the planet earth and some human populations will survive, there is much less optimism about the quality of life and the quality of the global environment in the future. Ecological principles indicate that if we do not limit runaway growth, natural mechanisms will. Drastic epidemics of infectious disease remain a realistic possibility (Garrett, 1994); a pandemic flu could be caused by a chance mutation of a virulent airborne virus. Millions of deaths could occur before modern medicine could deal with such an event. Mass mortality could also be caused by serious weather anomalies, which could deprive people of food and water, disrupt communications, and destroy homes. Each of these possibilities is related to population and environment, and each becomes more dangerous as global population increases and environmental quality decreases.

In brief, ecologists believe that we are using earthly capital for deficit financing of ourselves and our world. We are running hell-bent toward an impoverished future. Our options for corrective actions are shrinking every year.

THE WORLD ACCORDING TO ENTREPRENEURAL ECONOMICS

Many business leaders and entrepreneural economists feel that the dire predictions of ecologists are nonsense—the kind of ecological doom that ecologists have been preaching for more than 50 years. They maintain that the world is getting more prosperous, and more people are leading better lives now than at any time in history. Even China and India, those classic cases of poverty and crowding, have an emerging middle class in which people have better food, housing, education, and higher living standards than ever before.

It is easy to reminisce about the glorious past, but the facts are different. If anyone doubts this, he or she might like to compare London today with London in the time of Charles Dickens. Dickens's London was a filthy, polluted, crime-ridden city, with deathly air quality and water from neighborhood pumps contaminated with cholera, dysentery, polio, and tuberculosis.

Factual comparisons of many cities and many areas show that people are better off today than in the past. Modern Shanghai, with its crowds and polluted air and water, is a far better place than Shanghai of the 1920s, when it was swept by famine, disease, and warfare. Calcutta, still a prime example of a dysfunctional city with sprawling slums, is free of the epidemics, starvation, and conflicts that devastated it in the past. Even in African countries with serious social and environmental problems, progress in discernible. Most African and Asian nations have set aside national parks for wildlife conservation. Most have developed rural and urban health programs including efforts at family planning and prenatal care. Most have undertaken programs to control debilitating infectious and parasitic diseases such as schistosomiasis, trypanosomiasis, malaria, oncocerciasis, and leprosy. Obviously, these programs have a long way to go, but 50 years ago such efforts had not even been started.

According to many entrepreneurs, ecologists fail to recognize progress when it occurs; they are impatient with development, and unable to see opportunities for improvement. In their view, planet earth is poised to enter a glorious era of prosperity. Expanding knowledge and technologies are opening up incredible new horizons in agriculture, business, communications, education, health, housing, transportation, and other areas of human activity. Despite the pessimism of environmentalists, the world is facing its greatest boom in history (Morris, 1989; Naisbitt, 1994; Tiglao, 1990).

The scientific advancements since World War II are just now coming into full use. Computer technology, space exploration, global communications, the information superhighway, molecular biology, and genetic engineering are just now beginning to pay off in practical terms. All the world will benefit in terms of better food, better housing, better products of all kinds, better health care, better pollution control, better environmental management, and, in the long run, better conservation. We will learn to grow crops more efficiently and more productively. We will control disease more effectively and develop new methods of preventive medicine. We will enhance education, we will increase opportunities for all the world's peoples. Human and environmental prospects are indeed bright.

A BRIEF REVIEW OF THE POSITIVE EVIDENCE

World health is improving dramatically. Life expectancies have increased throughout the world, especially in China and India. Massive epidemics have been avoided. Smallpox has been eradicated, and remarkable strides have been made in the control of cholera, polio, river blindness (oncocerciasis), schistosomiasis, and trypanosomiasis.

Food supplies have improved in most of the world. Grain production in 1992 increased 3 percent over 1991 (Brown, 1993), and soybean production in 1994 increased 14 percent over 1993 (Brown, 1995). Meat production increased, and overall per capita food showed improvement after some instabilities in the late 1980s. Nutritional diseases have been reduced in many countries with the exception of a few African nations with famine conditions.

Birth rates declined markedly in some of the world's most populous nations during the 1970s and early 1980s. In China, for example, the fertility rate declined from 6.4 children per woman in 1968 to 2.2 children per woman in 1980. In India, the fertility rate fell from 5.8 children per woman in 1966 through 1971 to 4.8 in 1976 through 1981 (Starke, 1993). In Brazil, the fertility rate fell from 5.8

children per woman in 1970 to 3.3 in 1990 (Jacobsen, 1992). In Taiwan, the fertility rate has fallen to 1.7 (Weeks, 1992). Contraceptive use has increased from 31 to 50 percent in Tunisia, from 30 to 53 percent in Mexico, and from 20 to nearly 70 percent in Thailand and the Philippines. These improvements have occurred because of vigorous and farsighted leadership in these countries.

World economic development is showing favorable trends with particularly strong growth in world trade (Kane, 1993). A number of Asian and Latin American countries have increased productivity remarkably and are experiencing booming world trade. Some of the stellar financial markets in the world in recent years have been in Brazil, China, Taiwan, Korea, Thailand, Malaysia, Singapore, and Hong Kong. India and Indonesia are now considered prime candidates for striking growth in world markets. Although the initial benefactors of such growth are investors, expansive economic activity also means more jobs, higher per capita incomes, and better living conditions for the general population. The momentum is upward, and the trends in human terms are favorable. Improved living conditions for people will ultimately benefit the global environment by enhancing education, reducing population growth, and relieving poverty-related pressures on the environment.

Recent environmental and conservation measures have also shown favorable trends. The destruction of the ozone layer has been slowed due to global agreements on banning chlorofluorocarbons. Pollution restrictions in industrial nations have improved air and water quality in North America, western Europe, and Japan. The fear of rapid global warming in the 1980s has lessened since data from 1992 and 1993 showed returns to cooler global temperatures (Brown, 1993), but emerged again in heat waves in 1995.

Many species of endangered animals have been saved from near extinction and are showing population increases. These include animals such as the gray whale, Antarctic fur seal, and Galapagos fur seal in the Pacific Ocean; the Bengal tiger, black buck, and nilgai antelope in India; the bald eagle, peregrine falcon, black-footed ferret, and American alligator in the United States; the gold lion tamarin, muriqui, and ocelot in Latin America; and the mountain gorilla in central Africa. Several new species of nonhuman primates have been discovered in recent years, and none has been known to have become extinct.

More land and natural habitat has been put into nature reserves and national parks in the past 50 years than in all of previous history. In the United States, new national parks include Florida's Everglades (1947) and Biscayne Bay (1980), Utah's Canyonlands (1964) and Capitol Reef (1971), the Caribbean's Virgin Islands (1956), California's Channel Islands (1980), Washington's North Cascades (1968), and Alaska's Gates of the Arctic (1980), Glacier Bay (1980), Katmai (1980), Kenai Fjords (1980), Kobuk Valley (1980), Lake Clark (1980), and Wrangell-St. Elias (1980). In many other nations, nature reserves have expanded greatly in number and area, especially in Costa Rica, India, and several African nations.

Apart from national and international reserves, private sanctuaries have also increased dramatically. The Nature Conservancy, a nonprofit, nongovernmental organization in the United States with a membership of 720,000 individuals now has 1.3 million acres of protected nature reserves under its care.

All of these are encouraging signs of worldwide environmental awareness and commitment. More governments, corporations, private organizations, and individuals are devoting attention to conservation measures than at anytime in history. The environmental movement does have extremists and radicals, and there are certainly backlash movements against conservation, but the main tide is toward greater ecological sensitivity.

Another remarkable advance of the past 50 years, often forgotten by gloom-and-doom-sayers, is the fact that the world has avoided World War III, and even the Cold War military buildup has been greatly reduced. Military tensions and tragic wars obviously exist in many parts of the world, but many of these are being controlled. Despite regional wars, the world has been free of large-scale international wars such as the Vietnam War and Korean War since 1975.

These encouraging signs suggest a bright global future, according to the entrepreneural outlook. Problems are abundant and formidable, but human ingenuity and compassion can solve these as they have in the past if we do not become bogged down by an overly pessimistic outlook.

Both the ecological and entrepreneural points of view bring us to the question, Where do we go from here? The following agenda is written from the ecological point of view because the author feels strongly that the greatest weight of evidence is on the ecological side. At the same time, entrepreneural economics, science, and technology will play a part in how we handle the world's problems.

AN AGENDA FOR PLANET EARTH

Despite some signs of progress, the overwhelming ecological evidence suggests that the world is in serious trouble. This case is stated succinctly by Professor A. J. McMichael of the University of Adelaide, Australia, in the introduction to his book *Planetary Overload: Global Environmental Change and the Health of the Human Species* (1993):

> The most serious potential consequence of global environmental change is the erosion of Earth's life-support systems. . . . Our burgeoning numbers, technology and consumption are overloading Earth's capacity to absorb, replenish and repair.

Garret Hardin, Professor Emeritus of the University of California, Santa Barbara, made a similar case in *Living Within Limits* (1993). He emphasized that we must

> dramatically change the way we live in and manage the world. . . . Our world is in the dilemma of the lifeboat: it can only hold a certain number of people before it sinks.

The following agenda is presented in outline fashion, with steps of action listed more or less in order of priority. In other words, if the first priorities are not achieved, succeeding ones will be of little value. These injunctions are based on the principles and data presented in all the preceding chapters. In the ecological view, these goals must be achieved if we are to maintain civilized human societies on earth. As stated before, planet Earth will survive; it is our own livelihood and that of the present biosphere that is in jeopardy.

ECOLOGICAL PRIORITIES

I. Prevent Nuclear War

The worst imaginable devastation of the earth would be from a moderate to large-scale nuclear war. The most important goal of all human activity must be to avoid such a holocaust. To do this, we must

A. Reduce and eventually eliminate nuclear weapons.

B. Prevent nuclear proliferation.

C. Promote education and research on the causes of war and routes to the promotion of peace.

With the end of the Cold War between the former Soviet Union and the United States and the encouraging progress of SALT (Strategic Arms Limitation Talks), the world is moving in the right direction of nuclear disarmament for the first time since the end of World War II. Consistent diplomacy will be required to continue this process.

The threat of nuclear war has not disappeared. Many nuclear bombs and missiles still exist and could be called into action. With the breakup of the USSR, more nations have access now to nuclear weapons, and the threat of nuclear proliferation continues in nations such as Iraq, India, China, Pakistan, North Korea, South Africa, and potentially many more with the technical capability to produce nuclear weapons.

The world must devote much greater attention to understanding and avoiding the causes of war and promoting the roads to peace. We have many official "war colleges" supported by national governments, where the history, strategies, and techniques of military conflict are studied. But the world has few "peace colleges," where the components of peace are analyzed and the avoidance of war is promoted.

II. Reduce and Ultimately Prevent Regional Conflicts

Local wars and terrorism have very damaging ecological effects for all the reasons outlined in Chapter 24. They create tragic losses in the human community and the broader environment, and they have the potential of escalating into larger wars.

Specific steps must be taken to

A. Understand the basic causes of these conflicts in ideology, religion, environment, competition, and so on.

B. Find ways to alleviate the causes of conflict and achieve compromise and cooperation.

C. Devote the resources that support conflict to productive social and environmental causes.

Consensus and control must be achieved. The real needs of people and the environment must be met.

III. **Stabilize World Population Growth**

All the social and environmental recommendations will be relatively ineffective if human population growth is not stabilized in a rational fashion. If this is not accomplished, natural control mechanisms—disease, poverty, starvation, conflict, and war—will most certainly prevail. This backlash will not necessarily occur in an apocalyptic fashion; it will more likely take the form of a series of dehumanizing tragedies that sap the quality of life and destroy both civilized societies and local environments. We've already seen such tragedies in countries such as Bosnia, Cambodia, Chechnya, Rwanda, and Somalia, to name a few, and there is no doubt that similar tragedies could occur elsewhere.

To avoid being overwhelmed by this erosion of human life, we must

A. Promote family planning and family health activity in all nations, including our own.

B. Promote world literacy and educational opportunities.

C. Reduce birth rates in all nations that have not yet achieved stable populations. It is vitally important to stabilize population in the United States, now projected to grow from 260 million to 350 million in the next 50 years.

D. Reduce high infant mortality rates which stimulate higher birth rates. Promote prenatal health programs and postnatal care.

E. Reduce poverty and raise the standards of living of all people. Poverty and rapid population growth are strongly linked.

F. Promote world economic development in a balanced, sustainable and more equitable fashion.

IV. **Promote Conservation**

The world must come to grips with the need for conservation of all natural resources, both renewable and nonrenewable. Conservation is not a nice option to follow; it is an absolute necessity. Successful economics must ultimately involve sound ecological principles.

Specifically, we must

A. Care for the earth's life-support systems, including soil, water, air, and biota, with the goal of sustainability. We must reduce the wasteful, extravagant use of resources in our own country, the United States.

B. Control air, water, and soil pollution more effectively. This involves attention to greenhouse gases, noxious air and water emissions of all kinds, and special attention to ground water, coastal waters, and the world's oceans.

C. Preserve and properly manage all natural resources. This includes sustainable use of renewable resources and technological discoveries of replacements for nonrenewable resources. In other words, we must maintain the quality of agricultural lands, waters, forests, and grasslands while developing alternative sources for nonrenewable energy.

D. Reduce present rates of deforestation and desertification (Figure 26.1). Special attention must be given to the conservation of biodiversity for ethical, aesthetic, and economic reasons.

E. Pursue technological advances to develop economically feasible solar energy, superconductivity, and fusion energy. Increase efforts to develop renewable, nonpolluting energy sources for industry, transportation, and domestic use.

The above agenda is ambitious in its scope. Some might say it is unrealistic. At the same time, such an agenda is inspiring because it shows that everyone can contribute to improving planet earth and human prospects. All manners of skills are needed; every vocation and every individual has a role in this enterprise, including students, teachers, homemakers, farmers, manufacturers, engineers, lawyers, public servants, physicians, and politicians. Whoever and wherever we may be or whatever we may do, we can make a positive contribution to our collective survival and security. Although the world is extremely diverse and complex, we need to think in global terms. The bumper sticker, "Think Globally and Act Locally," is a very sound idea so long as the global thoughts and local acts contribute to human and environmental well-being.

PROFESSIONAL RESPONSIBILITIES

It is not very difficult to picture different professional roles in various parts of the foregoing agenda. Teachers, from elementary school to the graduate and profes-

Figure 26.1 Many species of wildlife can coexist in or near urban habitats. Here mule deer graze at the Rocky Mountain Arsenal just northeast of Denver. By public demand, the arsenal will become a wildlife refuge after decades-old chemical contamination is cleaned up. Many species of wildlife thrive in the arsenal's grasslands, providing wildlife viewing for more than 50,000 bus visitors a year. (Photo by Wendy Shattil/Bob Rozinski)

sional level, can work to instill a sense of environmental awareness and responsibility in students. The public official can make sure public policies are environmentally and socially friendly, ethical, and sustainable. The business leader can work to improve the economic picture over the long term rather than simply allow short-term exploitation. He or she can provide jobs, improve products, practice recycling, and meet society's needs for goods and services without environmental damage.

The engineer can find practical solutions to local and global problems of water supply, housing, transportation, pollution control, and energy sources that are sustainable and nonpolluting. The scientist can find new answers to current problems for which we have no known solutions.

The lawyer can help formulate, promote, and defend legislation that protects and enhances environmental and social goals. The physician and health worker can contribute to human health, family planning, and fertility regulation. The clergy can promote a sense of ethics and morality toward Mother Earth as well as toward our fellow human beings. The farmer and rancher can use agricultural practices that enhance, rather than deplete, the productivity of the earth.

Every job, every worker, every profession has a vital role to play in this human enterprise. It is relatively easy to slip into destructive, exploitative, or dysfunctional roles; we need to move ahead into protective and creative roles.

PERSONAL RESPONSIBILITIES

Apart from our professional lives, we can support the ecological agenda as individuals and contribute positively to the world in many ways. In recent years numerous books have provided guidelines on how individuals can contribute to a greener world (Anderson, 1990; Earth Works Group, 1990, 1991; MacEachern, 1990).

As individuals, we can choose to be reasonable and conservative, rather than wasteful, consumers. If we insist on driving huge automobiles that deliver 8 to 10 miles per gallon, we are part of the problem. If we require expansive homes, or multiple homes, filled with rare tropical hardwoods, we are part of the problem. If we use 500 gallons of fresh water per day per person, we are part of the problem. If we never recycle daily consumer items such as paper, plastics, glass, and aluminum, we are part of the problem.

We can contribute to the ecological agenda by being well informed on global issues, aware of current events, and knowledgeable about the world's environmental and social problems. The time is past for narrow, provincial attitudes limited to one point of view.

We can improve global conditions by remaining in good health ourselves and contributing to the good health of others. This means avoiding smoking, excessive drinking, overeating, drug abuse, reckless and high-speed driving, excessive pesticide use, and so on. The time has come for us to recognize the close linkage between our behavior and the health of ourselves and our environment.

Finally, we can contribute to the global agenda by broadening our ethical horizons, our understanding, and our tolerance. Our ethics must be expanded to include all of nature, our understanding must include recognition of social and environmental interrelations, and our tolerance must be broadened to encompass the world's diversity.

SUMMARY

We live in a world today that presents unprecedented problems and opportunities. According to the ecological view, the global environment is deteriorating seriously, threatening our life-support system and the health of all nature, including ourselves. The human population remains out of control, much to the detriment of the quality of life on earth.

According to the view of entrepreneural economics, there is no population problem and no serious environmental problem that we cannot handle with better management, technology, and science. In fact, the modern world offers expanding markets and unprecedented growth opportunities. More people are now living in better conditions than at any time in history, and we can continue to improve as free markets and capitalism spread around the world, according to this economic philosophy.

The opinion of this book is that there are elements of truth in both viewpoints but that the preponderance of ecological evidence points to increasing problems ahead. Planet earth will survive despite whatever we do to her, but the quality of human life and all of nature is at stake. It is the quality of the biosphere and the quality of our own survival that we are endangering.

Although impressive statistics of economic growth and technological advances can be cited, our environmental and social systems are deteriorating in a downward spiral. The qualities of the natural world on which we depend are eroding. In most of the world there is a clear linkage between population, poverty, and the environment (Dasgupta, 1995).

To improve our chances for a better future, we need to adopt an ecological agenda with at least four major components: (1) prevention of nuclear war and nuclear proliferation, (2) prevention of regional conflicts, terrorism, and local wars, (3) stabilization of population, including improving family health, strengthening the status of women, increasing literacy and education, and reducing poverty, and (4) conservation of soil, water, air, and wildlife, including reducing deforestation and desertification, protecting biodiversity, and properly managing both human-dominated and natural ecosystems. We need to develop sustainable sources of energy for domestic, industrial, and transportation uses.

Such an ecological agenda is discouraging because it is broad and far-reaching. It is encouraging because it offers every individual opportunities to contribute in positive ways both professionally and personally. The quality of human life and the future of the biosphere are at stake.

REFERENCES

Anderson, B. (ed.) 1990. *Ecologue: The Environmental Catalog and Consumers Guide for a Safe Earth.* Englewood Cliffs, NJ: Prentice-Hall.

Barney, B. O. 1980. *The Global 2000 Report to the President: Entering the 21st Century.* Washington, DC: Government Printing Office.

Bongaarts, J. 1994. Population policy options in the developing world. *Science* 263: 771–776.

Brown, L. R. 1993. Grain production rises. In *Vital Signs: The Trends that Are Shaping Our Future,* ed. L. R. Brown, H. Kane, and E. Ayres, pp 26–27. New York: W. W. Norton.

Brown, L. R. 1995. Soybean production jumps. In *Vital Signs: The Trends that are Shaping Our Future.* ed. L. R. Brown, N. Lenssen, H. Kane. pp 28–29. New York: W. W. Norton.

Dasgupta, P. S. 1995. Population, poverty and the local environment. *Scientific American,* 272: 40–45 (Feb. 1995).

The Earth Works Group. 1990. *Fifty Simple Things You Can Do to Save the Earth.* Berkeley, CA: Earth Works Press.

The Earth Works Group. 1991. *The Next Step: Fifty More Things You Can Do to Save the Earth.* Kansas City, MO: Andrews and McMeel.

Garrett, L. 1994. *The Coming Plague: Newly Emerging Viruses in a World Out of Balance.* New York: Farrar, Straus and Giroux.

Hardin, G. 1968. The tragedy of the commons. *Science* 162: 1243–1248.

Hardin, G. 1993. *Living Within Limits: Ecology, Economics and Population Taboos.* New York: Oxford University Press.

Jacobsen, J. L. 1992. Improving women's reproductive health. In *State of the World—1992.* ed L. R. Brown, et al. New York: W. W. Norton.

Kane, H. 1993. Trade continues steep rise. In Vital Signs: The Trends That Are Shaping Our Future, ed. L. R. Brown, H. Kane, and E. Ayres. New York: W. W. Norton.

MacEachern, D. 1990. *Save Our Planet: 750 Everyday Ways You Can Help Clean Up the Earth.* New York: Dell.

McMichael, A. J. 1993. *Planetary Overload: Global Environmental Change and the Health of the Human Species.* New York: Cambridge University Press.

Meadows, D. H., et al. 1972. *The Limits to Growth.* New York: Universe Books.

Meadows, D. H., D. L. Meadows, and J. Randers. 1992. *Beyond the Limits: Confronting Global Collapse, Envisioning a Sustainable Future.* Post Mills, VT: Chelsea Green.

Morris, C. R. 1989. The coming global boom. *Atlantic Monthly* 264(4): 51–64 (Oct. 1989).

Naisbitt, J. 1994. *Global Paradox.* New York: Avon Books.

Pimentel, D., et al. 1995. Environmental and economic costs of soil erosion and conservation benefits. *Science* 267: 1117–1123.

Postel, S. 1993. Water scarcity spreading. In *Vital Signs: The Trends That Are Shaping Our Future,* ed. L. R. Brown, H. Kane, and E. Ayres. New York: W. W. Norton.

Postel, S. 1994. Carrying capacity: Earth's bottom line. In *State of the World, 1994,* ed. L. R. Brown et al. New York: W. W. Norton.

Raven, P. H. 1994. Defining biodiversity. *Nature Conservancy* 44 (1): 10–15.

Starke, L. 1993. Population growth sets another record. In *Vital Signs: The Trends that Are Shaping Our Future,* ed. L. R. Brown, H. Kane, E. Ayres. pp 94–95. New York: W. W. Norton.

Tiglao, R. 1990. The Philippine paradox. *Far Eastern Economic Review* 149 (28): 31–34 (July 12, 1990).

Weeks, J. R. 1992. *Population: An Introduction to Concepts and Issues.* 5th ed. Belmont, CA: Wadsworth.

Chapter 27

PROGNOSIS

We know what has to be done, but can we do it? Do we have the honesty and insight to follow the right course? In one sense, we have no choice; if we do not take the proper steps, natural forces will make their own moves. The consequences will be grim. In another sense, we still have options to move in the right direction, and the choice of rational or irrational behavior remains with us.

Our society can send men to the moon, develop the information "superhighway," decipher genetic code, transplant hearts, and produce "smart" bombs, but at the same time we have great difficulty in dealing with each other. The ability to communicate and cooperate, however, is what will determine our future. Our relationship to the environment ultimately turns on behavioral issues and social relationships. As Aldo Leopold (1933) said more than 50 years ago:

> In short, twenty centuries of "progress" have brought the average citizen a vote, a national anthem, a Ford, a bank account, and a high opinion of himself, but not the capacity to live in high density without defouling and denuding his environment, nor a conviction that such capacity, rather than density, is the true test of whether he is civilized.

David Ehrenfeld (1993) brings these thoughts up to date on the dust jacket of a recent book:

> The twentieth century is drawing to a chaotic close amidst portents of unprecedented change and upheaval. The unraveling of societies and civilizations and the destruction of nature march together—linked—a fact whose enormous significance is often lost.

This final chapter deals with some hopes and fears related to the prognosis for humankind in the 21st century. Clearly, our human fate is closely tied to our global environment, and in particular to our social environment.

THREATS TO PROGRESS

The word *prognosis,* which usually has a medical connotation, refers to the course or outcome of a disease. It is derived from Latin and Greek words meaning "to know beforehand."

Applying a disease metaphor to the earth is appropriate. There is abundant evidence of pathologic processes in global ecology. Global warming or global cooling can be likened to fever or hypothermia. Deforestation and desertification represent wounds and scars on the earth's integument. Water and soil pollution are comparable to toxemia and septicemia. Conflicts and wars are analogous to the autoimmune process wherein elements of a system battle with themselves and self-destruct. Excessive population growth has been likened to neoplastic growth in an individual, a malignant process of uncontrolled or only partially controlled growth that eventually destroys the host (Gregg, 1955; Forrester, quoted in Gordon and Suzucki, 1991; Hern, 1993).

These are distinctly unfavorable and unpopular views of human population growth—no one wants to be compared to a malignant cell or an auto-immune process. But the comparisons are powerful, and if they provide insight to current events, they can and should be a rude awakening.

If we are to take any clues from the study of ecology, we have learned that the demise of a population comes about through the breakdown of the environment and the community, usually a combination of both. We should learn that the greatest threats to human progress lie in social disintegration caused by environmental pressures and excessive competition. The results can be chaotic: violence to ourselves and our environment.

We need to recognize that global trends are taking us in the wrong direction, the direction of increasing environmental and societal problems. There is little doubt that we must turn these trends around to avoid destructive consequences that nobody wants.

As the richest and most powerful nation on earth, the United States must show leadership on global issues. Unfortunately, we fail to do so in many areas. We remain one of the most consumptive and wasteful societies on earth, we have record crime and violence, we are the number one arms dealer in the world, and we fail to take appropriate responsibility in family health and population planning.

Sadly, the political winds are blowing in different directions. At a time when cooperative and collective action is needed, the cry is for greater individual freedom, less international commitment, less public support of health and education, lower environmental standards, less money for environmental and social restoration, and even attacks on family planning. If these are interpreted to mean less acceptance of collective responsibility and less long-term planning, as they seem to, they will be counter-productive. They all seem to say, "I'll run my own show for my own benefit." Ultimately, they will lead to greater population pressure, more job competition, less security, and fewer options to solve our problems.

WHAT IS NEEDED?

Modern scholars who share the ecological point of view feel that nothing less than a quantum leap in human understanding and behavior is necessary. Business as usual, with only mild recognition of population and environmental problems, will not suffice.

The historian, physician, and biologist David Ehrenfeld, in a powerful book *Beginning Again,* (1993) makes a case for beginning the twenty-first century with a totally new outlook. If we continue on the track of the twentieth century, he cautions,

> sooner or later we will find that it ends in global chaos—upheaval and breakdown on a vast scale.
>
> A likely cause of upheaval is the disintegration of the extremely complicated and finicky economic, industrial, social, and political structure that we have put together in the decades since the Second World War. This structure has been supported by resources, especially petroleum, that are waning, and by an environmental and cultural legacy—soil, vegetation, air, water, families, traditions—that we foolishly took for granted, squandered, and lost. The most terrifying thing about this disintegration for a society that believes in prediction and control will be the randomness of its violent consequences. The chaotic violence will include not only desperate, ruthless struggles over the wealth that remains, but the last great rape of nature.

Ehrenfeld feels that the only alternative for the next millennium is a new beginning, a transformation of human ambitions from

> a love of quantity and consumption, waste, and the idiot's goal of perpetual growth to one of honesty, resilience, appreciation of beauty and scale, and stability—based in part on the inventive imitation of nature.

The eminent geneticist Bruce Wallace concurs. In "The Re-making of the Modern Mind: The Case for Prudence" (1995), Wallace states that "the natural world and human civilization as we know it are in grave danger." He concludes that continued existence requires a new ethical system based on an understanding and appreciation of nature. In considering the beauty and fascination of the living world, Wallace notes, "The biosphere, finite as it is, cannot provide for peoples who are intent on global struggle."

Nor can it provide for peoples who give up all hope of constructive change. Vice President Al Gore in his 1992 book *Earth in the Balance,* states:

> When considering a problem as large as the degradation of the global environment, it is easy to feel overwhelmed, utterly helpless to effect any change whatsoever. But we must resist that response, because this crisis will be resolved only if individuals take some responsibility for it. By educating ourselves and others, by doing our part to minimize our use and waste of resources, by becoming more active politically and demanding change—in these ways and many others, each one of us can make a difference. Perhaps most important, we each need to assess our own relationship to the natural world and renew, at the deepest level of personal integrity, a connection to it.

In the end, we return to the thoughts of the philosopher Ernest Schumacher (1973):

Wisdom demands a new orientation of science and technology towards the organic, the gentle, the non-violent, the elegant and the beautiful. . . . We must look for a revolution in technology [and the human spirit] which will reverse the destructive trends now threatening us all.

REFERENCES

Ehrenfeld, D. 1993. *Beginning Again: People and Nature in the New Millennium.* New York: Oxford University Press.

Forrester, J. 1991. Quoted in A. Gordon and D. Suzuki. *It's a Matter of Survival.* Cambridge, MA: Harvard University Press.

Gore, A. 1992. *Earth in the Balance: Ecology and the Human Spirit.* Boston: Houghton Mifflin.

Gregg, A. 1955. A medical aspect of the population problem. *Science* 121: 681–682.

Hern, W. M. 1993. Is human culture carcinogenic for uncontrolled population growth and ecological destruction? *BioScience* 43: 768–773 (Dec. 1993).

Leopold, A. 1933. *Game Management.* New York: Scribners.

Schumacher, E. F. 1973. *Small is Beautiful.* New York: Harper and Row.

Wallace, B. 1995. The Re-making of the Modern Mind: The Case for Prudence. Blacksburg, VA: Virginia Polytechnic Institute and State University. Unpublished ms.

GLOSSARY

This is an abbreviated list of terms used in the text. An extensive list of ecological terms with thorough definitions is given in *The Concise Oxford Dictionary of Ecology,* edited by M. Allaby (New York: Oxford University Press, 1994).

abiotic without life; nonliving; absence of living organisms

abyssal of or relating to ocean depths

acid rain or acid precipitation rainfall or precipitation with a low pH (less than 7.0 and usually in the range of 3.0 to 6.0); forms in polluted air when oxides of sulfur and nitrogen dissolve in atmospheric precipitation

aeolian related to the wind; produced or blown by the wind

aerobic utilizing or requiring oxygen for life processes

aggression hostility or conflict, overt or suppressed

agroecosystem agricultural croplands, pastures or gardens dependent on soil and water and requiring subsidized inputs such as cultivation and fertilizers in the case of crops and gardens

AIDS acquired immune deficiency syndrome, a viral infection of lymphocytes that impairs the immune function in humans and makes the body susceptible to opportunistic infections

albedo ratio of light striking a surface to that reflected; a measure of the reflectivity of the earth's surface

allergen substance producing a pathogenic sensitivity within an organism

alpha particle in physics, a type of ionizing radiation consisting of a helium nucleus containing two protons and two neutrons; cf. *beta particle, gamma rays*

alpine describing mountainous regions of high elevation; refers to environments near and above the treeline

ambient existing conditions; surrounding on all sides

anadromous referring to animals that live in the sea and breed in fresh water, such as salmon

anaerobic capable of living without oxygen

anarchy a state of social disorder, lawlessness, confusion, chaos

anthropogenic induced or influenced by human action

antibiosis a process fatal or injurious to any form of life

antibiotic usually refers to a drug or chemotherapeutic agent that kills bacteria

antibody a chemical produced in a living organism in response to a foreign substance entering the organism

antigen a foreign substance that produces an antibody response in an organism

atmosphere multilayered envelope of gases around the earth which provides the basic components necessary for life; moderates temperatures and serves as a protective shield from excessive ultraviolet radiation and cosmic rays from outer space

autoecology the study of individuals, or species, in response to environmental conditions, cf. *synecology*

autotrophic capable of producing organic compounds from inorganic chemicals by means of photosynthesis or chemosynthesis

bacteria single-celled, prokaryotic organisms of the Kingdom Monera; although many bacteria are pathogenic organisms, other bacteria play essential ecological roles in biogeochemical cycles

benthic referring to the bottom layer of any body of water and the organisms therein

beta particle in physics, a type of ionizing radiation consisting of an electron, such as the radiation emitted by strontium 90 or carbon 14

biochemical oxygen demand (BOD) the amount of dissolved oxygen required by microorganisms in water

biocide any chemical that kills living organisms

biodegradable capable of being decomposed by bacterial action

biodiversity the number and variety of species in a given area

biogeochemical cycles the cyclic flow of essential elements such as carbon, nitrogen, phosphorus, and sulfur between the living and nonliving components of ecosystems

biological clock the rhythmic occurrence of processes within organisms at periodic intervals

biomass the total weight of living organisms per unit area or unit volume

biome a large regional ecosystem with characteristic plant and animal communities, usually covering a broad geographic area; examples are grassland, desert, coniferous forest, deciduous forest, and tropical rain forest

biometry the application of statistics to biology

biosphere the portion of the earth and its atmosphere capable of supporting life; the portion of planet earth where living organisms exist; the global assemblage of all living organisms

biota all living organisms, including plants, animals, microorganisms and fungi, in an area

BPP (biospheric primary production) the primary production of all living organisms on earth by photosynthesis or chemosynthesis

calorie the amount of heat energy needed to raise the temperature of one gram of water 1° centigrade (small calorie, or g-cal) or one liter of water 1° (large calorie or kg-cal)

Calvin cycle the biochemistry of photosynthesis as described by Nobel Prizewinner Melvin Calvin of University of California

carcinogen a substance that produces or stimulates cancer

carnivore an organism that consumes animals for food

carrying capacity the ability of a landscape or ecosystem to support a given plant or animal species, population, or community; the term implies a numerical limit

CDC (Centers for Disease Control) a U.S. organization headquartered in Atlanta, Georgia, concerned with controlling human disease and promoting health

chemosynthesis the synthesis of organic compounds using chemical energy, independent of light energy

chlorinated hydrocarbons synthetic organic compounds containing chlorine, carbon, and hydrogen, many of which are effective as insecticides, such as DDT, endrin, aldrin, dieldrin, chlordane, heptachlor, and toxaphene

cholera an infectious disease of the digestive system caused by bacteria of the genus *Vibrio*

cirrhosis a chronic disease of the liver often caused by prolonged alcohol abuse or prolonged infectious hepatitis

climate the integrated result of long-term weather patterns

climax community the mature biotic community capable of self-perpetuation under prevailing climatic and topographic conditions

commensalism the living together of two species, usually with benefit to one of them

community all the different populations of living organisms in a definable area

competition the mutual utilization of limited resources; may be intraspecific, between members of the same species; interspecific, between members of different species

competitive exclusion the ecological principle that if two closely related species compete for a limited resource essential for the survival of each, one of the species will tend to be eliminated; cf. *Gause's principle*

consumer organisms organisms that utilize the organic materials manufactured by plants and chemosynthetic bacteria; they are unable to produce their own organic compounds for basic energetic and nutritive purposes

continental drift movements of the crustal or tectonic plates of continents; the theory that the continents of earth were once a single large landmass that broke into major sections and diverged, or drifted apart

coral reef a geological marine formation in tropical oceans and seas composed of hard coral (Phylum Coelenterata) and associated organisms

cosmic pertaining to the universe, especially that outside the earth

cosmos the universe

crude birth rate the number of births per year for each 1,000 individuals in a population

crude death rate the number of deaths per year for each 1,000 individuals in a population

crustal plates solid portion of the earth's outer surface consisting of continental plates and oceanic plates beneath the oceans

cybernetics the study of communication and control systems in living organisms and machines

data analysis the calculation of statistical relationships between variables

DDT chemical used to control pest insects; an insecticide consisting of dichloro-diphenyl-trichloroethane, one of a group of chlorinated hydrocarbons

decomposer organisms bacteria and fungi that break down organic compounds into simpler substances

defoliation loss of foliage; often refers to the use of herbicides to destroy vegetation; applied as a military strategy in Vietnam

deforestation the destruction of forests; may involve clear-cutting or selective logging

demography the mathematical study of populations

density-dependent factor an ecologic influence that varies in relation to population density

density-independent factor an ecologic influence that operates independent of density

desert dry region with less than 25 cm of rainfall per year

desertification conversion of arid or semiarid land to desert

detritivore an organism, which may be a scavenger, that utilizes dead organic material and extracts energy from the process of decomposition; organism that lives on detritus

detritus dead organic material in solid or particulate form

disparity a lack of equality; a difference, often a prominent difference

diversity difference, complexity, or variety; in reference to species diversity, a measure of the number of species in a biotic community in relation to the total number of individuals

diversity conservation the conservation of diverse ecosystems to ensure more natural, more complex, and more stable flora and fauna

diversity indices mathematical models of biodiversity; the number and statistical distribution of species in a given area

DNA (deoxyribonucleic acid) the basic genetic material of all living cells; composed of chains of nucleotides wound in a double helix and having four organic bases—adenine, guanine, cytosine, and thymine—bonded to a five-carbon sugar and a phosphate group; cf. *RNA*

dominance supremacy, control, or authority; may refer to behavioral, political, economic, or ecologic relationships

dominance, behavioral or social the condition by which an individual or group of individuals has primary access to environmental resources

dominance, ecologic the condition in a biotic community in which one or more species exert a major or controlling influence

ecology the study of the interrelationships of living organisms to one another and their environments; the study of the abundance and distribution of living organisms

economics the study of the production, distribution, and consumption of goods and services; the study of material welfare of human beings

ecosphere the biosphere interacting with all other components of planet earth; the global ecosystem

ecosystem a biotic community and its nonliving environment as an interacting system

ecotone a transition zone between two adjacent and differing communities

ecotype a group of living organisms adapted to a certain set of environmental conditions

endemic referring to a group of organisms native to a given region and occurring only in that region

entropy the loss or degradation of energy; that portion of heat energy that is unavailable for work

environment all the conditions and influences surrounding and affecting an organism or group of organisms; consists of physical, biological, and social factors

environmental resistance those factors that tend to limit the growth or distribution of a population

EPA (Environmental Protection Agency) an agency of the U.S. government charged with administering environmental laws and regulations

epidemic the occurrence in greater numbers than usual of a pathogen, parasite, disease, or pest

epilimnion the upper layer of warm water in a stratified lake

erosion the wearing away of the earth's surface by wind and water

estuary a coastal zone where rivers and bays meet the ocean with a mixing of fresh water and salt water; estuaries are usually very productive areas, with high biodiversity, but sensitive to pollution

ethnology a branch of anthropology that analyzes human cultures; the study of the divisions of human populations

ethology the comparative and biological study of behavior

etiology the causes or developmental history of a condition; in a medical sense, the causes of a disease

eukaryotic cell a living cell with a nucleus, chromosomes, and membrane-bound organelles; cf. *prokaryotic cell*

euphotic zone the surface or upper layers of a body of water into which light can penetrate to support photosynthesis

eutrophication the process of nutrient enrichment and aging in an aquatic ecosystem; a natural process accelerated by domestic and agricultural pollution

eutrophy a condition of nutrient enrichment in an aquatic ecosystem

extermination, biological biological extinction of a species that is caused or hastened by human activities

extinction the total loss by death of all members of a species, subspecies, or population of a living organism

extirpation the extinction or extermination of a species, subspecies, or population in a given area

famine a shortage of food in a population leading to extensive malnutrition and starvation

fauna a collective term for all the animals in a given region or geological period

fecundity the production of gametes or sex cells; capacity to produce offspring; cf. *fertility*

feedback the process by which a system returns toward a former condition; the supply of information to an automatic control system so that errors or variations may be corrected

fertility the production of offspring; in humans and animals, the fertility rate is the birth rate; cf. *fecundity*

filariasis a helminthic disease of the lymphatic system transmitted in tropical regions by mosquitoes

filter feeder an organism that acquires food by straining or filtering particles or organisms from water

food chain a series of organisms dependent on one another for food

fossil fuel a fuel formed from the altered remains of once-living organisms that is burned to release energy (e.g., coal, oil and natural gas)

fungus an organism in the Kingdom Fungi; typically a filamentous, multinucleate, eukaryotic organism that functions in ecological decomposition

Gaia hypothesis the idea that the earth has evolved as a living functional unit in which biological processes have interacted with and modified physical conditions to be favorable for the evolution of the biosphere

gamma rays in physics, a type of radiation consisting of high-energy electromagnetic waves with very short wave lengths, similar to x-rays; gamma rays readily penetrate living tissue; examples are the radiation from cobalt 60 and iodine 131

Gause's principle the concept that two species cannot occupy exactly the same ecologic niche within a community

GDP, GNP (gross domestic product or gross national product) the total sum of all goods and services produced by a nation or political unit, usually expressed per year and per person

genus a group of closely related species; the first part of the scientific Latin name for an organism

geronticide the killing of older individuals

global referring to the planet earth

global change all of the ecological changes—physical, biological, and social—occurring on planet earth

global ecology study of planet earth in physical, biological, and sociobehavioral terms; the study of the interactions of lithosphere, atmosphere, hydrosphere, and biosphere

Gondwana a large landmass in the southern hemisphere existing 135 million years ago

GPP (gross primary production) total organic synthesis in plants by photosynthesis and in microorganisms by chemosynthesis

greenhouse effect the warming of the earth as a result of solar radiation entering the atmosphere, striking the earth's surface, and reradiating heat energy, which is retained; the greenhouse effect is a natural process that enables life on earth, but it is enhanced by increasing CO_2 and other greenhouse gases produced by human activities

greenhouse gases atmospheric gases, primarily carbon dioxide, methane, oxides of sulfur and nitrogen, chlorofluorocarbons, and water vapor, that retain infrared radiation or heat in the earth's atmosphere

green revolution the modernization of agriculture as a result of new high-yield genetic strains and increased use of fertilizers, irrigation, pesticides, and mechanization

habitat the natural abode of an organism including its total environment

Hadley cells oval clusters of air movement originating in the tropics; characterized by the upward flow of warm moist air, generating rainfall as it cools, and descending of cool dry air at higher latitudes.

herbicide any chemical that kills or injures plant life

herbivore an animal that feeds on plants

hierarchy a system of ranked individuals or things; a social ordering of animals, as in a dominance hierarchy

HIV (human immune virus) a group of retroviruses known to be the cause of AIDS

homeostasis the tendency of a system to maintain equilibrium; maintenance of a relative condition of stability

hydrology the study of the distribution, circulation, and properties of water

hydrosphere the water on planet earth; that portion of the earth containing water in liquid, solid, or vapor form

hypolimnion the lower layer of cold dense water in a stratified lake

indicator organism an organism that indicates the presence or absence of certain environmental conditions

infanticide the killing of infants or newborn individuals in a population

infared longwave radiation (from 0.8 to 1000 microns) that generates heat

interdiction in warfare, the policy of massive bombardment using heavy munitions against landscape where the enemy is thought to be hiding; massive firepower to deprive the enemy of resources

invasive the tendency to invade or enter a new location or niche

irruptive the tendency to increase suddenly in numbers

isotopes forms of the same element that differ in atomic weight and organization of the atomic nucleus

J-curve a J-shaped growth curve describing exponential growth of a population

joule a unit of energy equivalent to 0.239 calories or 0.738 foot pounds

K-selection a pattern of reproduction and population ecology characterized by relatively low reproductive potential and slow population growth

kingdom in a biological sense, the highest category of living organisms; modern biology recognizes five Kingdoms: Monera, Protista, Fungi, Plants, and Animals

kwashiorkor a condition of malnutrition resulting from protein deficiency

land modification ecological changes of landscape, such as construction, cultivation, deforestation, erosion, earthquakes, and volcanism

landscape the combination of physical and biotic factors, especially topography and plant communities, that create scenery and habitat

Laurasia a large landmass in the northern hemisphere that formed approximately 135 million years ago prior to the separation of the existing continents

leaching the removal or downward movement by water of soluble chemicals from the soil or other materials

leishmaniasis a protozoan disease causing severe skin lesions, transmitted by biting flies

limiting factor the ecologic influence that limits or controls the abundance or distribution of a species

lithosphere the solid and molten rock portions of the earth; divided into the crust, mantle (molten rock), and core (thought to be molten iron with an inner solid core); also called the geosphere

littoral pertaining to the shore of an ocean, sea, or large lake

logistic curve an S-shaped growth curve describing the growth of an organism or a population; characterized by slow initial growth, rapid intermediate growth, and a slow final stage with a gradual approach to an upper asymptote

Malthusian referring to the doctrine of Thomas Malthus that a population of living organisms tends to increase faster than its means of support

mantle a large mass of molten rock underlying the crustal or tectonic plates

marasmus a starvation condition resulting from total caloric restriction

megalopolis a large continuous urban area formed by the confluence of two or more cities

mesic referring to a moist or wet environment; cf. *xeric*

mesosphere a portion of the atmosphere that extends 30 to 50 miles beyond the stratosphere; it has gradually decreasing temperatures and ozone concentrations; at its outer limits temperatures reach $-90°C$

metabolism the chemical processes occurring within an organism or biological system by which living cells are produced, maintained, and recycled

Milankovitch cycles long-term astronomical cycles of axial and orbital variation postulated as the causes of major climate changes including periods of glaciation

millennium 1,000 years

mimicry superficial resemblance of an organism to another that gives it a selective advantage against predation by appearing dangerous, inedible, or inconspicuous

monoculture a simplified biotic community dominated by one species

monsoon a seasonal period of increased rainfall, usually heavy; most prominent in Asia

mortality death

mutagen a substance that produces changes in genetic material

mutualism an interspecific relationship of benefit to two or more interacting species

natality the production of offspring by organisms; birth rate

natural selection the process of evolution by which some organisms give rise to more descendants than others

neritic pertaining to the area of the sea over the continental shelf

net primary productivity organic synthesis in a plant in excess of that used in respiration by the plant

niche the role of a species in a community

non-point-source pollution pollution originating from widely dispersed sources, such as urban or agricultural runoff

NPP (net primary production) total organic synthesis by photosynthesis and chemosynthesis minus that consumed by respiration; total carbon fixation minus oxidation

nuclear winter a severe cold spell caused by reduced sunlight due to clouds of heavy smoke which would result from extensive nuclear warfare; smoke and dust clouds from nuclear detonations would theoretically impair photosynthesis and cause drastic reductions in the surface temperatures of the earth

nutrient a chemical substance that contributes to the growth and maintenance of an organism

oligotrophic referring to low-nutrient, relatively unproductive, and often cold lakes; such lakes usually have sparse plant and animal life, low organic content, and high oxygen levels

omnivore a living organism that consumes both plants and animals as food

organism an individual unit of life, constituted to carry on the activities of life

organophosphate any one of several synthetic chemical insecticides, such as parathion and malathion, that are very toxic but nonpersistent in the environment

ozone layer a layer at the outer edge of the stratosphere formed when molecules of oxygen (O_2), dissociated by solar radiation, recombine into ozone (O_3); ozone is a very reactive molecule that is an injurious pollutant at the earth's surface, but in the upper atmosphere (15–50 km, or 10–30 miles, high) it forms a protective layer that filters and reduces ultraviolet radiation

Pangaea a single large landmass on the earth that existed 200 to 250 million years ago, before the formation of existing continents

parasite a specialized consumer that feeds directly on plants and animals, thus deriving its nutrition over an extended period of time; parasites may be external (ectoparasites, e.g., ticks, mites) or internal (e.g., flukes, worms)

pathogen an organism or virus that causes disease

pathological a disease process in an organism or ecosystem causing a serious deviation from homeostasis

pelagic referring to the open ocean or seas, especially near the surface

pesticide a chemical agent that kills pests (e.g., insecticide)

photosynthesis the production of organic compounds in green plants from carbon dioxide and water using light energy with the aid of chlorophyll as catalyst

phylum a major division of living organisms in the animal kingdom and the kingdoms Monera and Protista; *division* is the equivalent term for organisms in the plant kingdom and fungi

phytoplankton autotrophic plankton, composed of algae, certain Monera and Protista, and some floating plants

pioneer an organism that first appears in a newly exposed or altered environment

plankton free-floating or suspended aquatic organisms, small or microscopic; see also *phytoplankton, zooplankton*

point-source pollution pollution coming from a specific point such as a smokestack or drainpipe; cf. *non-point-source pollution*

pollution an unfavorable or unwanted alteration of the environment; contamination by some undesirable agent

population a group of individuals, members of the same species, in a definable area

PPM (parts per million) a measure of concentration, the same as micrograms per milliliter or milligrams per liter

predator an organism, usually an animal, that lives by killing and consuming other animals or protozoa

prediction a statement of belief about what will happen; cf. *projection*

primary production the process by which organic compounds are formed through photosynthesis or chemosynthesis; carbon fixation; see also *GPP*, NPP

producer organisms green plants and bacteria that synthesize organic compounds; all green plants, algae, and cyanobacteria (blue-green algae)

projection a statement about what will happen if current trends, or certain assumed trends, continue

prokaryotic cell a bacterial cell lacking a nucleus and membrane-bound organelles; more primitive than a eukaryotic cell

protoplasm the matrix of living cells; composed of cytoplasm and nucleoplasm

protozoa eukaryotic, single-cell organisms in the Kingdom Protista, such as *Paramecium*

quadrat in ecology, a square or rectangular plot designated for the study of plant or animal populations

Quaternary period a geological period spanning the Pleistocene and Holocene or Recent epochs from 1.8 million years ago to the present; the Pleistocene was the great age of glaciation and human evolution; the Holocene of the past 10,000 years has been a period of global warming and the development of human civilizations

r-selection a pattern of reproduction and population ecology characterized by high reproductive potential and the possibility of explosive population growth

RAD (radiation absorbed dose) the amount of radiation absorbed by an object or organism, expressed as the amount of energy (in ergs) per gram of absorbing tissue or material

radioactive the property of certain chemical elements to spontaneously emit radiation from atomic nuclei in the form of alpha or beta particles or gamma rays

radioisotope an isotope that is unstable and disintegrates, emitting radiation energy

radon a natural radioactive gas resulting from the decay of uranium in soil and rocks; radon is odorless and tasteless and emits alpha particles

respiration the utilization of oxygen in living organisms to obtain energy; an analogous process in anaerobic bacteria in the absence of oxygen

restoration ecology a subdivision of ecology devoted to restoring damaged landscapes

RNA (ribonucleic acid) a type of nucleic acid containing the sugar ribose and the pyrimidine uracil; (cf *DNA*)

S.A.L.T. (Strategic Arms Limitation Treaty) The first arms control accord signed in Moscow in 1972 by the US and USSR; it limited antiballistic missiles of both nations, and prohibited the use of space for military purposes. cf. S.T.A.R.T.

Sahel a semiarid area in Africa south of the Sahara desert stretching from Senegal to Ethiopia, Uganda, and Somalia

salination the accumulation of salts in soil or water; same as salinization

saline salty; pertaining to chemical salts, especially sodium, potassium, magnesium, and calcium

saprophytic referring to an organism that lives on decaying organic material; describes many bacteria and fungi

savannah a grassland community with scattered tree growth; usually tropical or subtropical

schistosomiasis a parasitic infection and disease caused by a helminthic blood fluke of the genus *Schistosoma;* known as bilharzia or snail fever

sediment soil particles suspended in the water of rivers, lakes, seas, and oceans

sere a stage of an ecological community in a successional series; also called seral stage

slash and burn a type of agriculture in which plots are cut, burned, and cultivated on a temporary basis, then abandoned for a variable period of time

smog polluted air and water vapor

soil the aggregate of decaying organic material, living organisms, and weathered substrate

solar referring to the sun; solar energy is the energy of sunlight reaching the earth

species a specific kind of organism designated by a binomial scientific name; a species is a recognizable population that has some degree of reproductive isolation from other similar organisms

standing crop the amount of biomass present in a defined area, population, or community at any given time

S.T.A.R.T. Strategic Arms Reduction Treaty, a treaty signed in 1991 by the US and USSR to reduce nuclear arsenals 33 percent, especially to dismantle intercontinental ballistic missiles over a 7-year period

stochastic conjectural; unpredictable; occurring as a chance outcome

stratosphere a region of the atmosphere (10 to 30 miles high) that lies beyond the troposphere and is outwardly bounded by the ozone layer

stress a condition that forces a deviation from homeostasis

stressor a stimulus or factor producing stress

succession an ecological process whereby populations replace each other and gradually change community structure; ecosystems changing slowly over time

sustainable development an economic strategy designed to meet today's needs without compromising the ability of future generations to meet their own needs

symbiosis the living together of two or more species

synecology the study of the environmental relations of communities; cf. *autoecology*

synergism the interaction of separate agencies that together have a greater effect than the sum of their independent effects

system an aggregation or assemblage of objects united by some form of interaction or interdependence; a group of diverse units combined to form an integrated whole

systematics the taxonomy or classification of living organisms

systemic referring to the entire body of an organism

systems management a process of managing an ecosystem in which objectives are outlined, data are obtained on important variables relating to complex phenomena, and appropriate action is planned

systems measurement a detailed inventory of existing conditions, characteristics and identification of major influences on a system

systems modeling the use of functional or mathematical models to provide a theoretical basis for interpreting the variables

systems optimization a situation in which the best strategies for achieving system objectives are selected

systems simulation a process in which variables are manipulated mathematically to test a model and predict the consequences of changes within a system

tectonic plates crustal plates of the earth's lithosphere; may be continental or oceanic (beneath the oceans)

teratogen a chemical substance, such as thalidomide, that can cause developmental malformation in an embryo or fetus; a factor which causes birth defects

territorialism a system of partitioning spatial resources in an area held by an individual or a group

territory an area occupied by an animal or group of animals and defended against others of the same species; the concept of territory applies to human beings as well

thermocline the subsurface layer of a lake characterized by a significant temperature change

thermodynamics the study of the transformation of energy in physical and biological systems

thermosphere the outermost zone of the atmosphere which shows slightly rising temperatures ($-40°C$) as the atmosphere gradually merges with hydrogen and helium atoms in outer space

"third world" developing nations of LDC's, the lesser developed countries

threshold the level (duration or intensity) of a stimulus required to produce an effect

topography characteristics of the ground surface in regard to physical features

tort part of our legal code by which a person or an institution can be held responsible for an injury or wrong done to someone else

transpiration the loss of water vapor through the surface of leaves and other plant parts

trophic level one stage in a nutritive series including producers or various levels of consumers

trophic structure the organization of a community into a nutritive series; cf. *food chain, trophic level*

tropical referring to that portion of the earth's surface between 23° north latitude and 23° south; usually implies warm climatic conditions but geographically, the tropical regions also contain mountainous regions with temperate forests, alpine tundra, and permanent snow and ice

troposphere the portion of atmosphere closest to the earth and about 10 miles thick; contains 75 percent of all gaseous molecules in the atmosphere

trypanosomiasis an infectious disease caused by a protozoan of the genus *Trypanosoma;* known in Africa as sleeping sickness and in Latin America as Chagas disease

turbidity any condition in air or water that reduces its transparency; muddy water or smoky air is turbid

ultraviolet shortwave radiation (less than 400 angstrom units) emanating from the sun

UNEP (United Nations Environmental Program) A branch of the UN started in Nairobi in 1973 to coordinate a global environmental monitoring system (GEMS) and focus on problems of desertification, pollution, water use, and related ecological topics

UNESCO (United Nations Educational Scientific and Cultural Organization) A branch of the UN founded in 1946 in Paris to promote literacy, intellectual cooperation, cultural heritages, and international exchange programs

ungulates hoofed animals such as deer, horses, and cattle

UNICEF (United Nations International Children's Emergency Fund) a unit of the UN working to promote the nutrition, health, and welfare of children throughout the world

uranium a radioactive metallic element used in atomic and hydrogen bombs and as a fuel in nuclear power plants

vector-born infection pathogens transmitted by animals, often arthropods such as mosquitoes, ticks, lice, mites, or flies

viable capable of living

virulence the capability of a microorganism, such as a virus or bacterium, to cause disease; similar to pathogenicity

virus a microscopic particle of DNA or RNA, smaller than bacteria, that is parasitic on living cells; viruses are capable of replication, mutation, and infection of higher organisms but incapable of reproducing outside of living cells

vitamin an organic substance that is usually not synthesized in the body (except for vitamin D) and is required in the diet in small amounts for normal growth and metabolism

water pollution the unfavorable or unwanted alteration of water quality in aquatic ecosystems by natural or human actions

watershed the total land area contributing surface or ground water to a lake, river, or drainage basin

WHO (World Health Organization) an international organization headquartered in Geneva, Switzerland, that works to reduce disease and promote human health

xenophobic referring to the fear and hatred of strangers, often manifested as aggression toward strangers

xeric referring to a dry environment; cf. *mesic*

yield the part of primary production utilized by humans or other animals

yin and yang Chinese concept describing the earth as a complex entity with contrasting forces that can operate in balance or out of balance; if they are unbalanced the individual or the earth becomes ill

zonation arrangement into bands or zones

zoogeography the science dealing with the geographic distribution of animals

zoonoses (zoonotic infections) infections or diseases of animals that can be transmitted to humans, such as rabies, Lyme disease, encephalitis, Chagas disease, sleeping sickness, and yellow fever

zooplankton animal plankton

BIBLIOGRAPHY

In addition to the specific references at the end of each chapter, the following recent books and scientific articles are recommended as general reading. This list is by no means comprehensive, but it includes works with a wide range of opinions as well as reference works regarding the state of global environments and prospects for the future.

Bailey, R. (ed.). 1995. *The True State of the Planet*. New York: Free Press, Simon and Schuster.

Bartlett, A. A. 1994. Reflections on sustainability, population growth, and the environment. *Population and Environment: A Journal of Interdisciplinary Studies* 16: 5–34.

Berryman, A. 1994. The doomsday prediction. *Bulletin of the Ecological Society of America* 75(2): 123–124.

Brown, L. R., C. Flavin, S. Postel, and L. Starke. (eds.). 1995. *State of the World, 1995*. New York: W. W. Norton.

Brown, L. R., N. Lenssen, and H. Kane (eds.). 1995. *Vital Signs: The Trends that Are Shaping Our Future*. New York: W. W. Norton.

Brundtland, G. H. (ed.). 1987. *Our Common Future*. New York: Oxford University Press.

Cartledge, B. (ed.). 1995. *Population and the Environment*. Oxford and New York: Oxford University Press.

Colinvaux, P. 1993. *Ecology 2*. New York: John Wiley and Sons.

Corson, W. (ed.). 1990. *The Global Ecology Handbook: What You Can Do About the Environmental Crisis*. Boston: Beacon Press.

de Blij, H., and P. O. Muller. 1993. *Physical Geography of the Global Environment*. New York: John Wiley and Sons.

Easterbrook, G. 1995. *A Moment on the Earth: The Coming Age of Environmental Optimism*. New York: Viking Penguin.

Eblen, R. A., and W. R. Eblen (eds.). 1994. *Encyclopedia of the Environment*. New York: Houghton Mifflin.

Edwards, S. R. 1995. Conserving biodiversity. In *The True State of the Planet,* ed. R. Bailey. New York: Free Press, Simon and Schuster.

Ehrlich, P. R., and A. Ehrlich. 1991. *Healing the Planet: Strategies for Resolving the Environmental Crisis*. Reading, MA: Addison Wesley.

Few, R. 1994. *The Atlas of Wild Places: In Search of Earth's Last Wildernesses*. New York: Facts on File.

Freedman, B. 1994. *Environmental Ecology: The Ecological Effects of Pollution, Disturbance and Other Stresses*. Orlando, FL: Academic Press.

Glantz, M. (ed.). 1994. *Drought Follows the Plow: Cultivating Marginal Areas*. New York: Cambridge University Press.

Gotelli, N. J. 1995. *A Primer of Ecology*. Sunderland, MA: Sinauer Associates.

Goudie, A. 1994. *The Human Impact on the Natural Environment*. Cambridge, MA: MIT Press.

Kemp, D. D. 1994. *Global Environmental Issues: A Climatological Approach*. 2nd ed. London: Routledge.

King, J. J. (ed.). 1995. *The Environmental Dictionary*. New York: Wiley Interscience.

Krishnan, R., J. M. Harris, and N. R. Goodwin (eds.). 1995. *A Survey of Ecological Economics*. Covelo, CA: Island Press.

Likens, G. E. and F. H. Bormann. 1995. *Biogeochemistry of a Forested Ecosystem*. 2nd ed. New York: Springer-Verlag.

Mann, C. C., and M. L. Plummer. 1995. *Noah's Choice: The Future of Endangered Species*. New York: Alfred A. Knopf.

Myers, N. 1995. The world's forests: Need for a policy appraisal. *Science* 268: 823–824.

Myers, N., and J. Simon. 1994. *Scarcity or Abundance: A Debate on the Environment*. New York: W. W. Norton.

Nierenberg, W. A. (ed.). 1995. *Encyclopedia of Environmental Biology*. Orlando, FL: Academic Press.

Odum, E. P. 1993. *Ecology and Our Endangered Life Support Systems*. 2nd ed. Sunderland, MA: Sinauer Associates.

Orr, D. 1994. *Earth in Mind: On Education, Environment and the Human Prospect*. Covelo, CA: Island Press.

Perry, D. A. 1995. *Forest Ecosystems*. Baltimore, MD: Johns Hopkins University Press.

Pimentel, D., et al. 1995. Environmental and economic costs of soil erosion and conservation benefits. *Science* 267: 1117–1123.

Pollack, S. 1995. *The Atlas of Endangered Resources*. New York: Facts on File.

Primack, R. B. 1995. *A Primer of Conservation Biology*. Sunderland, MA: Sinauer Associates.

Ramakrishna, K., and G. M. Woodwell (eds.). 1995. *World Forests for the Future*. New Haven, CT: Yale University Press.

Raven, P. H. 1994. Defining biodiversity. *Nature Conservancy* (Jan.–Feb. 1994): 11–15.

Raven, P. H. 1995. Review of *A Moment on Earth: The Coming Age of Environmental Optimism,* by G. Easterbrook. *The Amicus Journal* 17(1): 42–45.

Raven, P. H., L. R. Berg, and G. B. Johnson. 1995. *Environment*. Philadelphia: Saunders College Publishing.

Rich, B. 1994. *Mortgaging the Earth: The World Bank, Environmental Impoverishment, and the Crisis of Development*. Boston: Beacon Press.

Ricklefs, R. 1990. *Ecology*. 3rd ed. New York: W. H. Freeman.

Sachs, W. (ed.) 1993. *Global Ecology: A New Arena of Political Conflict*. Halifax, Nova Scotia: Fernwood.

Sitarz, D. (ed.). 1994. *Agenda 21: The Earth Summit Strategy to Save Our Planet*. Boulder, CO: Earth Press.

Tudge, C. 1991. *Global Ecology*. New York: Oxford University Press.

Valiela, I. 1995. *Marine Ecological Processes*. New York: Springer-Verlag.

Weber, P. 1993. *Abandoned Seas: Reversing the Decline of the Oceans*. Worldwatch Paper 116. Washington, DC: Worldwatch Institute.

Wilson, E. 1992. *The Diversity of Life*. Cambridge, MA: Harvard University Press.

Wittwer, S. H. 1995. *Climate and Food: The Global Environment and World Food Production*. Boca Raton, FL: Lewis CRC.

Woodwell, G. M. (ed.). 1990. *The Earth in Transition: Patterns and Processes of Biotic Impoverishment*. New York: Cambridge University Press.

INDEX